The East Asian Computer Chip War

The semiconductor industry is a vital industry for military establishments worldwide, and the control, or loss of control, of this key industry has enormous strategic implications. *The East Asian Computer Chip War* focuses on the globalization of the strategic semiconductor industry and the security ramifications of this process. It examines in particular the migration of the Taiwanese chip industry to China as part of the globalization of production processes, and the extent to which such a globalization process poses security challenges to the United States, China, and Taiwan.

Transcending disciplinary boundaries between international political economy, security studies, and the history of science and technology, this multidisciplinary work provides an in-depth understanding of the globalization–security nexus, and disentangles the key policy issues connected to a potential explosive flashpoint in world politics today.

Ming-chin Monique Chu completed her PhD in international studies at the University of Cambridge, UK. She is a research fellow at St Antony's College and a postdoctoral research officer in Taiwan Studies, at the School of Interdisciplinary Area Studies, University of Oxford, UK.

Routledge Studies on the Chinese Economy

Series Editor
Peter Nolan
Sinyi Professor, Judge Business School,
Chair, Development Studies, University of Cambridge

Founding Series Editors
Peter Nolan, University of Cambridge and Dong Fureng, Beijing University

The aim of this series is to publish original, high-quality, research-level work by both new and established scholars in the West and the East, on all aspects of the Chinese economy, including studies of business and economic history.

Routledge Studies on the Chinese Economy – Chinese Economists on Economic Reform

The East Asian Computer Chip War

Ming-chin Monique Chu

Routledge
Taylor & Francis Group

LONDON AND NEW YORK

First published 2013
by Routledge

2 Park Square, Milton Park, Abingdon, Oxon OX14 4RN
711 Third Avenue, New York, NY 10017, USA

Routledge is an imprint of the Taylor & Francis Group, an informa business

First issued in paperback 2016

British Library Cataloguing in Publication Data
A catalogue record for this book is available from the British Library

Library of Congress Cataloging in Publication Data
Chu, Ming-chin Monique.
The East Asian computer chip war / Ming-chin Monique Chu.
    pages cm.—(Routledge studies on the Chinese economy)
    Includes bibliographical references and index.
    1. Semiconductor industry–Taiwan.   2. Semiconductor industry–China.
    3. Technology transfer–Taiwan.   4. Technology transfer–China.
    5. Computer security–Taiwan.   6. Computer security–China.   I. Title.
    HD9696.S43T23 2013
    338.7'6213950951—dc23
                                                                    2013020819
ISBN 978-0-415-56552-3 (hbk)
ISBN 978-1-138-20012-8 (pbk)

Typeset in Times New Roman
by RefineCatch Limited, Bungay, Suffolk

**Dedicated to the memory of my father
Chin-man Chu, 1932–2013**

Contents

Figures

Tables

Foreword

There is vast literature on globalization. A large part of this literature is abstract and "theoretical." However, most so-called "theory" is not theoretical at all, in the sense of making testable predictions about the real world. Much of what now passes for "theory" in the social sciences consists of complex words and phrases, often incomprehensible to most people, of little help in understanding the real world. However, if scholars quote each other sufficiently often what passes for "theory" establishes itself as a legitimate area of scholarly activity. The associated journals are able to achieve an imprimatur as "five star" journals and the leading theoreticians achieve high reputations in the academic community. A fast-growing area of the "theory" is the study of "how to research." "Methodology" now occupies a large fraction of academic activity in the social sciences. Students are urged to read so-called "theory" and spend large amounts of time learning "how to study" rather than encouraged to explore the extraordinarily interesting real world of globalization, using their own critical faculties and developing their independent analytical skills.

The world of globalization, which has unfolded at high speed since the 1970s, is based on real firms of all shapes and sizes. The firms that are the foundation of globalization mostly pursue profits, have real managers and directors, and real employees. They compete with other firms in order to survive and prosper. At the core of the globalization process is a relatively small number of systems integrator firms that sit at the core of their respective value chains. In almost every sector, from aerospace and automobiles, to beverages and confectionary, a handful of firms occupy a large fraction of the global market. These firms possess leading brands and technologies, and they have immense procurement power in relation to their supply chains. The relationship with the countries in which they have their headquarters has changed profoundly. The 'cascade effect' of pressure from leading global firms has stimulated comprehensive restructuring of global value chains in almost every sector. The global business system can be likened to an iceberg in which the leading firms are widely recognised, but beneath the water level a huge process of industrial transformation has taken place, which is invisible to most people. However, there is only a small amount of deep scholarly research on the transformation of global value chains. When people are informed about the extent to which the global value chains have been restructured in recent decades, the response is typically: "Why did no-one tell us?"

One of the reasons for the dearth of such deep empirical research within the value chains of global firms is that it is difficult for researchers to gain access. This is the case even for non-strategic industries. The impact of leading global firms extends far beyond the core systems integrator itself deep into the value chain. The supply chain of a leading global firm typically employs far more people than the core firm itself. Facilitated by the revolution in information technology, the very boundaries of the firm have become blurred during the recent era of globalization, with the evolution of a new form of separation of ownership and control. The way in which leading global firms interact with the surrounding value chains is a key part of competitive advantage and the interaction unavoidably involves power relationships that global firms typically prefer not to discuss in public or have investigated by scholars.

No sector is more important within the value chain of global firms than the semiconductor industry. The sector is centrally important within the information technology sector as a whole, which has transformed business and daily life in the modern world. It is critically important not only for commercial firms, but it also for firms in the military sector. Many of key parts of the semiconductor industry have dual use applications in both the commercial and the military sectors. It forms the foundation of new forms of warfare. The relationship between the semiconductor industry in the United States, the Chinese Mainland and Taiwan, is one of most sensitive and important parts of the global business system.

It is hard to imagine a sector in which research is more difficult. Through immense energy and skill, Dr Chu managed to penetrate deep inside the semiconductor industry in the United States, the Chinese Mainland and Taiwan. The result is a remarkable piece of research that sheds deep light on this little-studied but critically important industry. Her book is path-breaking. It is extremely helpful in deepening understanding of the modern era of globalization. It provides inspiration to scholars who are trying to deepen their understanding of global value chains. Her study will be read with great interest not only by scholars, but also by business people and by practitioners in the international relations community. It is hard to imagine a more worthwhile piece of research on the world of globalisation that we all inhabit.

Peter Nolan,
Chonghua Professor of Chinese Development, and
Director, Centre of Development Studies, University of Cambridge

Acknowledgments

Writing a thesis-turned-book on a complex subject has been a great challenge, analogous in many ways to taking the road less travelled described in Robert Frost's famous poem. Hence, I am grateful to all those who have put themselves out to help me complete this difficult journey.

In particular, I would like to thank Peter Nolan for his guidance and encouragement throughout my efforts to complete my PhD studies in international relations at the University of Cambridge. I would like to express my deep gratitude to the following sponsors of my studies: the Secretary of State for Education and Science in the United Kingdom; Cambridge Overseas Trust; Suzy Paine Fund at Cambridge; Taiwanese Ministry of Education; and the Chiang Ching-kuo Foundation for International Scholarly Exchange. I owe very special thanks to all of my interviewees in the United States, Taiwan, and China – most notably George Scalise, William J. Spencer, Klause Wiemer, Daniel Okimoto, Jackson Hu, and Yu Zhongyu. Their generosity in sharing with me their in-depth knowledge about the subject matter has enabled me to complete my work.

I very much appreciate the useful comments on earlier drafts of the book, either in the form of conference and journal papers or stand-alone chapters, by William J. Spencer, Ross J. Anderson, Stephen D. Bryen, Duncan Lennox, T. J. Cheng, Roger Cliff, Scott L. Kastner, Andrew Scobell, Robert Ash, Tse-min Lin, Robert Weatherley, John X. Przybysz, Tricia Gregg, Timothy Rich, Frank Muyard, and Jean-Pierre Cabestan. Khadidjah Mattar did a wonderful job in proofreading the manuscript. I wish to thank my editor, Peter Sowden, and his assistant, Helena Hurd, at Routledge for their patience and encouragement. Special thanks to colleagues at the School of Oriental and African Studies as well as the University of Oxford for their collegial support.

Over the years, Yu-shan Wu, Geoffrey Edwards, and the late Harvey Feldman have been important mentors to me, always willing to give me effective aid in all shapes and forms. I am also grateful for the help and encouragement of numerous friends in Cambridge and beyond in a variety of settings: Elvis Beytullayev, Maria Gaiyabu, Sara Tzinieris, Jim Kennedy, Bill Savadove, Mei Pan, Fion Chen, Kenneth Liu, and Iris Chyi.

As always, I owe most of all to my loving parents, brother and sister for their enduring love and support. During the dark days of my long journey, Guillaume

Bascoul provided me with constant shelter and warmth. While Guillaume was a guardian angel for my solitary intellectual quest on the road less travelled, our adorable twins Sophie and Olivia peppered my journey with joy and comfort in spite of their occasional efforts to sabotage the research monograph project.

Finally, this book is dedicated to my late father, Chin-man Chu, for his enduring love, extreme optimism, and confidence in me.

Ming-chin Monique Chu
London
September 2013

Abbreviations

A/D	analog-to-digital
ADGE	air defense ground environment
AGC	Apollo guidance computer
AMD	Advanced Micro Devices Inc.
APSCC	Asia Pacific Satellite Communications Council
ARM	Advanced RISC Machines
ASC	Actions Semiconductor Corporation
ASE	Advanced Semiconductor Engineering, Inc.
ASIC	application-specific integrated circuit
ASMC	Advanced Semiconductor Manufacturing Corporation
AVIC	Aviation Industry Corporation of China
AWR	Applied Wave Research
BCD	BCD Semiconductor Manufacturing Ltd.
BiCMOS	Bipolar Complementary Metal Oxide Semiconductor
BIDC	Beijing IC Design Center
C4	Command, Control, Communications and Computer Processing
C4ISR	Command, Control, Communications, Computers, Intelligence, Surveillance and Reconnaissance
C&T	Chips & Technologies
CAD	computer-aided design
CAGR	compound annual growth rate
CAS	Chinese Academy of Sciences
CASC	China Aerospace Science and Technology Corporation
CCD	charge coupled device
CCID	China Center for Information Industry Development
CCP	Chinese Communist Party
CEO	Chief Executive Officer
CETC	China Electronics Technology Corporation
CFO	Chief Financial Officer
CIA	Central Intelligence Agency
CIDC	China Integrated Circuit Design Center
CMC	Central Military Commission
CMOS	Complementary Metal Oxide Semiconductor

COCOM	Coordinating Committee for Multilateral Export Controls
COO	Chief Operating Officer
COSTIND	Commission of Science, Technology and Industry for National Defense
COTS	commercial off-the-shelf
CPU	central processing unit
CSIA	China Semiconductor Industry Association
CSIS	Center for Strategic & International Studies
CSIST	Chung-Shan Institute of Science & Technology
CSMC	Central Semiconductor Manufacturing Corporation
CTO	Chief Technology Officer
D/A	digital-to-analog
DARPA	Defense Advanced Research Projects Agency
DGI	Defense Group Inc.
DOD	Department of Defense
DPA	destructive physical analysis
DRAM	Dynamic Random Access Memory
DSB	Defense Science Board
DSCC	Defense Supply Center – Columbus
DSP	digital signal processor
DTICS	Defense Trusted Integrated Circuit Strategy
DTSA	Defense Technology Security Administration
EDA	electronic design automation
EEC	European Economic Community
EMP	electronic magnetic pulse
ENIAC	Electronic Numerical Integrator And Computer
FBI	Federal Bureau of Investigation
FDI	foreign direct investment
FPGA	field-programmable gate array
FSA	Fabless Semiconductor Association
GaA	gallium arsenide
GAD	General Armaments Department
GaN HFET	Gallium Nitride Heterostructure Field-Effect Transistor
GAO	General Accounting Office
GAPT	Global Advanced Packaging Technology Co. Ltd.
GDP	gross domestic product
GLD	General Logistics Department
GSD	General Staff Department
GSMC	Grace Semiconductor Manufacturing Corporation
HBT	hetero-bipolar transistors
HED	Huada Electronic Design
HEMT	high electron mobility transistors
IBM	International Business Machines Corporation
IC	integrated circuit
ICBM	intercontinental ballistic missile

IDA	Institute for Defense Analyses
IDM	integrated device manufacturer
IEDM	International Electron Devices Meeting
IEEE	Institute of Electrical and Electronic Engineers
IMP	Interplanetary Monitoring Platform
InP	indium phosphide
IP	intellectual property
IPO	initial public offering
ISR	Intelligence, Surveillance and Reconnaissance
IT	Information Technology
ITAR	International Traffic in Arms Regulations
ITRI	Industrial Technology Research Institute
ITRS	International Technology Roadmap for Semiconductors
IW	information warfare
JV	joint venture
KYEC	King Yuan Electronics Co.
LSI	large-scale-integration
MAC	Mission Assurance Category
MEMS	micro-electro-mechanical systems
MIL	SPEC military specification
mil-spec	military specification
MIT	Massachusetts Institute of Technology
MNC	multinational corporation
MOEA	Ministry of Economic Affairs
MOS	metal oxide semiconductor
MOTS	modified-off-the-shelf
MOU	memorandum of understanding
MP3	Moving Picture Experts Group Layer-3 Audio
MPW	multi-project wafer
NASA	National Aeronautics and Space Administration
NASDAQ	National Association of Securities Dealers Automated Quotations
NATO	North Atlantic Treaty Organization
NCW	net-centric warfare
NEC	Nippon Electric Company, Ltd.
nm	nanometer
NSA	National Security Agency
NSSI	Ningbo Sinomos Semiconductor Incorporation
NT	New Taiwan (dollar)
NYSE	New York Stock Exchange
OECD	Organisation for Economic Co-operation and Development
PC	personal computer
PLA	People's Liberation Army
PMOS	P-Channel Metal-Oxide-Semiconductor
PRC	People's Republic of China
QML	Qualified Manufacturers List

QPL	Qualified Parts List
R&D	research and development
RCA	Radio Corporation of America
RFMD	RF Micro Devices
RISC	reduced instruction set computing
RMA	revolution in military affairs
RMB	Renminbi
SATS	Semiconductor Assembly and Test Services
SCL	special comprehensive license
SCOSTIND	State COSTIND
SEHK	Stock Exchange of Hong Kong
SEMATECH	Semiconductor Manufacturing Technology
SEMI	Semiconductor Equipment and Materials International
SIA	Semiconductor Industry Association
SiGe	silicon germanium
SIM	Shanghai Institute of Metallurgy
SLBM	submarine-launched ballistic missile
SME	small and medium enterprise
SMIC	Semiconductor Manufacturing International Corporation
SOC	system-on-a-chip
SOE	state-owned enterprise
SOI	silicon-on-insulator
SPIL	Silicon Precision Industries Ltd.
SST	Silicon Storage Technology, Inc.
STM	STMicroelectronics
SWID	Southwest Integrated Circuit Design Co. Ltd.
TAPO	Trusted Access Program Office
TD-SCDMA	Time Division-Synchronous Code Division Multiple Access
TI	Texas Instruments Inc.
TSIA	Taiwan Semiconductor Industry Association
TSMC	Taiwan Semiconductor Manufacturing Corporation
UAV	unmanned aerial vehicle
UMC	United Microelectronics Corporation
US	United States
USA	United States of America
USSR	Union of Soviet Socialist Republics
VAT	value-added tax
VEU	validated end-user
VHSIC	Very High Speed Integrated Circuit
VIA	VIA Technologies Inc.
VLSI	Very Large Scale Integration
VP	vice-president
WSMC	Worldwide Semiconductor Manufacturing Corporation
WTO	World Trade Organization

1 Introduction

Modern nation-states are concerned about the consequences of international economic activities for the distribution of economic gains, and thus over time, as Robert Gilpin (2001: 80) predicts, "the unequal distribution of these gains will inevitably change the international balance of economic and military powers, and will affect national security." These concerns often focus on "the distribution of industrial power, especially in those high-tech industries vitally important to the relative power position of individual states." Hence, the "territorial distribution of industry and of technological capabilities is a matter of great concern for every state and a major issue in international political economy" (ibid.).

This book examines the changing territorial distribution of one such high-tech sector, namely the strategically important semiconductor industry which is crucial to the relative power position of states, and its impact on international security. The term semiconductor refers to a class of materials with electrical properties between those of conductors and those of insulators. Silicon is the most commonly used semiconductor material. The seminal invention of the world's first transistor as a semiconductor amplifier in 1947 marked the advent of the industry and the beginning of modern information technology (IT). Over the past six decades, semiconductors have emerged as key components, enabling electronic systems ranging from high-speed personal computers (PCs) to mobile phones to missiles. In 2009, worldwide semiconductor sales reached US$226.3 billion (Semiconductor Industry Association 2010), and the major products included integrated circuits (ICs), optoelectronics, sensors, and discrete components. The sector has been the "crude oil" of the information age because of its economic and defense significance, a subject explored in Chapter 2.

Transcending disciplinary boundaries between international political economy, security studies and the history of science and technology, this study investigates the globalization of semiconductor production activities and its security repercussions with specific reference to a neglected case; it explores the migration of the Taiwanese semiconductor industry, which is one of the most competitive global players, to the People's Republic of China (PRC), and the geopolitical implications of this migration for the triangular relationship among Taiwan, China, and the United States of America (USA). Specifically, the study intends to answer the following questions: (1) to what extent is the semiconductor industry relevant to national power and security which encompasses economic, technological, and defense security?; (2) why and how has the migration of the Taiwanese semi-conductor industry to China occurred, as measured by cross-border technology transfers, talent, and capital flows?; (3) what are the implications of this production globalization for the triangular security relationship among Taiwan, China, and the USA?; and (4) to what extent does this case study elucidate the impact of globaliza-tion on security?

The central arguments in this book are two-fold. First, the globalization of the industry in the context studied has deepened in recent years, contributing to China's indigenous semiconductor capability, which is central to Chinese power and security. This poses long-term economic, technological, and defense security challenges to the trilateral ties in a complex and contingent way. Second, globali-zation, as will be seen in the case study, affects security by changing its agency and scope, affecting the autonomy and capacity of the state, shifting the balance of power, and altering the nature of conflict. The case study thus serves as a basis for a reappraisal of the impact of globalization on security.

The remainder of this chapter defines the key concepts of globalization and security as used in the book, briefly reviews the existing literature, details the methodology involved, explains the significance of the study, and identifies the analytical framework and the structure of the book.

Definition of key concepts

This study embraces a broadly defined notion of globalization[1] and that of security[2] pertinent to the semiconductor industry as follows. The term globalization, as defined and used in the book, refers to the economic and military aspects of globalization involved in the industry. The former refers to production globalization, instead of trade or finance, through the cross-border activities of multinational corporations (MNCs) and concurrent or ensuing flows of technology, talent, foreign direct investment (FDI)[3] and other forms of capital. Production globalization refers to "the stretching of corporate activity and business networks across the world's major economic regions. In its most visible and institutionalized form, production globali-zation involves the operations of huge MNCs, organizing and managing cross-border business activities "through the ownership of plants, outlets or subsidiaries in different countries" (Held *et al.* 1999: 236–7). Moreover, MNCs outsource production to small and medium enterprises (SMEs) abroad, resulting in the creation of global production networks in which MNCs regularize contractual relationships. In Paul Krugman's phrase, MNCs exploit locational advantages across national borders as they "slice up the supply chain"; they break the production process into many geographically separated steps, adding bits of value in each stage. MNCs engage in cross-border activities including subcontracting, outsourcing, and inter-firm alliances in technological development (Krugman 1995: 332–3). Globalization of production spearheaded by MNCs can also drive the cross-border migration of people, espe-cially highly skilled expatriate managers. Hence, the transnational activities of MNCs have underpinned almost all aspects of the globalization processes, far beyond the mere production story (Salt 1997; Ietto-Gillies 2003: 140–4).

The military globalization of the industry is defined as follows. Because the industry in many countries supplies semiconductor components and technologies to national military end-users, it is part of these countries' defense industrial base. Its production globalization thus becomes a part of military globalization as defined by Held *et al.* (1999). They define military globalization as "the process (and patterns) of military connectedness that transcend the world's major regions as reflected in the spatio-temporal and organizational features of military relations, networks and interactions." Among the three indicators of military globalization they have proposed, the global arms dynamic, particularly the trans-nationalization of the defense industrial base, through which armaments production technologies and military capabilities are diffused globally, is linked to the chip industry. As semiconductor MNCs, which involve national defense production activities in many countries, globalize their production activities, semiconductor technologies and military capabilities are diffused across the borders, potentially affecting the defense security, autonomy, and national defense capabilities of states involved (Vernon 1998: 50–1; Held *et al.* 1999: 89).

This book subscribes to Barry Buzan's (1991: 20) belief that a precise definition of security should be directed towards specific case studies considering the fluid nature of the notion of security and the competing accounts of security in the study of international relations. Namely, "attempts at precise definition are much more suitably directed towards empirical cases where the particular factors in play can be identified." After having identified unique factors at play in the semiconductor industry, I define security directed towards the chip sector to include economic security, technological security and defense security. I have embraced a broadly based notion of security,[4] which some scholars have advocated in view of the increasingly extensive scope and agency of security threats beyond the military and the state because of deepening globalization.

In this study, the notion of economic security[5] is defined through the Realist perspective. The term refers to economic competitiveness and economic independence, and is seen as a direct contribution to the exercise of national power. A state's economic competitiveness refers to the degree to which it produces goods and services that meet the demand of international markets while expanding the incomes of its citizens. Moreover, the state's economic independence equips it with the flexibility to make decisions free from foreign dictates or economic coercion (Romm 1993: 78–80). The recognition of both military and economic means as the legitimate sources of power in international relations is well reflected in E.H. Carr's (2001: 109) words: "Power is indivisible; and the military and economic weapons are merely different instruments of power." The Realist approach to economic security thus emphasizes relative gains, power politics and interstate rivalry (Luttwak 1990; Friedberg 1991; Romm 1993). The concept of economic security as a direct asset to the exercise of national power is based on Borrus and Zysman's (1992: 9) definition of economic security. It refers to a nation's "ability to generate and apply economic resources to the direct exercise of power or to shape indirectly the international system and its norms." According to Nesadurai (2006: 8–12) and Lee (2006), it is appropriate to adopt a geo-strategic Realist approach to economic security to analyze the economic security of Taiwan because the Taiwanese economy is a vital means of empowering the vulnerable state and because any external manipulation by other states often makes Taiwan economically vulnerable. Thus, it is suitable to adopt a Realist approach to the analysis of Taiwan's economic security in the wake of the migration of its semiconductor industry, which is vital to its economy, to China.

The notion of technological security or techno-security refers to "a concept dealing with the perception and enhancement of the technological assets of a nation or a firm," which presumes that technology is an important element in national security (Simon 1997). According to Raymond Vernon (1998: 47–8), technology has been increasingly viewed as a security issue since the shining performance of the US military during the first Gulf War highlighted the significance of the technological edge in warfare. A nation's semiconductor industry indicates the nation's high-technology virtuosity. Its relative decline vis-à-vis that of its adversaries or competitors, which is often expressed in the notion of a shrinking "chip gap," has often triggered security concerns (Hanson 1982: 186–7; Central Intelligence Agency 1983; Friedman and Martin 1988: 106).

The notion of defense security refers to the traditionalist military-centric and state-centric definition of national security, whereby the state's territorial integrity is chiefly maintained through its military capability and alliances, which is seen as "the highest end" in anarchy – the prerequisite for the pursuit of other goals such as profit, power, and tranquility (Waltz 1979: 126). The semiconductor industry links to national defense security because it underpins the military power of a nation by supplying chip components and technologies for information-dependent military systems central to modern battlefield operations. According to Lawrence Freedman (1999), though greater firepower matters, advanced IT and "smarter" weapons – which benefit from semiconductor advances – have become more important than strictly physical military assets in shaping the conduct of modern warfare. Chapter 2 will further analyze the link between the industry and these three dimensions of security.

Literature review

The following strands of existing literature have deepened our understanding of issues pertaining to this study, although little has been studied concerning the case under investigation. They include: (1) studies on the impact of globalization on international security; (2) studies on the impact of globalization on security with reference to the semiconductor industry; (3) studies on the sectoral migration across the Taiwan Strait; and (4) studies on the globalization of production in the semiconductor industry.

First, the literature on globalization–security interconnections, which began to mushroom dramatically at the turn of the century (Defense Science Board 1999; Held *et al.* 1999: 102–4, 138; Allison 2000; Cha 2000; Hughes 2000; Hoffman

2002; Mansfield and Pollins 2003; Rudolph 2003; Brooks 2005; Adamson 2006; Kirshner 2006a; Smith 2006), has partially informed the study. Most recently,[6] Kirshner (2006) and Brooks (2005) respectively have conducted qualitative studies of the impact of globalization on security. Both pieces of work helped me formulate my analytical framework for the case study under investigation. By exploring different issue areas, Kirshner argues that globalization influences security in three major ways: (1) by changing the relative capacity and autonomy of the state vis-à-vis non-state actors; (2) by affecting the balance of power between states; and (3) by altering the nature of conflict. Brooks' study concentrates on the security repercussions of production globalization, which he views as the most critical feature of global commerce that has been left out of the debate on the influence of international commerce on war and peace. He examines how the globalization of production could influence security affairs with what he describes as "an open mind about how powerful an effect it might have." He reaches 25 findings on the globalization–security interconnections, sparking continuous debates on the subject matter (Brooks 2007; Caverley 2007; Gholz 2007; Kirshner 2007).

Brooks' finding on the growing globalization of US defense production and the defense industrial base – including the semiconductor industry – in the last two decades of the Cold War is most relevant to my inquiry of semiconductor production globalization. He argues that the US forwent its going-it-alone policy in defense production in order to enhance quality and reduce costs; such an approach contrasted with the autarkic defense production of the Union of Soviet Socialist Republics (USSR), thereby contributing to the technological gap between the two superpowers. He concludes that the US and Soviet cases reveal that great powers can no longer afford to pursue autarkic defense production. In delineating the globalization of the US defense industrial base, Brooks identifies key defense-related technologies in which globalization production would offer significant advantages to the US military. One concerns semiconductor materials and microelectronic circuits. The Pentagon identified Japan and North Atlantic Treaty Organization (NATO) countries, as US allies, where the offshore production of these military supplies to the US armed forces would benefit Washington. Because Brooks' primary focus is studying the impact of production globalization on security, the semiconductor industry is not a focal point in his book. This feature also characterizes the works of Buzan and Herring (1988: 42–6), Held *et al.* (1999: 89, 138), Kirshner (2006: 21), and Vernon (1998: 47–51) concerning the globalization of the defense industrial base.

Although the existing literature on globalization–security interconnections has partially informed my study, it has not yet facilitated a comprehensive understanding of the impact of globalization on security. For instance, in their survey of why and how cemented economic exchanges influence the outbreak of interstate military conflicts through various case studies, Mansfield and Pollins (2003) conclude that such a relationship is complex and contingent. While some contributors to the edited volume find a strong and significant relationship between open trade and peace, others find that the trade–peace tie is a contingent one. For example, Gelpi and Grieco (2003) provide empirical findings that the relationship is stronger for jointly democratic pairs of states than for other dyads, and that the tie is inverted for pairs of autocracies. The many untested hypotheses raised by the contributors suggest that much academic research remains to be done in order to understand the impact of globalization on security.

The second strand of literature relevant to this study involves empirical studies on the impact of globalization on security, with special reference to the semiconductor industry in distinctive historical contexts. Albeit different in degree, studies of the interplay between the chip industry, economics, and security have been conducted in the contexts of the USSR–US confrontation during the Cold War (Bucy 1980–1; Crawford 1993), the Japanese ascendancy in the global chip race during the late 1980s and the early 1990s (Okimoto *et al.* 1987; Moran 1990; Kanz 1991; Flamm and Reiss 1993; Flamm 1996), and the current and ongoing shifts of US and Taiwanese semiconductor production to China (General Accounting Office 2002; Howell *et al.* 2003; Lieberman 2003; Yang and Hung 2003; Cheng 2005; Defense Science Board Task Force 2005).

In this strand of literature, debates over the security impact of semiconductor globalization center on three themes. The first concerns the redefinition of traditionalist national security in terms of the agency and scope of security. The agency of cross-border chip technological transfers is predominantly non-state actors, namely MNCs. The scope of security involved goes beyond conventional military security to embrace economic (Flamm 1985; Brown and Linden 2005), technological and defense security. The second theme concerns the erosion of the autonomy and capability of the state because of the globalization forces at play, which include the increasing globalization of chip production activities and the related defense industrial base led by MNCs, and the ease of technological transfer via media and Internet. Hence, states become vulnerable because of their dependency on the offshore supply of chips to their military and intelligence systems. The third theme involves the balance of power, namely, the extent to which the globalization of the industry changes a nation's relative capability vis-à-vis that of its adversaries. These themes helped me formulate my analytical framework for the case study in question, though they did not dictate my research findings. After all, I began my project with an open mind about whether and how related production globalization forces might influence the trilateral security relations in question, with the aim of analyzing the case based on empirical evidence gathered from the field.

A major gap in this strand of literature, nevertheless, concerns a systematic analysis of the security implications of the Taiwanese migration across the Strait, although some scholars have identified economic security issues that concern actors in Taiwan's domestic politics because of the migration (Yang and Hung 2003; Cheng 2005). Although the US Defense Science Board (DSB) in 2005 published an important report examining the security risks Washington faces because of the continuous shift of chip manufacturing to China, the study remains US-centric. It has little, and even erroneous, reference to the Taiwanese dimension. Nor does it analyze China's domestic factors in terms of policies, motivations and capabilities, which I consider essential to the assessment of potential security risks the US may face because of semiconductor globalization. Like many studies of the economic aspect of the chip production shift across the Strait, which forms the third strand of literature discussed below, the DSB study is inadequate in that it is not based on first-hand field research. I consider first-hand research essential to the understanding of a fast-moving high-tech industry because related information concerning the industry dynamics is either lacking or incomplete in secondary materials. Worse yet, the DSB study's argument on the pertinent security risks is based on unexamined Realist vulnerability assumptions, which have not been adequately assessed against empirical evidence in any existing literature.

The third strand of literature concerns the economic dimension of the sectoral migration across the Strait. These studies (Chung 1997; Howell *et al.* 2003; Chase *et al.* 2004; Naughton 2004; Cheng 2005; Fuller 2008) have shed light on the industry dynamics in question. However, their primary focus is on the economic factors at play, and some of these studies are marred by a lack of substantial first-hand research. The fourth strand of literature refers to the study of globalized production in the semiconductor industry (Flamm 1985; Henderson 1989; Leachman and Leachman 2004; Brown and Linden 2005; Ernst 2005; United Nations Conference on Trade and Development 2005: 173–7; Hung *et al.* 2006; Dicken 2007: 317–45; Semiconductor Industry Association 2009), a sub-field of the study of the internationalization of production in the larger field of international political economy (Vernon 1966; 1979; Dunning 1988; Cantwell 1995; Ietto-Gillies 2003; Dicken 2007). This branch of literature provides an analytical framework for the case study in question, especially the economics of the semiconductor industry and the driving forces behind semiconductor globalization. In particular, some studies (Flamm 1985; Henderson 1989; Brown and Linden 2005; Ernst 2005) have elucidated on the extent of production globalization in the different production stages of the semiconductor supply chain in a diverse array of historical contexts. None of them, however, focuses on the Taiwanese migration to China. Although the aforementioned literature has partially informed my study, a major gap in the literature exists concerning a systemic and updated analysis of the Taiwanese migration to China and its security implications. This study attempts to fill such a void.

Methodology

After establishing a gap in the existing literature which needs to be investigated, how do I define the method of investigation? This theme will be explored in the following section. I will explain the methodological choice for the research undertaken, and specify strategies to meet various methodological challenges.

Methodological choice

A qualitative single case study approach was adopted, supplemented by quantitative data for this study because it is the most suitable for the research undertaken in view of topic sensitivity, and the paucity of existing knowledge about the chosen area of study.

First, the sensitive aspects of my research make qualitative methods more appropriate than large-*n* type quantitative ones. The sensitive nature of the subject matter relates to illegal Taiwanese semiconductor operations in China, the security ramifications of the migration, and the link between the Chinese military and defense microelectronics. My attempt to examine the complicated issues pertaining to the aforementioned sensitive topics would presumably invite discursive and complex answers, making qualitative interview methods a suitable methodology for data collection. Quantitative survey methods were rejected because it would be unlikely that issuing surveys to Taiwanese firms that had illegally moved their production activities to China would entice these respondents to disclose such sensitive information to outsiders. Besides, issuing on-site surveys to experts in China asking questions about China's defense security by a visiting researcher could be considered sensitive by the authorities to the point that this method would be unfeasible. Even if surveys were issued, to study such a sensitive topic through surveys could easily lead to a low response rate, or returned questionnaires with simplistic or even fabricated answers.

Second, the paucity of existing academic literature on the subject matter, as reviewed earlier, has further driven me to use qualitative case study methods. As Gerring (2004: 345) argues, "Case studies often tackle subjects about which little is previously known or about which existing knowledge is fundamentally flawed." As discussed earlier, existing literature on the Taiwanese migration and its security impact is rather flawed. Some have based their analysis of the migration on inadequate secondary materials instead of primary data. Others have based their study on imbalanced data by depending on materials gathered from a single rather than multiple geographical locations. Still others have examined one dimension of the topic. Given the complexity involved in the understudied multidisciplinary research, the utilization of the existing literature alone cannot enable a comprehensive study of the topic. In view of the aforementioned two factors, a qualitative single case study approach has been adopted, supplemented by the introduction of relevant quantitative data, in this research.

The merit of the book as a single case study of the globalization–security link is as follows.[7] In general, a single case study allows researchers to focus on marshalling facts in order to offer a holistic description of the complex case in question and to retain the meaningful characteristics of real-life events (Yin 1994: 2). Through fact-finding and valid description, a case study can help grasp the background, process and influence of the whole issue under discussion (Odell 2001). The use of a single case study is suitable in this instance because what little previous knowledge there is about the subject is problematic. Because in-depth case studies are essential for description, they are important to social science. One cannot explain what has not been described with a reasonable degree of precision. A careful description that focuses on important events is better than a bad explanation of anything (King *et al.* 1994: 44–5). In particular, a sector-based case study of the impact of globalization on security contributes to the international relations literature because, as discussed earlier, relevant studies of the globalization–security linkage have not facilitated a comprehensive understanding of the subject matter.

My aforementioned methodological choice has further guided the selection of appropriate research methods for the project, including in-depth interviews, the collection of primary and secondary materials, and on-site observations. Because results from these interviews are combined with findings from a vast variety of untapped sources to form the basis of the analysis in the book, it is important to briefly summarize the nature of these materials.

More than 160 rounds of elite interviews with officials and industry insiders from 65 semiconductor firms[8] were conducted in the USA, Taiwan, China and England between 2004 and 2005 as well as in 2009. After screening these interviews by weighing their relative significance to this research, I selected 143 interviews to form the basis of a statistical overview of the interview data (Figure 1.1).[9]

More importantly, many of my interviewees were influential players in their chosen profession (Table 1.1). My key interviewees in the US government included Pentagon officials in charge of defense technology export controls. Prominent interviewees in the US semiconductor industry included presidents and chief executive officers (CEOs) of various representative agencies, such as the Semiconductor Industry Association (SIA). I also interviewed senior executives at International Business Machines Corporation (IBM), Texas Instruments Inc. (TI), Advanced Micro Devices Inc. (AMD), among others. In Asia, senior industry executive input was gained through my interviews with top executives from, for instance, seven of the top eight chip-makers in China and Taiwan.[10] My interviewees also comprised market analysts and defense industry players with first-hand experience in defense microelectronics.

Primary and secondary materials include primary and secondary sources in mandarin Chinese and English, many of which are untapped in the existing academic studies. Primary mandarin Chinese sources include official statistics, government reports and yearbooks, internal circulars also known as *neibu* documents, corporate

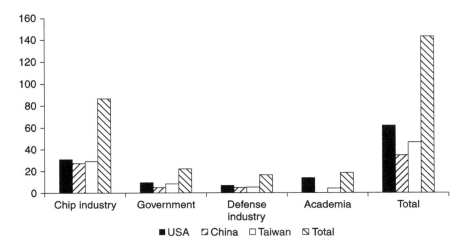

Figure 1.1 A profile of selected interviews.

Table 1.1 Representative high-ranking interviewees

Name	Affiliation	Name	Affiliation
Scalise	President, SIA	Tseng	Vice-Chairman, TSMC (largest foundry in the world)
Polcari	President and CEO, SEMATECH	Hu	CEO, UMC (second largest foundry in the world)
Myers	President and CEO, SEMI	Chang	CEO, SMIC (third largest foundry in the world; China's No. 1 foundry)
Donofrio	Senior VP, IBM		
Rhine	CEO, Mentor Graphics (third largest EDA company in the world)	Shyu	President, HeJian
		Wang	Senior VP, Applied Materials (largest semiconductor equipment company in the world)
Bryen	President, Finmeccanica North America, Inc.; founder of DTSA,[1] DOD, US	Zou	Chairman, GSMC
		Yu	President, CSIA
		Lin	Director General, Ministry of National Defense, Taiwan
Minnifield	DTSA official, DOD, US		
Borman	Deputy Assistant Secretary of Commerce, US		

Note: 1 DTSA stands for Defense Technology Security Administration, the Pentagon agency that reviews applications for the export of high technology to foreign countries. Bryen founded DTSA in his capacity as Deputy Under-Secretary of Defense. At the time of the interview, he was also a commissioner on the US–China Economic and Security Review Commission.

reports, those from the websites of firms and government-run defense agencies. Secondary mandarin Chinese sources include books, journal articles in the areas of business, industry, defense, science, and technology, and newspapers. English sources include journals in the areas of international trade, business, industry, defense, science, and technology, corporate annual reports, corporate websites, investment bank reports, and the statistical data and economic indicators compiled by major international organizations, notably the World Trade Organization (WTO).

Methodological challenges and solutions

Over the course of the discursive research journey, I have developed the following strategies in order to meet four major methodological challenges, which involved interview access, interview process, data reliability and data representativeness.

Access

Topic sensitivity and unequal power relationships inhibited interview access. Topic sensitivity hindered entry on numerous occasions. The fact that I intended to examine the strategic implications of the migration alerted many prospective interviewees. Several firms refused to be interviewed on the ground that any disclosure of their illegal China operations to outsiders could potentially get them into trouble. Semiconductor Manufacturing Technology's (SEMATECH) communications chief declined my request to interview the institution's CEO on the ground that her agency had nothing to do with "security" or "China." In reality, the consortium had been established with financial support from the Pentagon in view of potential security threats the US would face because of Japanese ascendancy in the global chip sector. Topic sensitivity also drove some of my prospective interviewees to suggest I change my research topic. Worse yet, unequal power relationships between me, then a doctoral student, and targeted elite interviewees in powerful institutions further inhibited entry in numerous cases.

To overcome obstacles to access, I deployed several strategies. They included: (1) "snowballing" techniques; (2) "guerrilla warfare" tactics; (3) continuous negotiations with gatekeepers; (4) direct negotiations with senior executives at various agencies bypassing gatekeepers; (5) the choice between a general or specific description of my project when requesting an interview; and (6) agreed compromises.

The first strategy was snowballing, which means that the researcher "begins from an initial set of contacts before being passed on by them to others, who in turn refer others, and so on" (Lee 1993: 65). Although snowballing is not without critics, it was the most useful technique to overcome entry obstacles. Some heavyweight intermediaries in the industry introduced me to their counterparts in the circle to help me gain access. The strategy was useful in this research because of the prior existence of a relatively organized professional referral network among a majority of my targets, especially those in the global chip industry. Such a network was established through partnerships at inter-agency or inter-individual levels in the global supply chain. Although not every referral worked, the use of the technique enabled me to interview numerous key players that I would have otherwise stood slim chance of gaining access to under normal circumstances. For instance, intermediaries as social insurance guarantors helped me complete some of the most important interviews I conducted in China. The head of the US-based SIA introduced me to his Chinese counterpart to ensure entry. I interviewed a Taiwanese senior executive who was involved in a legal dispute with Taipei because of his firm's operations in China, and access to him was made possible through a Taiwanese intermediary who was connected to the executive.

The second solution referred to what I described as a "guerrilla warfare" technique that I used on the sidelines of international semiconductor conferences.[11] I used this improvised technique, which I learnt along the way, to approach identified participants in these conferences in order to gain entry. At these venues, senior executives, research and development (R&D) chiefs, or mid-level managers in semiconductor and defense agencies would be largely free from the "protection" of their agencies' communications chiefs as gatekeepers. Hence, invaluable input was gained through interviews conducted on the sidelines of these conferences. They included some of the most unexpected respondents, such as Chinese and American defense industry players. The third strategy involved continuous negotiations with gatekeepers. Examples abound. To interview a leading American semiconductor expert at one of the world's top-notch chip companies, I re-negotiated with the company's communications chief numerous times regarding entry. Both sides made compromises, and agreed upon terms and conditions for the proposed interview, which eventually took place in Dallas.

The fourth strategy was to bypass gatekeepers, after they initially denied entry, in order to bargain directly with top management of the organizations by email over terms of entry. My interview with Pentagon officials in charge of defense technology export controls was made possible after I requested a US Deputy Under-secretary of Defense for an interview by email after being denied entry by her secretary because I was not a US national. The same strategy also facilitated my interviews with at least two important American CEOs. In 2005, the president of a Taiwanese chip firm considered the strategy successful. The job of a responsible gatekeeper in a corporation, as he put it, would be to "protect" the CEO from any unnecessary outside exposure, which would normally include the acceptance of a research interview on sensitive issues.

The fifth method involved deciding between a detailed or general description of my project when requesting entry. According to Lee (1993: 103) and Brannen (1988: 553), a researcher should decide whether or not the topic of the interview should be described in detail at the outset. They contend that a preferable way is to allow the issue to emerge in the interview. In my case, a general or even partial description of the project was more effective in gaining entry than a detailed one, given the sensitivity of the topic.

The sixth method involved compromises. In some cases, I made compromises in order to gain access after deciding that the benefits of the planned interviews would outweigh the costs of the compromises on my part. The compromises I made were numerous. They included my acceptance of prospective interviewees' suggested restrictions on the methodology used during interviews, such as tape recording. They also included permitting prospective interviewees to read interview questions in advance, and allowing them to confine the scope of the interviews. In some instances, I had to accept prospective interviewees' suggestion that they could check interview transcripts once they became available, or I had to ensure interviewees that their names would not appear in any published work as a result of my research.

In sum, my use of the aforementioned strategies enabled me to complete a relatively large number of elite interviews, which form the valuable backbone of the book. However, granted entry would not automatically ensure success in any interview; it had to be combined with the use of meticulous interview strategies.

The interview

Major challenges to my interviews were three-fold: (1) how to ask sensitive questions; (2) how to win the trust of any interviewee; and (3) how to tackle the issue of objectivity and bias. To entice interviewees to answer sensitive questions, I deployed the following techniques. The first technique involved the sequence of questions. Questions could be arranged from the general to the specific, or vice versa. They could also be sequenced from the sensitive to the least sensitive, or vice versa. Time constraints and the nature of the interviewees in question would affect the way I arranged the order of questions. Often such a decision was made prior to an interview, but unexpected circumstances sometimes forced me to change the order of questions on the spot with the most care possible to maximize my gains from the interview. The second useful technique involved framing sensitive questions in a relatively broad way, namely, to package the core of sensitivity with a general blanket. For instance, if the word "security" might be seen as taboo by interviewees in China, why not ask them to comment instead on the "importance" of the chip industry to the country's "development?" The third technique was to first mention a third party's controversial comments on sensitive issues and to ask the interviewee to respond to these remarks. This helped me to avoid presenting myself as a hostile interviewer in the eyes of the interviewee. It also enabled me to maintain a certain level of trust established earlier, which in turn would entice the interviewee to respond to sensitive questions.

To establish trust between the interviewer and the interviewee to encourage the latter to disclose general or sensitive information, I found the use of mature communication skills helpful. Eye contact to show sincerity, a demonstrated curiosity about what the interviewee had said earlier, and expressed respect for the person in front of me – all did help. A framework of trust could also be provided by my guarantee of anonymity when using the interview materials and my demonstrated non-judgmental attitude towards the respondent's revelation of his or her law-breaking behaviors.

As for attempts to tackle the issue of objectivity and bias, often raised by critics of qualitative research in general and the interview as a relevant method in particular, part of the answer is to admit that absolute objectivity is unattainable in any research interview, whereas relative objectivity is. As Devine (2002: 205–6) argues, "the relationship between an interviewer and interviewee is not aloof . . . since the interviewer participates in the conversation." More importantly, the inter-personal tie in question should not become "distant" if the interviewer intends to gather confidential information from the interviewee. Besides, it was natural that a certain amount of subjectivity would creep into the interview process from the side of the researcher and/or that of the respondent. For instance, US military strategists, Chinese government officials and Taiwanese businessmen as the project's respondents may be biased towards their cause. This is a potential pitfall that the researcher has to acknowledge. I would stress, however, that one's role as a relatively objective researcher is still attainable and even desirable in ensuring the success of any interview. I tried to reach this objective by allowing the respondent to play the key role over the course of any interview. Often I acted as a conversation facilitator instead of a dominant and subjective interrogator. I would become "invisible" by remaining silent in order to allow the respondent to continue his or her monologue if I considered such a strategy to be conducive to the success of the interview. Although at times I engaged the respondent in a seemingly subjective way by elaborating on my views on certain issues, I did so to get the respondent to resume his or her pivotal role during the interview.

Data reliability

Data reliability served as the third methodological challenge for this research. A common pitfall in any qualitative research results from the possibility that information gathered – through either interviews or document analysis – is unreliable, and mine is no exception.

The triangulation of different types of data, which forms the basis of the empirical analysis that follows, was the optimum strategy I used to cope with the challenge of data reliability.

> [Triangulation] involves data collected at different places, sources, times, levels of analysis, or perspectives, data that might be quantitative, or might involve intensive interviews or thick historical description. The best method should be chosen for each data source. But more data are better. Triangulation, then, refers to the practice of increasing the amount of information brought to bear on a theory or hypothesis.
>
> (King *et al.* 2004: 192)

The verification of interview materials, especially on past events, is an integral part of good research because people normally choose to remember what they want to remember (Burnham *et al.* 2004: 37). In this research, I discovered that some interviewees tended to overstate their importance to their organizations. In one instance, a former CEO of a Chinese semiconductor company recalled what he described as his contribution to the firm's market ascendancy with a flamboyant gesture. I then cross-checked his claim by consulting related materials, which contradicted his showy remark. Data triangulation thus enabled me to make a balanced assessment of the case in question.

Triangulation sometimes took place within an interview. This could be done by asking follow-up questions or requiring the interviewee to clarify unreliable earlier statements. Repeated interviews with the same target also enabled me to cross-reference certain information. Sometimes a one-time interview was not fruitful because I could not ask follow-up questions for data clarification. A follow-up interview enabled me to remedy the above shortcomings. Moreover, a second encounter with the same respondent would often generate a fuller and deeper account if a frame of trust had been established during the first interview. My interviews with two experts from the US defense industry illustrated the utility of repeated interviews. The first interview was superficial, whereas the second was in-depth especially their analysis of semiconductor–defense link.

In short, triangulation enables a good researcher to bring diverse kinds of evidence to bear on a problem, and to balance the strengths and weaknesses of each type of datum used. The process requires the researcher's careful and open-minded evaluation of all of the available data, no matter however discrepant, inconsistent or contradictory they are. It also requires a researcher to avoid the risk of favoring only materials that support a pre-existing or preferred interpretation (George and Bennett 2004: 99). It was a key strategy that I used to meet the challenge of data reliability.

Data representativeness and the issue of generalizability

The fourth methodological challenge concerned the question of whether the data were representative and the issue of generalizability. As Lee (1993: 60) points out, "The aim of sampling is to select elements for study in a way which adequately represents a population of interest ... both in relation to the purpose of the research, and at reasonable cost." Neither of these aims of sampling (i.e. cost-effectiveness and representativeness) can be achieved easily when one researches a sensitive topic using qualitative methods.

In my case, both aims could have proved difficult to attain. It was costly to travel extensively in order to conduct interviews to improve data representativeness. Besides, the sampling of the Taiwanese chip firms which had illegally spearheaded the migration and China-based companies which had supplied the Chinese military was challenging. The sampling of the American giant companies was also expected to be tough because, as an honest industry player cautioned, these firms would avoid any candid discussions on China, as few in the US would know how to deal with the country.

Despite these caveats, I used the following tools to cope with the challenge of data representativeness and eventually to present a relatively satisfying and sufficiently representative sample. Apart from traveling extensively to conduct interviews as a way to improve data representativeness, I used the methods of list sampling, network sampling (i.e. snowballing), and relational outcropping (i.e. "guerrilla warfare" tactics) to generate my sample in the most systematic way possible.

The first step was, by using related revenue rankings, to create a list sampling of the top 10, 20, or even 30 chip firms in various sub-sectors of the chip industry in the USA, Taiwan, and China. I also added national semiconductor industry associations, major government agencies and academics in the three regions to the list. Although negotiations over access to these samples were not completely successful and the sample obtained in this way was not entirely representative especially in the industry sector, it served as a basis for the implementation of the next two strategies.

As discussed earlier, snowballing worked from the outset. My initial interlocutors in the US industry forwarded my interview request to their colleagues in the sector, and their introductions served as what Lee (ibid.: 124) dubs "a form of social insurance which spreads the risk of being studied." Such a strategy was further justified by the fact that the samples thus generated turned out to be quite diverse and that interviewees introduced by my intermediaries did not necessarily offer identical views on the same questions I asked. This overturned the conventional reservation on the utility of the technique, which claimed that snowballing would tend to produce relatively homogeneous samples.

As Lee (ibid.: 68) suggests, a researcher should "maximize sample variability and the theoretical utility of snowball sampling." I made three such efforts to implement the technique. I first used as wide a variety of starting point contacts as possible in order to cover the study population as extensively as I could. The growing referral chains of contacts were regularly monitored and examined. I frequently assessed whether the increasing chains of contacts could help me reach certain types of respondents. Background checks on those inside and outside the referral chains sometimes turned out to be useful. Finally, I used the initial list as a reference guide and asked my growing number of intermediaries to help me gain access to agencies on the list where entrée had not yet been granted. Consequently, I covered an extensive part of the list which I originally produced for sampling purposes.

The final strategy, "relational outcropping" (ibid.: 69), refers to my efforts to approach participants at major industry gatherings. These conferences attracted the attendance of a fairly representative sample of the industry. To improve data representativeness, I tried to interview people from what could be a sufficiently representative pool of the industry population at these gatherings. For instance, members of the Chinese defense chip industrial base became my samples when approached on the sidelines of a major industry event in Shanghai, and they even revealed their business with the Chinese military. My goal to seek variation in my sample was thus realized.

In short, my interviewees were not chosen randomly and few, if any, statistical generalizations can be made on the basis of my findings. The resultant interview pool is far from a completely representative sample. In fact, my concern was not to have a randomly derived sample; I was concerned with variation. Mine is a purposive sample intended to gain as much information as possible about the migration and its security repercussions. The result was that the interviewees accumulated are, at best, sufficiently representative of the study population. Because the sample covers high-ranking heavyweights in various relevant sectors, the strength of the claims that can be made about the data should not be undervalued. Although I have to be tentative about extrapolating too much from my interview data to the population at large, the value of these data is not diminished considering the severe challenges that I faced throughout the research.

Significance of the study

The significance of this study is two-fold. First, this research fills a gap in the existing literature concerning the security repercussions of the fusion of the strategically significant sector which unfolds against the backdrop of the uneasy trilateral relationships among the USA, China, and Taiwan, as depicted below.

The Taiwan Strait has remained a flashpoint in world politics since the 1950s. The conflict between Taiwan and China began when Chiang Kai-shek's National Party, defeated by Mao Zedong's People's Liberation Army (PLA), at the end of the Chinese civil war on the Chinese mainland, retreated to Taiwan at the end of 1949. From 1949 until the 1980s, the two sides engaged in a bitter military and ideological struggle that precluded any type of cooperation. Whereas Chiang aspired to retake the mainland by force, Mao sought to "liberate" Taiwan. The USA was caught in the middle. The heightened military confrontations between the two sides in the 1950s even prompted Washington to once consider using nuclear weapons to end the conflict (Bush 2005: 19; Jackson and Towle 2006: 55). In 1995, Washington granted Lee Teng-hui, the then president of Taiwan, a visa to facilitate his visit to his alma mater, Cornell University, resulting in the 1995–96 missile crises. Beijing engaged in various displays of military power such as the launch of ballistic missiles at targets outside Taiwan's ports. In turn, Washington dispatched two aircraft carrier battle groups to skirt the Strait in order to deter Chinese aggression (Garver 1997; Ross 2000). Those episodes demonstrate that the China–Taiwan dispute might have erupted into war, potentially escalating into a devastating armed conflict involving the world's No. 1 military clout and other regional powers.

Today, the US and Taiwan maintain what some US scholars, such as Alastair Iain Johnston (2003), dub a "quasi-military" alliance relationship. This is in spite of the fact that the bilateral ties have been informal since Washington switched its diplomatic recognition from Taipei to Beijing in 1979. Nevertheless, the Taiwan Relations Act authorizes Washington's arms sales to Taipei to ensure that the island's military has a "sufficient self-defense capability." It also states that it is US policy to consider "any effort to determine the future of Taiwan by other than peaceful means, including by boycotts and embargoes, as a threat to the peace and security of the Western Pacific area and of grave concern to the United States" (cited in Bush 2005: 22).

As for Sino-US relations, the geopolitical links between the two countries make them far from allies, especially in security issue areas, partly due to the widely perceived role of the US as a guarantor of Taiwan's security. Moreover, China's rise within a relative short span of time, from its extraordinary economic growth to its expanding military muscle and political influence in world affairs, has prompted some in the US to perceive Beijing as Washington's "strategic competitor" (Nolan 2004: 34). As debates over the rise of China and its impact on the US hegemony continue, scholars from the Realist camp, contrasting with the engagement and liberal school of thought, often regard Beijing as the strategic challenge of the future for Washington. They further predict that Sino-US confrontations are inevitable and that Washington and its allies should contain China by forging alliances to hedge China's ascendancy (Mearsheimer 2001: 4, 362).

Additionally, the unsolved sovereignty dispute and security tensions between China and Taiwan continue to complicate the trilateral ties. Until the inauguration of Ma Ying-jeou as Taiwan's president in May 2008, which results in the obvious thaw in cross-strait relations, Taipei and Beijing over the past 15 years have pursued hedging and deterrence policies which have fed the insecurity spiral. Taiwan has been threatened by China's military capability, the growth of which continues unabated and is closely watched by the Pentagon (Department of

Defense 2004; 2005; 2006; 2007; 2008; 2009; 2010). Beijing has feared that any pro-independence move by Taipei will hurt its national interests. Paradoxically, the Taiwanese and Chinese economies are becoming increasingly integrated (Bush 2005: 28–35; Cheng 2005; Steinfeld 2005; Kastner 2009).

The migration of the Taiwanese semiconductor industry to China contributes to the increasing economic ties between the two sides, which occurs in the afore-mentioned sensitive geopolitical context in East Asia today. The importance of the migration has been accentuated by the fact that today Taiwan is one of the world's semiconductor powerhouses. It is second only to the US in highly skilled IC design, which is central to future weaponry systems, and is the global leader in pure-play foundry operations.[12] Given the strategic significance of the industry to national power and security, to what extent does this semiconductor capability shift from a technologically strong Taiwan to a technologically weak China poten-tially influence inter-state security relations, culminating in the East Asian computer chip war, as the title of the book suggests? This question may only be properly addressed through rigorous interdisciplinary empirical research. The importance of this study thus becomes clear.

Second, the sectoral approach this study adopts towards the examination of the impact of globalization on security may shed light on the globalization–security nexus, which is still not well understood. A sector-based analysis of the intercon-nections between globalization and security focusing on the semiconductor sector is novel, compared to the existing studies of globalization–security interconnec-tions, and such an approach is worth pursuing for two reasons. First, this approach allows the researcher to identify the details of the different types of interactions in the sector in question because it "confines the scope of inquiry to more manage-able proportions by reducing the number of variables in play." To justify their sectoral logic of security, Buzan *et al.* (1998: 8, 127) argue "security means survival in the face of existential threats, but what constitutes an existential threat is not the same across different sectors."[13] Thus, this approach allows us to analyze the main players closely in the given sector, the reasons they lead the sectoral globalization, and why the globalization force, which is often spearheaded by non-state actors, may, under certain conditions, engender unique threats and vulnerabilities in respect to national survival for the states involved and alter inter-state security ties.

Besides, the study of a fraction of the contemporary globalization processes is favored by some scholars (Held *et al.* 1999; Dicken 2003), because such an approach helps to facilitate a vigorous analysis of the specific phenomena in ques-tion. After all, changes in the various dimensions of globalization do not neces-sarily take place concurrently. Moreover, the multi-dimensional facets and complex outcomes of globalization processes are often difficult to comprehend fully if analyzed in a lump sum fashion. Held *et al.* argue that a sectoral approach to the study of production globalization, rather than measures of FDI flows, offers us the only reliable clues for understanding the magnitude and geographical scale of contemporary patterns of global production. Sectoral studies of production globalization are not rare. Dicken (ibid.: 5, 315–506) recognizes the advantage of slicing up globalization into smaller pieces when he conducts sectoral studies of the globalization of production in order to grasp specific factors at play. Brooks (2005: 28–30) also highlights sectoral differences when he argues "MNC production is *not dispersed equally across all industries*." The geographical dispersion of MNC production appears to be most salient in sectors of manufacturing characterized by "high levels of research and development and significant firm-level economies of scale." These sectors include electronic components and computers (Dunning 1988: 344–6; World Bank 1997: 42).

Considering the merits of a sectoral approach to the study of security and that of globalization, it follows that similar advantages can be gained from a sectoral approach to the study of the impact of globalization on security. If one defines security as survival in the face of existential threats, assuming that what constitutes an existential threat differs across sectors, and if one confines the scope of inquiry to more manageable portions of the semiconductor industry, this approach may help us understand the nature of threat and the means of survival which are embedded in the sector. This would be a preliminary step before dissecting the ensuing spill-over effects, if any, on inter-state security relations. Moreover, as the semiconductor industry is arguably strategic and is one of the most globalized sectors today, the study of the security impact of semiconductor globalization is an important case study of the impact of globalization on security.

Analytical framework and plan of the book

The analytical framework of the study contains a three-level analysis which derives from the empirical research findings (Figure 1.2). The first step is to examine the extent to which the semiconductor industry is related to national power and security, first at an aggregate level and then in the context of China. The second level of analysis focuses on the production globalization of the semiconductor industry in general and that in the single case study in particular. The third level of analysis concentrates on the security implications of the globalization force in the context studied. This security analysis will first proceed along the economic, technological, and defense dimensions of security. It will then touch upon the four aspects of security, as identified in Figure 1.2, in the concluding part of the book, where I explore globalization–security interconnections based on the results of the case study.

The aforementioned analytical framework thus dictates the structure of the book. Following the introductory chapter, the remainder of the book is divided into six chapters. Chapter 2 analyzes the relevance of the semiconductor industry to national power and security. It traces the major technological evolution of the industry, and examines the link between the industry and economic power and security, and that between the sector and defense power and security. Chapter 3 is devoted to outlining key features of the globalization of semiconductor production; it specifically examines how and why defense and commercial semiconductor production activities have become globalized over time. In the final portion of the chapter, I argue that the web of the production of semiconductor production

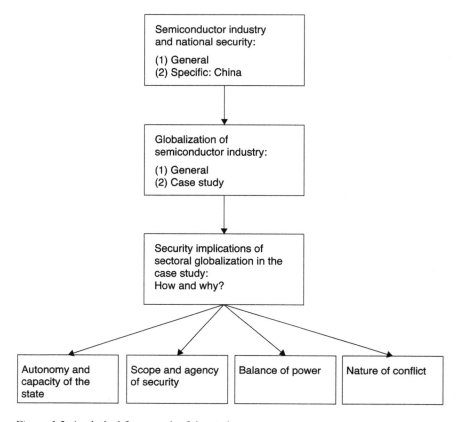

Figure 1.2 Analytical framework of the study.

is complex. While some of the manufacturing operations in the supply chain have become deeply globalized, others have remained homebound.

Chapter 4 shows the link between semiconductors and security in the context of the PRC. The chapter begins by identifying national security motivations behind China's resolve to upgrade its semiconductor capability from the Chinese standpoint. The second portion of the chapter analyzes the evolution of Chinese defense chip industrial base, arguing that such an indigenous industrial base has been an asset as well as a liability for China's national security. The third portion of the chapter is devoted to outlining the evolution of Chinese commercial chip industrial base amid the growing trend of sectoral globalization and its mixed national security implications.

Chapter 5 summarizes my empirical findings on the understudied process of sectoral migration across the Strait. It first explores the direction, scope, and causes of the migration, and then presents firm-level case studies in the IC design and fabrication sub-sectors to deepen our understanding of the migration. The chapter concludes by highlighting the impact of the migration on the autonomy and capacity of the Taiwanese state – the first aspect of security impact of the migration – and on the development of the Chinese semiconductor industry.

Chapter 6 analyzes other security implications of the migration. It examines the extent to which the globalization forces in question have threatened or may threaten the states involved. It first explores the impact of the migration in economic security terms. What follows is an examination of three aspects of the technological and defense security repercussions of the geographical shifts. They include: (1) the Chinese chip industrial base, the PLA modernization, and the balance of power; (2) technological risks resulting from the narrowing chip technology gap and China's semiconductor-targeted information warfare (IW) attacks; and (3) foreign dependency vulnerabilities. Chapter 7 summarizes the major findings of the research, examines the impact of globalization on security based on the case study in question, and identifies contributions and limitations of the study.

After identifying my research puzzle, establishing a gap in the existing literature, and defining my method of investigation, the next step is to examine the extent to which the semiconductor industry is related to national power and security. This theme will be explored in Chapter 2.

Notes

1 Many scholars have viewed the concept of globalization as controversial. See Held *et al.* 1999: 1; Hirst and Thompson 1999: 17; Perraton 2003. Some have embraced a broad definition of globalization. See, for example, Held *et al.* 1999; Keohane and Nye 2000: 16–17.

2 Some scholars have viewed the notion of security as ambiguous. See, for example, Wolfers 1952; Buzan 1991; Baldwin 1997: 10–12; Smith 2006: 33–55. Realism, Liberalism, and Constructivism have defined the notion of national security differently. See Lippman 1943: 51; Waltz 1979; Keohane and Nye 1987: 731–3; Luttwak 1990; Nye 1990; Wendt 1992; Romm 1993: 109; Cable 1995; Tickner 1995: 177; Krause and Williams 1996: 243; Baldwin 1997: 21–2; Buzan *et al.* 1998; Ruggie 1998; Walt 1998; Cha 2000: 126; Carr 2001; Booth 2005: 14–15; Mearsheimer 2005: 146; Smith 2005: 55–8; Nesadurai 2006.

3 FDI refers to "ownership of or investment in overseas enterprises in which the investor plays a direct managerial role." See Held *et al.* 1999: 191.

4 Followers of such an approach have been described as "wideners." They broaden the security agenda by claiming security status for issues and referent objects in the economic, technological, energy, and environmental sectors. See Nye and Lynn-Jones 1988; Tickner 1995; Buzan *et al.* 1998; Vernon 1998: 46–53; Klare 2004; Zweig and Bi 2005; Chu 2008: 54–5.

5 The economic elements of national security became underlined after the oil shocks of the 1970s. The end of the Cold War and the accelerated pace of globalization further intensified related debates. See Tetsuya 1988; Huntington 1993; Romm 1993; Tickner 1995; Green 1996; Kahler 2004: 485–6.

6 For earlier works, see Vernon 1971; 1998: 47–51; Cox 1996.

7 It is important to evaluate qualitative research on its merits instead of conducting the assessment test through the lens of a statistician. For related debates, see King *et al.* 1994; Munck 1998; Brady and Collier 2004; George and Bennett 2004.

8 These companies included fabless firms, integrated device manufacturers (IDMs), foundries, packaging, and testing firms, semiconductor equipment suppliers, mask makers, electronic design automation (EDA) companies, among others. Fabless firms refer to chip merchants that outsource their fabrication needs to the foundries. An IDM is a semiconductor company which designs, manufactures, and sells IC products. Foundries refer to contract manufacturers performing fabrication. Mask refers to the device used to shape desired geometries on the surface of the wafer.

9 All the interviews were semi-structured and the average duration was about one and a half hours. The majority of interviews conducted in English were taped; however, a few interviewees were unwilling to be recorded. The majority of these interviews were conducted face to face, but four of them were conducted by phone.

10 This is based on a 2004 *IC Insights* revenue ranking. See (Jones 2005). Senior executives include the chairman, the vice-chairman, the CEO, or president of the firms in question. These firms included Taiwan Semiconductor Manufacturing Corporation (TSMC), United Microelectronics Corporation (UMC), Semiconductor Manufacturing International Corporation (SMIC), HeJian Technology (Suzhou) Co., Ltd., Huahong NEC, Central Semiconductor Manufacturing Corporation (CSMC), and Grace Semiconductor Manufacturing Corporation (GSMC, or Grace).

11 These include the International Electron Devices Meeting (IEDM) and the Semiconductor Equipment and Materials International (SEMI) China.

12 For a discussion of how semiconductor technology can be used as a national defense strategy by Taiwan against Chinese attack, see Addison 2001.

13 However, I do not subscribe to the Copenhagen School's securitization approach which belongs to the constructivism camp; mine is a modified Realist approach to security.

2 The semiconductor industry and national power and security

Semiconductors are the backbones of all these ... You can't do net-centric warfare, you can't do sensors to shooters, and you can't do decentralized command and control ... without electronics. And the heart of these is small microprocessors, and other types of processors like gate arrays, digital-analog converters, and all of these things, which are unfortunately for us also in the commercial world.

(Stephen D. Bryen, President of Finmeccanica North America, Inc.[1])

If you had a trusted ally that you can rely on [for] electronic components, then having a strong chip industry at home will be relatively unimportant. But if you didn't have a trusted ally that you can rely on [for] these components, then having a strong chip industry at home is just as important as having the ability to integrate the system and the software.

(A Pentagon official in charge of defense technology export control[2])

Introduction

Since its inception, the semiconductor industry has often been viewed as a strategic sector, analogous in many ways to the steam engine of the first industrial revolution. As William J. Spencer, chairman emeritus of International SEMATECH, put it, "It was a foundation for both economy and security."[3] This chapter examines the relevance of the semiconductor industry to national power and security on the premise that both economic and defense elements are essential to the making of national power and security. The chapter thus lays the foundation for our subsequent analysis of the globalization of semiconductor production activities and its impact on US–China–Taiwan security relations. This chapter is composed of four sections. The second section traces the major technological evolution of the semiconductor industry providing a basis for our analysis of the semiconductor–security linkage. The third section analyzes the important tie between the semiconductor industry and economic power and security. The fourth section examines the complex linkage between the sector and defense power and security, which revolve around four inter-related arguments. The final section summarizes the major conclusions of the chapter.

Historical overview of the semiconductor technological evolution

The second section reviews major technological milestones in the industry and identifies the growth pattern of the sector following "Moore's Law." It is argued that the invention of the transistor as a scientific discovery did not lead to the burgeoning development of the then nascent industry. Instead, it was through twists and turns in semiconductor product and manufacturing process innovation that the industry established itself as an important force today churning out increasingly miniaturized semiconductors that permeate our modern life.

From transistor, IC, microprocessor to system-on-a-chip

Major breakthroughs in the industry included the invention of the transistor, planar process technology, IC, the metal oxide semiconductor (MOS) process technology, the microprocessor, and the present-day system-on-a-chip (SOC).

The advent of the industry, marked by the invention of the transistor by John Bardeen and Walter Brattain at the Bell Laboratories in 1947, has revolutionized the electronics sector. The world's first transistor was a three-terminal electronic device which amplifies or switches an electrical signal through a solid crystal of germanium, outputting information in binary form (Ross 1997: 6; Warner 2001: 2458). The pair succeeded in inventing a solid-state replacement for the vacuum tubes, which had permeated a variety of electronic equipment for half a century. It was viewed as "the major invention" (Braun and Macdonald 1982: 33) of the twentieth century or as "one of the most important technical developments" of the century (Ross 1997: 3). It took a few years for the transistor to achieve technological parity with the vacuum tube, gradually replacing it as the key electronic component (Hanson 1982: 78, 82; Chang 1998: 95; Lecuyer 2006: 133,141). While duplicating the ability of vacuum tubes to amplify and switch, transistors operate without their predecessors' bulky, fragile glass envelope, or power-hungry heater (Tilton 1971: 10; Hanson 1982: 38, 77; Misa 1985: 262; Berlin 2005: 24; Lecuyer 2006: 141). Transistors consume less power and can be packed together in a small space without excessive heat generation, making them ideal for logic applications. The significance of the transistor was that it could do things the tube could never do. Gradually, the era of semiconductor supremacy in the electronics industry would come into being.

The next watershed was the invention of the first IC in 1958. An IC is one in which the functions of several discrete components are performed in a single chip of semiconductor material. In an industry that had previously linked vast quantities of separate components to perform useful functions, it was a logical step to supply complete circuits (Braun and Macdonald 1982: 88). It marked the advent of the rapid growth of the sector "from the cradle of the research lab to become the largest value-added manufacturing industry" in the USA (Wessner 2003: 10).

The birth of the first IC addressed the "tyranny of numbers" problems confronted by the industry at the time (Reid 2001: 120). Engineers designed increasingly complex circuits containing hundreds or thousands of transistors and other components, which had to interconnect to form electrical circuits. It was time-consuming and costly to solder thousands of components by hand. Worse yet, every soldered joint became a point of failure, potentially destabilizing the circuit. Additionally, this construction resulted in unacceptable time delays as electrical signals had to propagate a relatively long distance in the circuits even though transistors were faster than vacuum tubes. These headaches became obsolete when J.S. Kilby invented the first IC in 1958 with a major contribution added from Bob Noyce in the following year. Both were given credit as the co-inventors of the IC. While Kilby's crude IC would not be manufacturable in any quantity, Noyce's silicon-based IC could (Braun and Macdonald 1982: 89; Berlin 2005: 108–9; Lecuyer 2006: 156).

Noyce's invention was enabled by the planar process technology, which Jan Hoerni introduced to address the tyranny of numbers problems (Braun and Macdonald 1982: 89; Hanson 1982: 97; Riordan and Hoddeson 1997: 262–3; Lecuyer 2006: 150–6). The planar process, by rapidly emerging as the standard process technology following its debut, helped to kick-start the assembly-line manufacture of silicon chips, expanding the industry in commercial terms. It also helped to reduce the size of semiconductors, enabling the construction of multiple circuits on a silicon chip.

It would take years for ICs to become ubiquitous. The introduction of MOS process technology by Radio Corporation of America (RCA) in 1962 fostered the growth of IC. The technology, advanced and improved over time, outperforms bipolar technology as it enables ICs to be fabricated with increasing densities and lower power consumption (Braun and Macdonald 1982: 111). Soon after MOS transistors hit the commercial market in 1964, MOS became the predominant process technology. Rapid advancements in complementary metal oxide semiconductor (CMOS) process technology further led to the fall of bipolar technology (Semiconductor Industry Association 2003: 3). Improved MOS technology also enabled the fabrication of low-cost and high performance memory chips. In 1970, Intel produced the first 1-kbit Dynamic Random Access Memory (DRAM) using MOS technology. By then, the IC had emerged as the "miracle chip" in the eyes of the public (Reid 2001: 120; Berlin 2005: 301).

The next milestone was the invention of the microprocessor in 1971 at Intel. A microprocessor refers to a specialized IC that can execute programs and store information. It is "the brain," or central processing unit (CPU), of a computer. Its debut addressed a serious IC design problem in the 1970s. As logic designs became more and more complex, design cost continued to grow, and the relative usage of each circuit continued to fall. The need to design a multifunction circuit thus emerged.

The 4004 microprocessor, as the world's first CPU, was a large-scale-integration (LSI) single-chip processor, a "micro-programmable computer on a chip" (Noyce and Hoff 1981: 8). This gadget could be programmed to operate as the basic control for almost any electronic product. Measuring 1/8 inch times 1/6 inch with 2,300 transistors, it carried the same computing power as the Electronic Numerical

Integrator And Computer (ENIAC), the electronic computer that had filled 3,000 cubic feet with 18,000 vacuum tubes. The birth of the microprocessor took the industry further down the path of pervasiveness and miniaturization. It opened the door to inserting intelligence into many products for the first time (ibid.: 3; Braun and Macdonald 1982: 106), such as the world's first PC which hit the market in 1975. The miniaturization process is equally impressive as the 4004 microprocessor began the evolution of Intel's line of 386, 486, Pentium, Celeron, Xeon, Itanium, Core, Core 2, and Atom microprocessors. By the time the latest generation of the Intel Core 2 processors hit the market in February 2008, it had 820 million transistors, over 350,000 times as many as the 4004 CPU.

In the 1990s, another major milestone was reached, namely the ability to put large-scale electronic systems on a single chip, known as SOC ICs. An SOC IC "incorporates at least one processor, memory, and any number of other functions, such as protocol converters, signal processors, and input, and output controllers" (Linden and Somaya 2000). Continuous advances in chip production techniques have allowed the packing of diverse chip elements in the same silicon chip, enabling the SOC to realize its potential in achieving cost, size, and functionality goals (Rincon *et al.* 1999). SOC designers and manufacturers intend to advance SOC progress by reducing their costs, improving their operating speeds, and scaling down the power consumption of electronic products inserted with SOC devices (Linden and Somaya 2000: 547).

The recent surge of SOC-based systems responds to emerging market demands such as the need for small and low-power hand-held devices. For instance, CMOS digital camera SOC devices, originally designed for the National Aeronautics and Space Administration (NASA), holds great potential for commercial applications because of surging market demands. The long sought-after SOC prize is the world's first SOC supercomputer, the IBM BlueGene/L. Its prototype made its debut in 2004. At the time, it was the world's most powerful supercomputer and had a computational power 50 times greater than any of its existing competitors. Its use of a cost-effective SOC approach with higher levels of system integration makes it a milestone of high-performance computing, with applications in materials science and nuclear stockpile simulations (Almasi *et al.* 2002; Guizzo 2005; Moreira *et al.* 2007).[4] The development in SOC technology promises to further drive growth in related end-user sub-sectors, thereby increasing the SOC's share of the chip market (Linden and Somaya 2000: 547–9).

Growth following Moore's Law

In recent decades, the industry has been characterized by rapid growth[5] and decreasing component cost. The ability to increase device capability and decrease device cost has been the foundation of the growth of the industry, a pattern noted by Gordon E. Moore in 1965. Moore (1965) extrapolated then that the number of transistors on a chip would double every year for the next ten years, thus opening up the route to low-cost electronics. In 1975, he updated his prediction, expecting the IC progress to change from doubling every year to doubling every two years

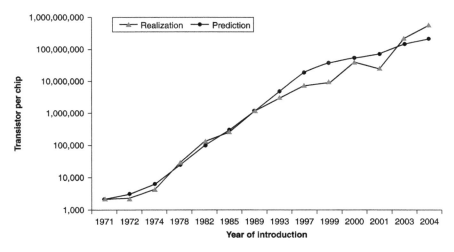

Figure 2.1 Moore's Law: prediction and realization for Intel microprocessors.

Sources: Company reports.

(Moore 1975). His prediction, known as Moore's Law, has dictated the pace of technological change in the industry for more than 40 years in spite of predictions of its demise (Figure 2.1).[6]

Moore's Law has far-reaching implications for the industry. First, because of advances in the fabrication technology, the expense per transistor was halved with each doubling in transistor density because the doubling is not accompanied by an increase in cost. This rapid increase in density per chip has been driven by the shrinkage in the minimum transistor feature size, which has been achieved since 1960 due to continuous improvements in lithographic and fabrication techniques that allow higher resolution feature definition.[7] Second, higher levels of integration mean that greater numbers of functional units can be placed on the chip, and more closely spaced devices can interact with less delay. Consequently, "the advances gave users increased computing power for the same money, spurring both sales of chips and demand for yet more power" (Hutcheson and Hutcheson 1996: 40).

Many have nevertheless questioned the sustainability of Moore's Law due to foreseen physical constraints, technical barriers in semiconductor manufacturing processes, and intensive capital required to sustain chip fabrication (ibid.: 40–6; Lee 2002: 51; Borsuk and Coffey 2003; Reuters 2007). Despite this, efforts to sustain Moore's Law have continued. Nanotechnology, for one, has helped conventional chip processes and tools produce smaller and denser circuits.[8] In 2007, Intel and IBM respectively announced a major breakthrough, which is expected to sustain silicon dominance for much of the next decade (IBM 2007; Intel 2007a). It refers to a transistor process technology using a new material to make smaller transistors. When the world's first 45nm microprocessor, fabricated using the new technology, made its debut in late 2007, it took the industry further down the road of miniaturization: about 400 of these transistors could fit on the surface of a single human blood cell. Moore described the breakthrough as the "biggest change in transistor technology since the introduction of polysilicon gate MOS transistors in the late 1960s" (Intel 2007a).

As the *International Technology Roadmap for Semiconductors* (ITRS) (2007), the industry's bible, announced its estimates of the sector's future technology development in 2007, it envisioned a continuous miniaturization process for the next 15 years amid challenges ahead.[9] In spite of talk of a moribund semiconductor sector in the imminent future, the industry has become a pivotal player in economic, technological, and defense terms today.

The semiconductor industry, economic power, and security

The third section examines the relevance of the industry to economic power and security in four inter-related aspects. First, semiconductor technology is a general-purpose technology which has spurred breakthroughs in computer and information-based technologies often associated with economic growth and productivity improvement. Second, the semiconductor sector has facilitated the development of many other large end-user industries that are important parts of various national economies. Third, a strong national semiconductor capability, as evidenced in the case of the US, has contributed to national economic power because of its key role in economic productivity and growth, industrial competitiveness, and employment. Fourth, a healthy national chip industry is the cornerstone of national economic security.

An enabling technology

First, semiconductor technology is a general-purpose technology and an enabling technology central to breakthroughs in computer and information-based high technologies. A general-purpose technology, as defined by Lipsey *et al.* (1998: 43), refers to "a technology that initially has much scope for improvement and eventually comes to be widely used, to have many uses and to have many Hicksian and technological complementarities." Chip technology is a general-purpose technology in civilian and non-civilian terms[10] because of evolutionary improvements in semiconductors and the resultant ubiquity of chips in our modern society. The industry carries an importance far beyond its specific trade, employment and revenue figures.

Semiconductor technology is also an enabling technology, which has helped to facilitate innovation in computers and information-based high technologies, as will be elaborated later at the level of respective industries. In the eyes of the semiconductor industry players, chip technology "enables you to do things that you couldn't do before or enables you to do things better than what you could do before."[11] It also helps "push the envelope of technologies."[12]

An enabling industry

Second, the semiconductor sector has been an enabling industry as its technological advances have helped develop a wide array of end-user industries, old and new, such as computers, communications, and automobiles (Wessner 2003: 18). Given the role of chip as an enabling technology, an examination of the trends in the semiconductor end-user market enables us to preliminarily assess the impact of semiconductors on the larger economy (Jorgenson and Wessner 2002: 72). In the US semiconductor market alone, the early diffusion of semiconductors centered on defense systems, hearing aids, portable radios, and computers, whereas telecommunications and automobiles were late adopters (Helpman and Trajtenberg 1998: 111–17). By the mid-1960s, the computer and communication industries surpassed the US military as the predominant markets for semiconductors. Whereas the military purchased 100 per cent of the chips in 1962, it bought only 55 per cent of those made in 1965. Since then, commercial applications have dominated the semiconductor market (Macher *et al.* 2000: 249; Berlin 2005: 137). On a worldwide scale, reports since 1991 have shown that the computer industry has accounted for about half the semiconductor end-user market (Wessner 2003: 14). Since 1999, the communication industry has remained the second largest chip market (21–5 per cent), followed by the consumer electronics sector (16–19 per cent). Most of the remaining market has been shared by the industrial, automotive, and government (especially military) sectors (Aizcorbe *et al.* 2007). As the worldwide chip market reached US$269.5 billion in 2007 (Figure 2.2), the computing application accounted for 37.6 per cent of the total market, followed by communications (25.4 per cent), consumer (18.6 per cent), industrial/military (10.9 per cent), and automotive applications (7.5 per cent) (PricewaterhouseCoopers 2007: 10).

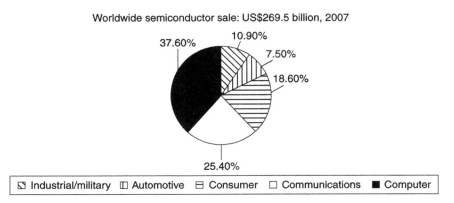

Figure 2.2 Distribution of world semiconductor sales by end-user, 2007.

Source: PricewaterhouseCoopers 2008.

The above analysis shows that chips have become ubiquitous in modern electronics systems in various sectors of the economy over the past few decades. Without advancements in chip technology, all of the industries and their products would have been non-existent. Mark Melliar-Smith, former president and CEO of International SEMATECH, summed up the significance of the sector as an enabler of other industries:

> The industry is a strategic industry because it provides the sort of the building blocks for many other industries, and it has continuously managed to facilitate other significantly large industries . . . It facilitated the PC industry, it facilitated modern telecommunications, it facilitated consumer electronics, and it is beginning to facilitate automobile electronics . . . It is a foundation block for many other industries.[13]

A prime example is the computer industry, one of the key backbones of the IT age. Improvements in semiconductor technology have fostered the development of the computer sector, thereby sustaining improved productivity and creating new business opportunities in the IT era.

The birth of the transistor marked the beginning of modern IT. As transistors replaced vacuum tubes in mainframes, the size of computers shrank, the performance improved, the price dropped and the scope of users widened. Moreover, the debut of the microprocessor gave birth to PCs, as described previously. A major milestone in the deployment of IT, the debut of the PC fostered the trend of personal computing (Hanson 1982: 215; Ruttan 2001: 334). As advancements in semiconductor technology have enabled a swift decline in the price of computers and other related IT goods, economic incentives for firms to quickly substitute IT-based equipment for other forms of capital and labor services increase. This injection of new capital contributes to the rise in productivity, which I shall elaborate on later.

The pervasive use of PCs has widened the use of the Internet, which grew 100 per cent per year during the 1990s. The increasing use of the Internet not only has created significant potentials for increasing productivity, but also has generated new business opportunities, such as e-commerce.

Aside from its distinctive contributions to the development of the computer industry, semiconductor technology advances, combined with advancements in other areas, have fostered the rise of new industries. These include, for instance, micro-electro-mechanical systems (MEMS) and the biochip sector. The combination of silicon-based microelectronics with micro-machine technology has made the realization of MEMS possible. Appreciation of the potential of very small machines predated the availability of technology that could make them. MEMS became practical once they could be fabricated using modified semiconductor manufacturing technologies (Amato 1998; Rothstein 1999; Zolper 2005).[14] The biochip sector is a sphere where electronics, molecular biology, MEMS, and other technologies and fields intersect. The continuous exploration of the interface between computers and biology would have been impossible without advances in semiconductor technology (Reuters 2007).

Finally, the economic importance of the industry tends to be underestimated because it is difficult to quantify how chip progress has helped to develop the majority of these end-user sectors aside from the computer industry. The current semiconductor knowledge gaps in most of its end-user sectors could potentially hinder an in-depth understanding of the industry's economic significance (Jorgenson and Wessner 2002: 72–3). Ruttan (2001: 359) argued that the best estimates concerning the contribution of semiconductors to the larger economy had failed to capture completely the full utility generated by the adoption of chips. Despite this, advances in semiconductor technology have enabled the chip industry to facilitate a wide array of sectors underpinning today's economy through its externalities and spillover effects onto other industries.

A contributor to growth, productivity, competitiveness, and employment

Third, the semiconductor industry contributes to economic power because of its important link with economic growth, productivity, competitiveness, and employment. This argument is substantiated by the following analysis of the case of the USA, the world's leading semiconductor player.

Various studies have shown that the US chip sector is important to US economic growth and productivity both through its performance and its contribution to productivity gains in its customers (Wessner 2003: 20; Semiconductor Industry Association 2009: 3–12).

The value-addedness of the US industry, as measured by its contribution to the US gross domestic product (GDP), has been impressive. In 2002, the industry was the largest value-added manufacturing sector in the US economy, accounting for nearly 1 per cent of US GDP (Jorgenson and Wessner 2002: 86–7). According to Bob R. Doering, TI senior fellow and one of the three US representatives to the ITRS International Roadmap Committee, such a distinctive feature of the chip manufacturing industry in the US context is one of the major benefits any country can enjoy because of having a strong indigenous sector.[15]

Moreover, various studies have concluded that there is a strong link between the US chip industry and various high-growth US manufacturing sectors. Flamm discovered that semiconductor production was one of the most technologically dynamic sectors of the US economy during 1958–76 and 1959–73. Both the chip industry and its principal customers were in the rank of the most rapidly growing US sectors during the given periods. "This indicates that the manufacture of semiconductor devices is not only a highly important growth industry, but is also one with important linkages to other growth industries," he concludes (Flamm 1985: 45–7).

Jorgenson's study (2004) on industry contributions to the US economy from 1977 to 2000 indicates that five semiconductor-related industries had significantly contributed to the value-added growth of the US economy in three separate periods, 1977–89, 1989–95, and 1995–2000 (Figure 2.3).

Furthermore, a number of quantitative studies identified investment in IT applications, driven largely by semiconductor advances, as one of the primary

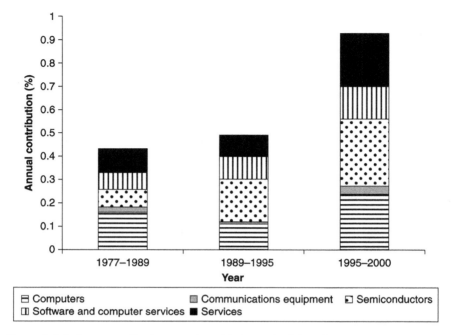

Figure 2.3 Industry contributions to value added growth of the US economy.

Source: Adapted from Jorgenson 2004.

explanations for the unexpected US economic growth revival since 1995 (Jorgenson and Stiroh 1999; Jorgenson *et al.* 2000; Mowery 2000; Jorgenson 2001; Jorgenson and Wessner 2002; Wessner 2003: 1; Jorgenson *et al.* 2006).

Jorgenson's (2001) study concluded that the development and deployment of chips had underpinned the resurgence in American economic growth since 1995. The decline in IT prices, which witnessed a steep acceleration in 1995, was triggered by a much sharper price decline of semiconductors in 1994, which resulted from semiconductor technology advances. Since semiconductor technology drove the spread of IT, "the impact of the relentless decline in semiconductor prices is transmitted through falling IT prices" (ibid.). As chip technology found its broadest applications in computing and communications equipment, the decline in semiconductor prices also helped to reduce the cost of a wide array of related products.

The study by Jorgenson *et al.* (2006) on the post-1995 growth resurgence of the US economy suggested that rapid technological progress in the IT-producing sectors had primarily accounted for the resurgence of US economic growth. It also noted that a consensus had emerged among economists that the resurgence in the US economic growth through 2000 could be traced to the production and use of IT equipment and software.

Lewis *et al.* (2002) traced the role of the US chip industry in productivity improvement, concluding that the sector alone contributed 0.20 percentage points to the 1.33 per cent jump in productivity in the US economy from 1995 to 1999. Furthermore, the study identified managerial and technological innovations in six highly competitive industries as the most important drivers behind production growth in the US economy during the same period. Aside from the chip sector (i.e. the fourth-largest contributor), the six industries also included computer manufacturing and telecommunications, which were two key semiconductor end-users. In total, the six sectors accounted for 99 per cent of net productivity acceleration over the period studied.

According to various studies, the US semiconductor industry has also contributed to the US industrial competitiveness. The National Research Council contended that advances in the industry had accounted for the resurgence of at least half the US industrial competitiveness over the past 15 years. It also discovered that many US industries had regained competitive positions relative to their foreign counterparts over the same period, over half of which had been transformed by the use of IT, which was underpinned by developments in semiconductors and software (Mowery 2000; Wessner 2003: 22).

The US semiconductor industry has also contributed to the US economy because of its role as a high-wage generator. According to preliminary 2006 data by the US Bureau of Labor Statistics, the industry employed about 474,115 people in 2006 at an average annual salary of US$93,526.[16] According to Wessner's study (2003: 17–19), employment had increased steadily by more than two-and-a-half times since 1972 in the US semiconductor and related device industries. Over the same period, real average hourly earnings in these industries had risen a remarkable 50 per cent: they increased from about US$14.50 in 1972 to US$21.50 today in 2001 dollars. In contrast to the impressive increase in hourly earnings in these semiconductor-related industries, real wages in the overall US manufacturing sectors witnessed stagnation and then decline over the same period. From 2001 to 2006, average weekly and average annual wages in the semiconductor industry had shown steady and slight increases (Office of the Governor 2007: 4).

In a nutshell, a healthy domestic semiconductor industry, as evidenced in the case of the US, would contribute to all of the key measurements of economic power, which include traditional macroeconomic indicators such as GDP growth, real wages and domestic productivity (Green 1996: 8).

The cornerstone of economic security

Fourth, and finally, the industry is the cornerstone of economic security, which is construed to include economic competitiveness and economic independence.

A country's healthy domestic chip sector contributes to its economic competitiveness, as evidenced in our analysis of the US case. Moreover, a healthy industry further helps to ensure the country's economic independence. It does so by supplying semiconductor goods and technologies to meet the needs of the domestic market, thereby reducing the prospect of undesirable foreign dictates or coercion due to dependency on chip supplies from foreign sources.

Since a country's strong IC capability helps to advance its economic power and security, any underperformance of the country's strategic sector can generate economic security concerns for the country in question, which, if realized, can erode national economic power and diminish a major pillar of the national might. In particular, as the country's semiconductor firms begin to globalize their production activities, the cost of globalization often generates a sense of economic insecurity for the given country. This often revolves around the fear of competitiveness loss, job reduction, and the erosion of economic independence (Flamm 1985; Brown and Linden 2005).

To conclude the third section, the semiconductor industry, a vital upstream segment of the crucial information industries, is strategic in economic terms. Advances in semiconductor technology have helped to spur technological innovation in computer and information-related industries. The sector has also contributed to the growth and development of its end-user industries. These sectors range from the computer industry to communications sector – all pillar sectors of various national economies, particularly developed ones. Moreover, the industry, as shown in the case of the US, has contributed to corresponding national economic power through its important links with the given economic growth and productivity, industrial competitiveness, and employment. Hence, a country's strong IC capability helps to advance its economic power and security; conversely, as noted above, underperformance in such a strategic industry triggers economic security concerns pertaining to economic competitiveness and independence.

The semiconductor industry, defense power, and security

Aside from its economic relevance, the semiconductor industry is strategic in defense terms. The fourth section explains how and why the industry is linked to defense power and security by exploring the following four inter-related aspects: (1) the role of semiconductors as the building blocks of information-dependent defense systems; (2) the relative importance of semiconductors to the making of defense power compared to other factors and the issue of strategic supply; (3) semiconductors and IW; and (4) the extent to which the defense significance of the sector is complicated by pertinent technological changes, i.e. the dual use nature of semiconductors and the dominant trend of spin-on.

A building block of defense capability

Given the role of chips as building blocks of information-dependent systems that underpin national defense, it is argued that a strong semiconductor industry at home has been and will continue to contribute to national defense power and security by its provision of semiconductors for end-uses in national defense and aerospace systems. Revealing evidence of military semiconductor end-uses in the early and recent years of the industry history and their contributions to the corresponding improvements in military capability is presented to support the argument. Moreover, it is contended that so long as the move towards IT-driven revolution in military affairs (RMA)[17] intensifies, the importance of chip-level advancements to advances in military power and security will increase accordingly, further ensuring the relevance of a strong indigenous chip capability to national defense and security.

The two-car analogy

To begin with, the defense importance of the industry can be illustrated by the two-car analogy. To explain the link between the industry and national security, the CEO of an IC design house used a car comparison analogy to illustrate his point. Car 1 has limited capability and is equipped with components that are larger and heavier than those in Car 2. Thus, Car 1 requires a bigger engine, which consumes much more gas and performs badly. In contrast, Car 2 has a control system composed of small, light electronic elements, which improves its capability over Car 1. It does not need the bigger engine of Car 1 to accomplish a similar performance, and it is more agile and flexible than Car 1. "If you have the semiconductor industry, you will be able to get Car 2. It helps you move faster, [be] agile, [be] flexible, and [be] less energy-/power-consuming," concluded the senior industry player, who had previous experience in US military chip design.[18]

Compelling evidence of military semiconductor end-uses further substantiates the strategic importance of the sector in defense terms, thereby echoing the two-car analogy to a substantial extent.

Early-day chip applications in defense and aerospace systems

In the first two decades of the US semiconductor industry, one major development[19] concerned the immediate exploitation of semiconductors by the US military to advance its defense programs. Washington's competition with Moscow during the Cold War ignited its interest in microelectronics. Moscow's launch of Sputnik in 1957 sent a shock wave through Washington. "The Soviets obviously now had the rockets they needed to lob their dreaded H-bombs onto U.S. targets," noted Riordan and Hoddeson (1997: 252). To close the perceived "missile gap" between Washington and Moscow, the USA expanded its defense and space projects, of which microelectronics were an integral part (Braun and Macdonald 1982: 95; Hanson 1982: 92). Of the many US defense and space systems developed in the post-Sputnik years, two of the most prominent users of ICs were the Apollo Program and the Minuteman intercontinental ballistic missile (ICBM) II of the US Air Force.

The Apollo Program was the most publicized military end-user of ICs in the 1960s (Reid 2001: 145). The size, weight, and reliability requirements for the rockets' navigation and guidance systems were achievable only with miniaturization (Manners and Makimoto 1995: 38). Real-time calculations would be required to navigate successfully a swift-moving rocket from a fast-orbiting planet through two different atmospheres and two distinct gravitational fields to a precise landing on a quick-moving satellite. Only a competent computer could shoulder this task. Such a computer on a spacecraft would have to be smaller, lighter, more energy-efficient and more reliable than any computer in existence. The computer in question would have to be built from ICs (Reid 2001: 146). The birth of the Apollo guidance computer (AGC) provided the solution. The project had demonstrated "the reliability of integrated electronics by showing that complete circuit functions are as free from failure as the best individual transistors" (Moore 1965: 114).

The Minuteman II guidance and control system was the second prominent defense use of ICs in the post-Sputnik era, marking the first time that ICs were employed in a major missile subsystem. As early as 1956, the US Air Force championed the digitalization of avionics equipment to bolster the reliability and capabilities of its weapon systems, including missiles and aircraft. Until that time, these systems had been primarily controlled by analog computers, which were driven by failure-prone vacuum tubes. The computers failed, on average, every 70 hours. They were not suitable for use in the ICBM guidance systems, which required the utmost reliability. The solution lay in the use of digital computers in aircraft and missiles because digital computers outperformed their analog counterparts in terms of computational accuracy, speed, functions, and reliability. The Air Force's increasing use of digital computers in its weapon systems opened a lucrative market for high-performance digital devices. Autonetics, supplier of the guidance and control system of the Minuteman I, purchased transistors from various chip makers (Texas Instruments 1964; Berlin 2005: 121–2; Lecuyer 2006: 149). In 1962, Autonetics decided to move from discrete transistors to ICs for the on-board computer of the Minuteman II. The decision enabled Minuteman II to provide large early markets for IC-based computers (Frost & Sullivan 1981: 23).[20] More importantly, the decision achieved the most dramatic performance enhancement in guidance that took place between the two generations of the missile. "The resulting greater power of the computer meant greater accuracy through the use of more refined trajectory computations," argued Mackenzie (1990: 206–7). TI (Texas Instruments 1964), whose ICs performed more than 90 per cent of the electronic functions of the guidance computer, concluded that the use of ICs in the Minuteman II computer had enabled the system to outperform its predecessor (Table 2.1).[21]

In sum, the increasing use of semiconductors in the US defense and space programs in the 1960s[22] highlighted Washington's recognition of the defense relevance of semiconductors. More importantly, the use of US-made semiconductors by systems designers during this period further showed that progress made in the US sector helped improve the performance of these systems. The case of the USA thus demonstrates that a strong indigenous chip industry contributes to national defense power and security.

Contemporary chip applications and revolution in military affairs

The military, to a large extent in developed countries and to a smaller extent in developing countries, continues to transform itself into a more nimble and

Table 2.1 Impact of ICs in the world's first major missile subsystem

	Weight	Space	Number of parts	Power consumption
Guidance computer for Minuteman I, Discrete Transistor-Based (MI)	70 lbs	1.5 cu ft	14,711	350 watts
Guidance computer for Minuteman II, IC-Based (MII)	36.5 lbs	0.915 cu ft	5,510	195 watts
Improvement from MI to MII	−48%	−39%	−63%	−44%

Source: Texas Instruments 1964.

information-aware fighting force in an era characterized by the RMA. High-demand and computer-intensive military systems need an increasing amount of processing power in ever-smaller spaces. This relies on the supply of advanced chips as backbones of these systems. As will be argued in this section, the microchip revolution has driven and will continue to drive the military system revolution during the RMA era, thereby enabling far-reaching progress in military capability and modernization in almost all high-tech countries. In the same way that semiconductors enabled the Apollo Program and the Minuteman II in the 1960s, they have further supported precision forces, Command, Control, Communications, Computers, Intelligence, Surveillance and Reconnaissance (C4ISR) and net-centric warfare (NCW), which forms the backbone of RMA today. Hence, the defense relevance of a strong domestic semiconductor industry has remained intact because it provides cutting-edge ICs critical to advances in RMA, and such importance will continue to grow so long as the RMA moves ahead.

The concept of RMA, as defined by military transformation advocates, refers to changes in military fighting due to the systemic introduction of advanced IT, with the emphasis on the contributions of data and analysis to the effectiveness of war fighters and their weapons (Dombrowski and Gholz 2006: 4, 6). Information and IT have changed the way the military fights (as illustrated by various major wars since the 1991 Gulf War) and are expected to further influence the future of warfare, with the pursuit of information superiority as a paramount objective in force planning and operations. In recent years, the rhetoric of RMA has remained pervasive and the concept of RMA as a military doctrine has become influential among military planners in many countries (e.g. the USA, the United Kingdom, and China). *Joint Vision 2020: America's Military – Preparing for Tomorrow*, for instance, foresaw the dominance of the US armed forces in all aspects of military operations by exploiting information superiority (Department of Defense 2000).[23] Although I do not intend to survey RMA in detail here, this research does not regard technology as the sole driving force behind RMA (since I am not a follower of technology determinism), nor does it embrace the rhetoric of RMA without any scrutiny.[24] All things being equal, what I intend to analyze below is the relationship between the introduction of semiconductor and IT technologies and the realization of RMA in the areas of precision force, C4ISR, and NCW. I also plan to explore the extent to which a future semiconductor technology paradigm is likely to transform further the face of modern military fighting.

The increasing electronics content in information-dependent modern military systems has indicated the defense relevance of semiconductors. In 2003, as much as 60 per cent of the content of many weapons systems was composed of electronic components (Gain 2003). More importantly, military electronics has been a growing market, even when the US defense budget is stagnant. Computers and electronics are estimated to account for more than one-third of US defense procurement spending, and this proportion is steadily increasing (Andrews 2003). To explain the contribution of semiconductors to defense capabilities, a number of interviewees have also highlighted the considerable electronics content in modern weapons systems. According to the CEO of an American chip association, semiconductor technology is a significant part of the fundamental technology in defense applications though the military market is small today.[25] As another senior American industry player put it:

> Defense systems use more chips than ever. However, defense systems use a smaller percentage . . . I think [the] semiconductor is just as strategic to defense as it always has been. It's no less strategic to defense. It's just that other components have caught up. It's become very important to commercial in addition to being strategic for defense . . . Defense systems count on chips to work without a doubt, and they will never go backward and not count on chips.[26]

A few interviewees, however, have played down the defense security significance of the chip industry because the military consumes only a small fraction of worldwide chip production today. As a senior vice-president of Applied Materials argued, "Although a majority of chips were for defense applications in early years, a large proportion of chip are for consumer products today. So there is little connection to security."[27] A director at AMD also diluted the security link of the chip industry:

> Fifteen or twenty years ago, there were a lot of security concerns, defense concerns. And that's just not the case anymore . . . What's actually driving and shaping the industry absolutely is economic concerns to this point . . . Semiconductors no longer play an important role in defense and security . . . Chips themselves have over time turned into commodities – driven by the consumer-oriented applications.[28]

Despite the dissenting views above, it is generally agreed among the majority of the interviewees in the USA and Asia that semiconductors remain the backbone of modern weapons, communications, navigation, space, and battle management systems, all of which act as force multipliers in modern military affairs, transforming modern warfare (Frost & Sullivan 1981: 117–25). I shall pinpoint below three critical elements in RMA – precision force, C4ISR, and NCW – and explain why advances in semiconductors affect these areas of RMA.

First, the case of improvement in accuracy and reliability of precision-guided weapons as a function of advances in semiconductors and IT is the most obvious area in which IT-related technological progress has transformed the face of modern warfare.

Barnaby's study (1984: 47) tracked the impact of microelectronics on war. He argued: "The most dramatic consequences of advances in military electronics are those that make a nuclear world war more probable. Such advances include, in particular, improvements in the accuracy and reliability of strategic missiles." These missiles included cruise missiles, ICBMs, and submarine-launched ballistic missiles (SLBMs).

In the case of cruise missiles, for example, Barnaby (ibid.: 48–52) contended that the modern cruise missile was a strategic nuclear weapons system whose development "depends almost entirely on microelectronics." Renewed interest in cruise missiles in the USA occurred in the 1970s because of two major technological advances related to semiconductors. They involved the miniaturization of computers "in terms of volume and weight for a given power output" and the availability of an improved database that described the coordinates of potential targets. Both advances enabled small but accurate cruise missile guidance systems to become a reality.

A DSB study (Defense Science Board 1987b: 91) also identified strategic bombing as a major illustration of the impact of electronics on warfare. "Modern bomb/navigation systems permit precision location of aircraft and target position as well as supporting flight at extremely low altitudes. These capabilities have been achievable in substantial part through advancements in solid state electronic technology."

Arguably, the constantly improved precision-guided weapons had been much more dependent on microprocessors and memory chips than earlier systems (Wessner 2003: 18), and various interviewees with in-depth knowledge of military semiconductors have largely endorsed this view. Some have done so by providing concrete qualitative evidence to explain why chips matter to precision force operations.

Stephen D. Bryen, President of Finmeccanica North America, Inc., used a mistake made by the US military during the war in Yugoslavia to illustrate his point. The case showed how the crude electronics embedded in a US guided weapon reduced the weapon's discrimination, resulting in the death of 50 children on a school bus on a targeted bridge. When the weapon was launched, nobody was on the bridge. However, when it hit the bridge, there was a school bus on it, and it killed 50 children. Although the weapon's crude electronics were adequate to guide the weapon to hit the target, they were never intended to be so sophisticated as to determine who was on the bridge and then transmit a picture back to the launch operator to enable the operator to divert the weapon at the last few seconds before impact.[29] "If you took a more advanced image processing chip with better resolution, you might improve that weapon on a simplistic basis so that you would not face that mistake again," Bryen concluded. "And that's all semiconductor technology . . . That's just one example. But that's a very real one."[30]

A senior American chip industry player also gave telling examples of the linkage between semiconductors and bombs. He did so by recalling his involvement in the development of smart bombs as early as the 1970s, and his experience fabricating application-specific integrated circuits (ASICs) for missiles at TI, as part of the TI-Very High Speed Integrated Circuit (VHSIC) project in the 1980s.[31] The TI veteran viewed chip technology as the single most important technology involved over the course of his involvement in TI's endeavor to improve Paveway laser-guided systems for bombs. In his view, the ability to build smaller chips with a higher level of integration and increasing functions into the missile electronics would enable the missile to achieve its ultimate mission.[32] Similarly, he contended the prospect of the critical role advanced chips might play in enabling the "fire and forget" missiles to accomplish their tasks had underpinned the TI-VHSIC program. The chips in question were digital signal processors (DSPs). DSPs outperform multi-purpose microprocessors because of their capabilities in processing an enormous amount of signals, including image and sound, into digital information in nano-seconds (Fong 2000: 167). These devices would enable the "fire and forget" missiles, once launched, to continue to process a rapid and continuous series of incoming radar and infrared signals in the search for a moving target.

By the time of the First Gulf War, when bombs guided by Paveway III kits were in action, they had reached an improved 90 per cent accuracy rate. Advancements in pertinent IC technology, as the TI veteran recalled, had helped to improve the accuracy rate of these bombs. The weapon's improved precision rate partly accounted for the US-led allied victory. As former US Air Force historian Richard Hallion said of the precision-guided weapons: "It was as revolutionary a development in military air power as, say, the jet engine." By 2004, TI (and after 1997, Raytheon) had delivered more than 250,000 Paveway systems – all equipped with better electronics than their predecessors. Their precision continues to be seen on televised news reports from the battlefields of Afghanistan and Iraq.

A historical comparison of the number of air strikes required to take out a target across various modern wars would shed further light on the tie between electronics and precision force, according to William J. Spencer, chairman emeritus of International SEMATECH.[33] Secondary data indicate that during the Vietnam War, TI put Paveway I into operation, thereby successfully reversing the US Air Force's large margin of errors in its bombing rate. While four jets armed with Paveway I kits destroyed a critical bridge in a single run, the US Air Force was unable to accomplish a comparable mission with several hundred unsuccessful sorties, including the loss of ten planes (Broad 1991).

In short, the constantly upgraded precision-guided weapons, which form an important part of the march towards RMA, have benefited from the improved computerized guidance systems on-board that are underpinned by advanced chips.

Improvements in C4ISR and NCW, the second and third aspects of RMA under discussion, have also benefited from advances in semiconductors and IT. C4ISR is the acronym used to represent the group of military functions defined by command, control, communications and computer processing (C4), intelligence, surveillance and reconnaissance (ISR), which enable the coordination of operations. Command and Control is the "exercise of authority and direction by a properly designated commander over assigned forces in the accomplishment of his mission." The addition of communications to the grouping reflects the fact that communications are required to enable this coordination. The inclusion of computer processing in the grouping means that computer processing is involved in the process of the coordination. Thus, at the heart of the C4 system is "a complete, accurate, and timely information set on which the commander and his staff base their decisions," rendering information a critical driver of warfare (Panel on Information in Warfare 1997: 7–8). Adding to the grouping is ISR, which describes methods of observing the enemy and one's area of operations (Williams 2001: 48).[34]

NCW is the acronym used to describe "the combination of strategies, emerging tactics, techniques, and procedures and organizations that a fully or even a partially networked force can employ to create a decisive warfighting advantage." Such increased combat power is generated by "networking sensors, decision makers, and shooters to achieve shared awareness, increased speed of command, high tempo of operations, greater legality, increased survivability, and a degree of self-synchronization" (Office of Force Transformation 2005: 3–4). The US Office of Force Transformation defined the operation of NCW as follows:

> [It] translates information advantage into combat power by effectively linking friendly forces with the battleforce, providing a much improved shared aware-ness of the situation, enabling more rapid and effective decision making at all levels of military operations, and thereby allowing for increased speed of execution. This "network" is underpinned by information technology systems, but is exploited by the Soldiers, Sailors, Airmen, and Marines that use the network and, at the same time, are part of it.
>
> (ibid.: 5)[35]

Both in real-world experience and in futuristic visions, the implementation of C4ISR and NCW would have been impossible without the introduction of semi-conductor and computer technologies. Examples abound. James Mulvenon, Deputy Director for Advanced Studies and Analysis at Center for Intelligence Research and Analysis at Defense Group Inc. (DGI), regarded semiconductors as the basis for all the modern systems pertinent to C4ISR and NCW.[36] Hua-ming Shuai, retired Taiwanese army lieutenant-general, regarded semiconductor mini-aturization as an enabler of the implementation of C4ISR. In the 1980s when the US military invited him to visit Washington to study the idea of C4ISR, computers were too bulky to be effectively used in real-world battlegrounds, and only a lab simulation of C4ISR was possible. As chips continued to shrink in size, leading to a reduction in computer size, the roadblock to implementing C4ISR was removed, enabling it to emerge as one of the major features of modern high-tech warfare. The improvement of the US C4ISR thus contributed to the US victory in the First Gulf War and beyond.[37]

Joe Yu-wu Chen, former president of Chung-Shan Institute of Science & Technology (CSIST), highlighted his experience in building Taiwan's air defense ground environment (ADGE) in order to illustrate the centrality of semiconductors to modern command and control systems. The operations of ADGE were under-pinned by "the integration of sensors and weapons for command and control purposes, a system based on IT whose building blocks are ICs."[38]

Viewing NCW as one of the major qualitative changes in Western forces, which began with the First Gulf War and has gathered momentum since then, Bryen of Finmeccanica stressed that semiconductors were building blocks of this war-fighting matrix. In his words,

> Semiconductors are the backbones of all these . . . You can't do net-centric warfare, you can't do sensors to shooters, and you can't do decentralized command and control . . . without electronics. And the heart of these is small microprocessors, and other types of processors like gate arrays . . . [and] digital-analog converters.[39]

Leading-edge US semiconductor firms and weapons designers have shared Bryen's excitement. They have projected an information advantage that the modern military force can exploit using cutting-edge semiconductor materials and gadgets in next-generation weapons systems which form an integral part of the C4ISR and NCW.

For instance, any duo-core or multi-core CPU, which enables a microprocessor to perform concurrently two or more tasks, is expected to improve the speed and performance of a variety of next-generation electronic defense systems, including unmanned aerial vehicles (UAVs), radar, and sonar control (O'Hanlon 2000: 40; Carlston 2007; Roberts 2007; Thryft 2007).

A telling example involves the case of UAVs. UAVs are increasingly popular sensor platforms capable of getting closer to the battle action in order to provide detailed information for combat operations. At least ten different types of UAVs were reportedly used during the 2003 Iraq War with success (Betz 2006: 522). Two Pentagon authors (Borsuk and Coffey 2003) used a hypothetical example, involving an image processor flying on a mini-UAV, to show that projected silicon technology under Moore's Law would inhibit US military capability in the future. They charted the data detection and identification requirements for the processor, and then discovered that silicon IC technology advances following Moore's Law would be unable to meet these requirements. This would inhibit the Pentagon's ability to reach its stated objective of information superiority in future warfare. Despite their prediction, various attempts are underway to further improve UAVs. Carlston (2007), from Intel, envisaged that a rugged and light-weight dual-core Intel Core Duo microprocessor would enable a UAV to perform concurrently multiple computation-intensive tasks. The cutting-edge CPU is expected to become a suitable choice for the high-performance and low-power system. Experts at BAE Systems further identified five new technologies that would drive the UAV transformation. They included semiconductor materials and MEMS (Fulghum 2005). These R&D efforts, if realized, will further improve next-generation UAVs, contributing to the US military's pursuit of information superiority in future wars.

Another illustration involves the case of future radar systems. The director of the US Defense Research and Engineering predicted that the continuous use of gallium arsenide (GaAs) semiconductors might lead to a several-fold increase in capability and perhaps a doubling in any given radar's detection or tracking range (O'Hanlon 2000: 40).

In sum, the semiconductor component revolution has driven the military system revolution.[40] Continuous semiconductor advancements, often combined with other technological improvements, are expected to enable the birth of advanced military systems in ways beyond our imagination so long as non-technological drivers behind RMA remain strong.

To conclude, the defense importance of the semiconductor industry lies in the function of semiconductors as the enabling building blocks of military systems as evidenced in the early and recent years of the industry. As Lorber (2002: 210) pointed out, the military applications of ICs involve the "enabling smaller and lighter equipment, faster processing in computers, cheaper mass production, and decreased power consumption with its attendant savings in weight and volume." The analysis in the section underscores his point. The defense significance of a strong indigenous semiconductor capability remains if it supplies much-needed ICs to the national military in question to contribute to its system-level modernization. Such an importance is likely to grow so long as the march towards RMA continues. Hence, the two-car analogy introduced at the beginning of the section will continue to be viable: a country's strong semiconductor sector[41] is ready to supply the national military with miniaturized chips with strong computational capabilities, thereby empowering the armed forces to seek information superiority over its enemy, which is a key, albeit not the only, factor to the victory in future warfare characterized by RMA.

Semiconductors, software, and system integration, and strategic supply

Even though the semiconductor sector is important in defense terms because it supplies chips for end-use in most information-dependent military systems today, the extent of such an importance becomes controversial when viewed from two different angles. According to the first school of thought, even though chips are the building blocks of modern military systems, semiconductors alone cannot modernize a military force nor enhance the armed forces' capability because other factors (e.g. software and system integration capabilities) are also critical to defense modernization. It follows that even if a country has a strong chip industry, there is no guarantee that it will automatically enhance the country's military capability, especially when other factors which are crucial to the making of national military power are inadequate. Hence, a country's semiconductor capability, at best, contributes rather than determines its defense power. According to the second school of thought, the extent to which any country's indigenous chip industry is important in defense terms depends on the issue of strategic chip supply. Any country's indigenous chip sector is as strategically important to that country as its indigenous software and system integration capabilities are, if the country lacks trustworthy offshore supplies of military chips.

The Tour de France analogy

A senior avionics engineer at a leading American defense company, who supported the first school of thought, used a Tour de France analogy to make his point. To explain the relative defense significance of semiconductors vis-à-vis that of other capabilities intrinsically significant to the making of national military power, he said:

> There are many, many things involved in producing a modern, state-of-the-art, high-performance military system, and one of those is the chip. It is just like that lightweight carbon fiber bicycle is one of the things that will help you win the Tour de France. Think of all the rest of the things involved in that . . . For the last 40 years, I have seen what's possible. I know how long it takes. I know how much money it requires . . . Having that one little thing [referring to the chip] in your hand isn't the miracle piece, like the carbon bike.[42]

Based on his 40-year experience in avionics systems design, he saw system engineering and system integration as two key capabilities that would make up a modern military system:

> If I look at any one given military system, the integrated circuit itself may be an enabling capability, but there are numerous other capabilities that make up the total system that you have to account for, such as the ability to put all those things together. System engineering and system integration are two really key things which require investment, require a highly trained staff with also lots of experience.

His colleague further stressed that even though ICs would contribute to national defense because certain military systems would not be able to function without them, some countries will never be able to make these systems with advanced ICs if they lack system integration capabilities.[43]

According to these two senior American defense industry players, a high-performance military capability requires elements other than chips. This is analogous to a victory in the Tour de France: the eventual trophy requires a confluence of various factors, which includes not only a good carbon fiber bicycle, but also a physically strong cyclist, a sufficient amount of financial investment, professional training, and teamwork.

Supportive arguments

Other interviewees in the chip and defense industries supported the view embodied in the Tour de France analogy, adding that chips are not magic wands in defense modernization.

When asked to analyze the linkage between defense and chips, George Scalise, President of SIA, stressed the importance of system integration capability: "The ability to get hold of that kind of components isn't where the issue lies. The issue lies in the ability to put the components into a system that is capable of whatever you want to have happen."[44] William A. Reinsch, former Under-Secretary of Commerce, made a similar observation:

> While chips and high-end electronics are essential to the production of sophisticated military equipment and weapon systems, they are by no means the only elements . . . the most critical element in developing a complex weapons platform is system integration, which is a capability that is very difficult to develop and hard to transfer . . . One person told me once, "You know what is a F15? An F15 is 80,000 parts flying in close formation." The point is close formation.[45]

A senior manager at AMD also stressed that if he were from a developing country with an ambition to help leapfrog his country's military modernization, it would be software and system capabilities that he would steal from developed countries rather than silicon technology. In his reasoning, "The software that goes into these things is much more valuable than the semiconductor components that would go into these systems because software has become very complex."[46]

In a similar vein, former CSIST president Joe Yu-wu Chen emphasized that any country's ambition to increase its military capability today would depend on its system integration capabilities, as they are essential to IT-based complex weapons systems. "Whoever has stronger system integration capabilities has stronger defense might," he contended. Moreover, even if a country has a strong chip sector, this alone would not necessarily guarantee that the country would possess viable system integration capabilities which are critical to its military modernization.[47]

Refutational arguments

The second school of thought does not embrace the Tour de France analogy wholeheartedly. Some contend that even though having a high-quality carbon fiber bicycle would not guarantee victory in the famous race, it would at least increase the probability of success. Others argue that if no guaranteed and trustworthy supply of good-quality bicycles were available, the ability to design and manufacture one's own bike would be critical to the prospect of victory.

Thomas Howell, attorney at Dewey Ballantine LLP, and an author on the globalization of the chip sector, was only partly convinced by the Tour de France analogy. On the one hand, Howell agreed that chips are not magic wands: "Just because you can make an advanced semiconductor doesn't mean you can build an aircraft carrier or any other sophisticated system that needs system integrators to do that kind of thing." On the other hand, he judged these systems would not work without ICs as their fundamental components: "If you take the chips out of the systems, the systems don't function. And you can say that, to a greater degree, that the chips are almost the other aspect of the systems."[48] More importantly, Howell stressed that it would be strategically important to establish a strong indigenous chip sector when a trustworthy external supply could not be guaranteed: "If you lost the ability to make these things, or couldn't get them somewhere else yourself, you had a hard time making the advanced systems work . . . So in that sense, that's strategic."

A Pentagon official in charge of defense technology export controls echoed Howell's emphasis on the strategic importance of a solid domestic chip industry in the absence of a trustworthy offshore IC supply for critical national military systems. While acknowledging the importance of system integration capability to the advancement of national defense capability, the official said:

> If you had a trusted ally that you can rely on electronic components, then having a strong chip industry at home will be relatively unimportant. But if you didn't have a trusted ally that you can rely on these components, then having a strong chip industry at home is just as important as having the ability to integrate the system and the software.[49]

To sum up, even though a solid indigenous semiconductor industry does not necessarily lead to a strong national defense capability, it increases such a probability. A country's powerful semiconductor industrial base can supply its military with leading-edge chips, which are indispensable building blocks of information-dependent weapons systems today. Although a strong chip sector on the home soil is not a sufficient condition for the country's military modernization, it is a necessary one.[50] Most important of all, possessing a powerful indigenous chip capability is of paramount strategic importance to any country whose military modernization objectives have been inhibited by difficult access to trustworthy and guaranteed foreign supplies of chips for its mission critical military systems. China falls into such a category, as will be discussed in Chapter 4. Concisely, the analysis in both sections leads us to conclude that both the two-car analogy and the Tour de France analogy remain relevant to our understanding of the defense importance of the chip industry.

Semiconductors and information warfare

A third aspect of the importance of the industry in defense terms relates to the link between semiconductors and information security. It is argued that a country's strong indigenous chip sector can contribute to its security by increasing its defensive and offensive capabilities in the arena of semiconductor-related IW attacks.[51]

Information warfare as a double-edged sword

To begin with, information handling lies at the heart of RMA. On the one hand, information superiority over one's adversary constitutes military victory, and exploiting advanced semiconductors contributes to such a pursuit. On the other hand, because information and IT are double-edged swords in modern warfare operations, what is equally important to military victory is the ability to shield the information infrastructure from any IW attacks launched by enemies and to initiate similar attacks against one's foes. Any IW attacks, which engender threats and vulnerabilities associated with RMA, thus contain defensive (Cobb 1999; Hayward 2000: 129; Bishop and Goldman 2003; Anderson 2008b)[52] and offensive (Cobb 1999: 143; Denning 1999; Bishop and Goldman 2003: 117) elements.[53] Among a vast array of IW attacks identified in the literature, two of them target semiconductor-dependent electronics systems: electronic magnetic pulse (EMP) strikes and "chipping."

Semiconductor-targeted information warfare: EMP strikes

Semiconductors, particularly unprotected ones, are susceptible to EMP strikes, and a confluence of trends, which contributes to the progressive semiconductor evolution, has ironically increased the prospect of such attacks against any semiconductor-dependent high-tech country.

Both nuclear denotations and non-nuclear devices can generate EMP, causing the disappearance of all or most computing capabilities of semiconductor devices. Nuclear-generated EMP was first observed during the early testing of high altitude airburst nuclear weapons in the 1950s, suggesting a new class of destructiveness (Glasstone 1964: 503, 510–19). Non-nuclear devices, such as a new generation of directed-energy weapons including high-power microwave weapons, can also generate destructive EMP. Dubbed as the "mother of all weapons" by *IEEE Spectrum*, high-power microwave weapons generate an intense blast of EMP in the microwave frequency band which is strong enough to overload electrical circuitry (Abrams 2003; Maggio and Coleman 2007).[54] These EMP generators can be either placed near a desired target, or carried by a UAV, a cruise missile or an aerial bomb (Washington 1995; Walling 2001: 97–100; Lorber 2002: 214; Abrams 2003).

Unprotected semiconductors, particularly silicon-based rather than GaAs-based ones, are most vulnerable to EMP for the following reasons. Since semiconductors operate by regulating a flow of electrons, a large amount of EMP, when applied to unprotected semiconductors, can induce more electrons to flow than the components are built to handle. This may potentially degrade, upset or destroy the devices (O'Hanlon 2000: 58–9; Walling 2001: 93; Abrams 2003; Anderson 2008b: 584–6). EMP strikes can hit the core of these components unless they are protected through "shielding" which enables them to be radiation-hardened. Radiation-hardening protects systems and applications from radiation, such as cosmic rays, which degrades the reliability and performance of conventional electronics. More importantly, silicon-based semiconductors, which are pervasive in our modern society, are more vulnerable to EMP attacks than GaAs-based ones because the radiation resistance of GaAs is at least ten times greater than that of the silicon (Hansell 1988: 11).

Ironically, a confluence of inter-related trends, most of which contribute to the semiconductor evolution, may exacerbate the prospect of EMP strikes against any country whose business and defense electronics systems rely on pervasive chips.

First, since the majority of semiconductors which form the backbone of electronics systems in any high-tech country are not radiation-hardened, EMP strikes may "deafen" and "blind" an entire country. While semiconductor advances have enhanced US economic and defense power, the very pervasiveness of semiconductors in the American society makes the country most vulnerable in the event of EMP attacks (Anselmo 1997; Schiesel 2003; Wilson 2005: 3–5; Anderson 2008b: 586).

Second, while chips have replaced EMP-resistant vacuum tubes, enabling a vast array of modern electronic systems, the much-lauded era of semiconductor dominance has an inherent vulnerability: the danger of electronic meltdown generated by EMP attacks has grown accordingly because unprotected ICs are more susceptible to EMP strikes than vacuum tubes (O'Hanlon 2000: 58–9; Anderson 2008b: 585).

Third, the smaller semiconductors are (e.g. MOS devices), the more vulnerable they become in the face of EMP strikes (Barbe 1980: 20; Anselmo 1997). Although semiconductor miniaturization has contributed to any high-tech economy and defense infrastructure, it has deepened the vulnerability of the very infrastructure to any EMP strike. The fourth trend concerns the increasing insertion of commercial off-the-shelf (COTS) semiconductors in military systems whose radiation hardness tends to decrease as chips become smaller and smaller. COTS chips are generally less resilient than most tailor-made defense chips, so the increasing use of COTS semiconductors in military systems may exacerbate the susceptibility of these gadgets to EMP strikes (O'Hanlon 2000: 59).[55] This vulnerability is compounded by the fact that the radiation hardness of COTS chips drops as technology plumbs the deep sub-micron regime (Dellin *et al.* 1998).[56]

The fifth and final trend concerns the decreasing level of investment on radiation hardening of military systems in countries such as the USA. As the Pentagon has reduced its emphasis on radiation hardening in recent years, the susceptibility of the US armed forces to EMP threats may worsen accordingly. One estimate in 1997 indicated that Pentagon expenditures for hardened semiconductors had decreased from US$1 billion to US$100 million annually. If such a trend continues, it will deepen potential vulnerabilities in future US military forces: while shielded military systems may emerge unscathed, other systems would be harmed by EMP strikes (O'Hanlon 2000: 59–60, 174–5). Arguably, other countries' military forces may face greater levels of EMP threats than those faced by their US counterpart for reasons given below. In theory, the US military has a stronger capability to invest in radiation hardening of its military systems than its counterparts elsewhere in order to reduce its vulnerability to any EMP attacks because the USA is the world's largest military spender. Since the Pentagon has decreased its investment in hardened semiconductors for its military systems, it is logical to infer that its counterparts elsewhere may have scaled down their investment in the same area, according to their proportionally lower level of investment. If this holds true, it is plausible to conclude that these countries' military forces may face more serious EMP threats than the USA.

O'Hanlon's study (ibid.: 59) concluded that the increasing use of COTS in military systems and the Pentagon's reduced investment on hardened semiconductors might jointly increase the amount of EMP threats the US military faces. "As such, it works at cross-purposes with the popular notion that modern electronics – as well as commercial off-the-shelf technologies – will enable inexpensive improvements in defense capabilities," he lamented. As some military strategists and scholars have increasingly viewed the prospect of EMP weapons as a real threat rather than an intellectual exercise because of related technological advancements, such an attack became a real possibility when the US reportedly launched its first electromagnetic bomb against an Iraqi TV station during the Iraq War in 2003 (Abrams 2003: 24; Martin 2003). Ross J. Anderson (2008b: 586) argued that "there remains one serious concern: that the EMP from a single nuclear explosion at an altitude of 250 miles would do colossal economic damage, while killing few people directly." Arguably, continuous R&D activities in directed-energy weapons in various countries increase the prospect of the actual deployment of EMP gadgets on the high-tech battlefield.[57]

Semiconductor-targeted information warfare: "chipping"

Aside from EMP, "chipping" is the second major IW attack that directly targets semiconductors. In this section, chipping as an IW tactic is defined, its variety is summarized, and its prospects are analyzed.

The term "chipping" was coined by Schwartau (1994b: 164). Defined as the "Trojan horse of microelectronics," chipping involves nefarious acts performed at the hardware and/or software level. These attacks may involve the "modification, alteration, design or use of integrated circuits for purposes other than those originally intended by the designers."[58] Cobb (1999: 143) confined chipping to nefarious acts performed at the hardware and software level which involve making special inserts into ICs "at the time of manufacture to allow unauthorized access."

As economic and/or defense motivations can underlie chipping attacks, those driven by defense considerations may involve the following possibilities in different contexts (Schwartau 1994a; Cobb 1999; Cohen 2000; Pretorius 2003; Defense Science Board Task Force 2005; Livingston 2007).

First, chipping might be launched during peacetime when chips are made on an adversary's soil and be activated during wartime. Offshore production of chips in a foreign country deemed as a foe may create a window of opportunity for the enemy's Information Warriors to make special inserts into chips at the time of manufacture during peacetime. These inserts do not make chips come to an immediate halt but allow these Information Warriors' unauthorized access as a belligerent act when the two countries are at war (Pretorius 2003). In the second scenario, chipping may involve the supply of counterfeit chips to an adversary, thus inflicting damage on its military operations (Schwartau 1994b: 168; Cohen 2007; Livingston 2007),[59] and the "more common the chip, the easier it is to counterfeit and distribute" (Schwartau 1994b: 168).

Third, chipping may enable a military force to conduct important missions including tracking and surveillance. By inserting a chip that electromagnetically broadcasts a distinctive signal or pattern as a tracking device, the military would be able to gather information in an insidious fashion. The fourth possibility involves an attempt to enhance a relative military position vis-à-vis that of the adversary through chipping. In a hypothetical scenario, Schwartau envisaged that the Central Intelligence Agency (CIA) and the Pentagon might launch pertinent attacks against customers who would buy American weapons, especially those perceived as Washington's foes. These American agencies might modify chips in the related military systems in order to trigger their failure "in three months [*sic*] time, to shoot off course by three degrees, or to blow themselves up after two shots" – all with the intent of gaining a military advantage over their adversaries.

To what extent does chipping pose a real threat? Three observations follow. First, there had been neither substantiated reports of chipping nor numerous discussions of chipping in the public domain by the late 1990s.[60] In her study of IW strategies, Denning (1999: 266) concluded that there had been no substantiated reports of chipping. In the case of the USA, however, *Time* magazine alleged in 1995 that the CIA did run several chipping projects (Washington 1995). The report remains unverified. Even if the report were based on facts, it would be unclear whether chipping has already been deployed by the CIA against Washington's foes.

Second, the limited discussions of chipping in the public domain as of the late 1990s could result from the possibility that discussions of closely guarded chipping techniques had been confined to the US intelligence circle. The US intelligence agencies were reluctant to foster, through public discussions of chipping, the prospect of a foe doing unto them what they would do unto others (Schwartau 1994b).

The third observation relates to the recent surge of literature on chipping amid the growing globalization of chip production activities and the resultant research projects on chipping. As discussion of chipping has increased in the public domain in various countries (including China and the USA), the focus is largely on the growing chipping threats posed by potential adversaries as the globalization of semiconductor production activities deepens. These threat scenarios are at least two-fold when seen from the US perspective. First, cemented offshore chip productions on an adversary's soil may create a window of opportunity for the adversary to launch chipping attacks during the production process, and the insertion of these problematic chips into US military systems may cause damage rarely heard of when these productions are kept on home soil (Defense Science Board Task Force 2005; Cohen 2007; Adee 2008). Another threat may arise as foreign-made counterfeit chips find their way into US defense systems. While only recently studied, the piracy of ICs has become a growing problem for both the US military (Cohen 2007; Livingston 2007; Adee 2008; Derene and Pappalardo 2008) and the electronics industry (Clarke 2006). The 2005 DSB report, which identified these potential threats, sparked a new project under the Defense Advanced Research Projects Agency (DARPA), namely Trust Integrated Circuits Program. So far, several US university labs and private firms, such as the Massachusetts Institute of Technology (MIT) Lincoln Laboratory and Raytheon, have taken part in the project (Sharkey 2007).[61] This theme will be further explored in Chapter 6 in the case study under investigation.

Implications for the defense significance of the industry

In the context discussed so far, to what extent is the consideration of these semiconductor-related IW attacks connected with the defense importance of the semiconductor industry?

The answer is two-fold. First, a strong domestic chip industry should, in theory, help accumulate in-depth semiconductor-related knowledge and technologies on the home soil. The dissemination of these assets to the relevant national intelligence and defense agencies may contribute to the given national security by increasing the country's offensive and defensive capabilities in the arena of semiconductor-related IW attacks. It cements the country's capability to launch these attacks against its adversaries, which, in turn, increases the prospect of military victory over its foes. It also increases the country's ability to shield important semiconductors in order to minimize any damage that IW attacks may cause. Second, a strong indigenous semiconductor industry should, in theory, provide sufficient onshore IC design and fabrication services to designers of critical national electronics systems, thereby minimizing the prospect of chipping attacks which may occur at offshore production sites. As chip production activities continue to globalize, the prospect of chipping threats generated at offshore production centers increases accordingly. To minimize its susceptibility to these threats, a country needs to retain at least a viable amount of chip capability at home to supply classified ICs, designed and manufactured in a trusted domestic environment, to its military unless it can rely on its allies as trusted suppliers. Thus, when factored in the issue of information security and warfare, a country's strong IC industry is an asset to its national defense and security.

Semiconductors, spin-on, and dual-use technological trends

In the section below, I shall consider the impact of inter-related technological factors (e.g. dual-use and spin-on) on the defense importance of the semiconductor industry. The section will first define these technological trends. It will then identify their applicability to the semiconductor industry, with reference to four related initiatives in the US context. It concludes with an analysis of why these technological trends have complicated the defense significance of the chip industry in three main ways. It is argued that a country's mainstream commercial chip subsector is potentially important in defense terms if it manages to transfer resources, hardware, and know-how alike, to satisfy national defense needs. Moreover, a country's tangential defense chip sub-sector remains important to its security if it provides classified chips for critical national defense systems through relatively enclosed and yet trustworthy operations at home. Third, and finally, a country needs to adopt a dual-track acquisition strategy to ensure that it can acquire reliable chips for its military end-uses from both commercial and defense chip sub-sectors.

Definitions of spin-on and dual-use

A brief review of spin-on and dual-use follows. The discussion of spin-on cannot be separated from that of spin-off. Spin-on refers to a technology development trajectory in which technology diffuses from the civilian to the defense sector, whereas spin-off involves technology diffusion in the reverse direction (Samuels 1994: 18, 26). Both paradigms, as Samuels pointed out, "have their own logic and their own consequences for economic development. Each suggests different ways nations can develop technology and provide for national security."

Technology has been always inter-diffused. Neither spin-off nor spin-on is new as military production has enhanced the civilian economy, and civilian technological advances have, for centuries, informed military production (Van Creveld 1989: 220; Fong 2000: 162–4; Stowsky 2004: 263). In the mid-1960s, however, the paradigm began to switch to spin-on. As Lorell *et al.* (2000) argued, "The post-war paradigm of 'spin-off' is turning into 'spin-on': more and more defense technologies are driven by developments in the commercial world." In their respective studies of the Japanese case in the early 1990s, Vogel and Zysman also noted the growing importance of spin-on technologies. In Zysman's words (1991: 98), "Spin-on technologies become established in the commercial sector, and are directly or with minor modification the basis of more sophisticated products or advanced military systems."[62] Likewise, there is a consensus among the project's interviewees that the majority of technological innovation today comes from the civilian rather than the defense sector. As a senior AMD engineer put it: "If you look at the most advanced technology, it is not in defense systems. It is in my BlackBerry."[63] Under the dominant trend of spin-on, the majority of cutting-edge technologies crucial to military power today are expected to originate from highly sophisticated commercial technologies.

Dual-use is a concept pertinent to the discussion of spin-on. A technology is defined as dual-use when it is "developed and used both by the military and space sectors on the one hand and by the civilian sector on the other" (Cowan and Foray 1995: 851) and "when it has current or potential military and civilian applications" (Molas-Gallart 1997: 370). More importantly, dual-use technologies are protagonists for spin-off and spin-on. The process of spin-on becomes possible due to the dual-use nature of many modern technologies and the growing importance of these dual-use technologies for military applications (Moran 1990; Branscomb *et al.* 1992; Samuels 1994: 27; Brooks 2005: 84).[64]

Spin-on and dual-use in the industry

The increasingly dominant process of spin-on and the dual-use nature of most technologies also apply to the semiconductor industry (Okimoto *et al.* 1987: 33–4; Kanz 1991: 338; Samuels 1994: 28). Spin-on has overwhelmed a large segment of the sector (Morrus 1988: 86–107).[65] Major semiconductor technological innovations have all originated from civilian firms. According to William J. Spencer, chairman emeritus of International SEMATECH:

Certainly, the Department of Defense made the first purchases of the integrated circuits . . . in the middle 1960s, and some of their money was used to improve manufacturing. I would agree with that. But when I looked at the invention of the transistor, invention of the integrated circuits, ion implanter, electron beam mass makers, photo lithography equipment—all of these breakthroughs and all of these processes came from the industry.[66]

By the early 1990s, commercial firms' ICs had been either on a par with or ahead of comparable military devices in the vast majority of cases, and few areas of Pentagon leadership had remained (Slomovic 1991). Thus, the defense sector, often aided by pushes from the bottom and the top, has largely benefited from superior chip technologies of civilian origin by spurring technology transfers from the commercial to the military side. Spin-on has dominated the semiconductor industry, co-existing with the tangential trend of spin-off.[67]

Such co-existence and inter-dependence between spin-on and spin-off in the semiconductor arena were observed by Nicholas M. Donofrio, senior vice-president of IBM. For him, "Things have been going back and forth all the time across that interface" between the mainstream commercial IC inventions and the tangential military applications. While there are always efforts to use more standard commercial parts in military systems, there are also pushes "to pull out valuable assets and valuable parts and commercialize them." He identified DARPA and the National Science Foundation as the key US agencies that would sponsor projects that push the aforementioned "crossing back and forth because it ends up better for everyone." Bob R. Doering of TI (2004) also noted that members of the SIA had reached consensus on the inter-dependence of defense and commercial research and technology needs for current and future ICs. Despite his recognition of such inter-dependence between spin-on and spin-off, Donofrio acknowledged that spin-on had been dominant, as "the industry will probably do a better job . . . as the driving force."[68]

The dual-use nature of modern technologies also applies to the semiconductor industry, with chip components and technologies often being cited as primary examples of dual-use. As Branscomb *et al.* (1992: 4) argued, microchips that would make precise missile guidance possible could also turn up in children's toys and in automobiles, and "the theoretical knowledge and practical know-how that comprise the essence of technology are even more protean than the artefacts themselves."[69]

Large charge coupled device (CCD) arrays and GaAs-based chips have been cited as primary examples of dual-use IC goods by various industry interviewees. CCD is a specialized semiconductor technology mainly used for image sensing applications, which convert the image to an electronic signal. CCD arrays appear in both military reconnaissance systems and civilian digital cameras. Often military reconnaissance satellites with missions to look for images on the ground need imaging chips with higher resolution capability than those used in civilian end-uses. The basic semiconductor technology in both military and non-military applications, nonetheless, is very similar.[70] Another tangible dual-use semiconductor product is a

GaAs-based chip. As Doering of TI pointed out, "GaAs is a very germane dual-use technology."[71] Given its unique properties, GaAs has become a desirable device material for both military (e.g. radar systems) and civilian applications (e.g. cell phones). Its high electronic mobility compared to silicon makes it a desirable option for high-speed device applications. Additionally, its large radiation resistance and high temperature performance allow it to be used in much of today's space and military applications (Hansell 1988).[72]

More importantly, the semiconductor fabrication process technology used to make chips for military end-uses and that for non-military ones is similar. According to Michael Polcari, President and CEO of International SEMETACH, the underlying semiconductor manufacturing process technology used to make chips for consumer and military electronics is fundamentally the same.[73] Similarly, in the 1990s both Intel and Motorola admitted that there would be no difference in wafer fabrication process for chips rated for commercial and military end-uses. Both stressed that the only difference would occur in the final segments of operations when military products would require additional inspection, test, and finish tasks than commercial ones. In its 1990 military product data book, Intel stated:

> There is no distinction between commercial product and military product in the wafer fabrication process . . . Intel's military products have the advantages of stability and control which derive from the larger volumes produced for the commercial market. In the assembly, test, and finish operations, Intel's military product flow differs slightly from the commercial process flow, mainly in additional inspection, test, and finish operations.
>
> (Humphrey *et al.* 2000: 596)[74]

Initiatives pertaining to spin-on and dual-use in the US context

To further explain the dominant trend of spin-on and dual-use in the semiconductor industry, this section reviews four related initiatives in the US. While acknowledging the dual-use nature of semiconductors, all of these initiatives have involved resource transfers from the commercial chip firms to the defense side, further facilitating and even institutionalizing the actual process or prospect of spin-on. The majority of them are products of top-down policy changes masterminded by the Pentagon. They are: (1) technology transfers from Intel to Sandia National Laboratories since the 1980s; (2) the VHSIC program in the 1980s; (3) greater use of commercial chip components, practices, and standards in military systems primarily spurred by the 1994 Perry COTS Initiative; and (4) the Trusted Foundry program and complementary Accredited Trusted IC Suppliers scheme since 2004, which implemented the 2003 Wolfowitz memo on "Defense Trusted Integrated Circuit Strategy" (DTICS).

The first example involves technology transfers from Intel to Sandia since the 1980s (Table 2.2). It is a telling illustration of spin-on because Intel is the world's commercial chip giant and Sandia is the primary US government agency in charge of radiation-hardened chip design and fabrication for nuclear weapon end-uses.

Table 2.2 Spin-on and dual-use: from Intel to Sandia

Date	Spin-on from Intel to Sandia
1980s	Free-of-charge transfer of Intel 8085 and 8051 microcontroller technology to Sandia for the development of rad-hard processors for US space and defense applications. Gordon E. Moore directed the transfer of the first Intel microprocessor design to Sandia
1995	Intel used 9,200 Intel Pentium Pro processors to develop the world's fastest teraflop* computer for Sandia under government contract
1997	The redesigned rad-hard version of an earlier Intel chip by Sandia was inserted in the rover for NASA's July 1997 Mars Pathfinder mission
1998	Free-of-charge transfer of Intel Pentium processor technology to Sandia for the development of rad-hard processors for US space and defense applications

Source: Sandia National Laboratories 1998.

Note: *Teraflop means one-trillion-operations-per-second.

In 1998, Washington initiated the spin-on deal for the Pentium processor, one of the world's most popular computer chips at the time, in order to exploit existing commercial chip technology for cost and efficiency considerations (Sandia National Laboratories 1998; Schwartau 2000: 303; Sullivan 2003). Craig Barrett, Intel president and CEO, summarized the benefits Sandia would garner from the deal as follows: "The Pentium processor design will offer tremendous performance, flexibility, and reliability for critical government applications ... This agreement allows the government to apply the vast research and development activity that Intel has undertaken for the commercial market to its mission-critical needs" (Sandia National Laboratories 1998).[75]

The spin-on case benefited US defense and security in two major ways. First, the five generations of chips that Sandia had hardened by 1998 were used in Earth-orbiting satellites, the Galileo mission, missiles, nuclear weapons, and in other applications where radiation would degrade the reliability and performance of conventional electronics (ibid.). Second, the given technology transfers may potentially help Sandia identify measures to protect semiconductor-based civilian systems, which form part of the American national infrastructure, against any related IW attacks including EMP strikes (Schwartau 2000: 303). This helps to enhance the defensive capability of the USA, contributing to its information security. The case thus highlights the way a strong indigenous chip industry, by helping develop defensive measures against IW attacks, can contribute to the defense security of the country in question.

The second case involves the Pentagon's VHSIC program. Established in 1980, the project began with the prospect of actual or potential spin-on and spin-off benefits (Fong 1986: 270–7).

The Pentagon included commercial chipmakers in the project because it recognized the dual-use nature of semiconductors and the technological superiority of the commercial sector over the defense sector (Fong 2000: 165–8).[76] To meet the needs of the military for the mid-1980s and beyond, the Pentagon had to rely on technologically superior commercial firms. The program incentivized commercial firms to provide tailor-made, high-performance, silicon-based chip technologies for military systems, to help accelerate the insertion of VHSIC technologies into new and existing fielded weapons systems, and to help the USA maintain a qualitative lead over potential adversaries (Borkan 1982; Fong 1986: 268–91; Gansler 1988: 71; Sheppard 1990; Slomovic 1991; Fong 2000: 153, 166–8; Stowsky 2004: 261).[77] The project aimed to develop product and manufacturing processes that have commercial and military uses. This dual-use chip technology development, if successful, would benefit the Pentagon and participating commercial firms. Twenty-five US companies participated in the eight-year, US$1 billion scheme. They included ten of the top 15 American merchant semiconductor producers,[78] representing 63 per cent of the chip industry. Eight of them won the bid and became contractors or subcontractors in the program (Fong 1986: 277, 290).

In Phase I (1981–84), each contractor was required to develop a programmable chip set, to set up a pilot production line, and to develop a complete electronic brassboard subsystem that would meet a specific Pentagon system requirement. The contractors were also required to present a feasible demonstration of advances to 0.5 micron feature sizes, 1×10^{13} gate-Hertz/cm^2 and 100MHz. This would be carried into production during Phase II (1984–88) (Borkan 1982).

The outcomes of the project were mixed. On the one hand, the positive impact of spin-on seemed uncertain. By the time the program ended, none of the new VHSIC chips had been inserted into fielded weapons systems (Stowsky 2004: 261). Supportability issues were expected to further inhibit the insertion of VHSIC technology into fielded electronics systems (Murphy 1988). On the other hand, accomplishing cases of spin-on did occur. Some successfully integrated VHSIC technology into fielded military systems in the immediate years following the end of the project. For instance, IBM Federal Systems Company infused VHSIC technology into AP-102 avionics processors, enabling the delivery of the initial upgraded units in 1992. As an engineer at IBM observed, "VHSIC technology infusion into the AP-102 computer family achieved enhanced technical performance." These improvements included greater throughput,[79] increased memory capacity, reduced power consumption and weight, improved reliability, and reduced logistics support. Based on the case, the IBM engineer concluded that VHSIC technology did "allow weapons systems to effectively incorporate new functions and capabilities to improve mission success" (Stuart 1992: 188). To sum up, the VHSIC program's foundation was the trends of dual-use and spin-on in the chip sector, although the actual spin-on benefits resulting from the project were mixed.

The third initiative pertaining to semiconductor spin-on and dual-use is the 1994 Perry COTS Initiative, a major US military procurement policy change in the mid-1990s. The reform represented a top-down approach which aimed to institutionalize further the process of spin-on and the greater use of COTS chips in military systems.

Several important developments preceded the initiative. Numerous studies and industry lobbies in the 1980s and the early 1990s pushed for greater use of commercial products and practices in military systems for cost, efficiency, and performance considerations. Besides, a series of defense acquisition reforms, introduced in the 1980s, helped to foster the development of military equipment using commercial components in the USA, ultimately culminating in the Initiative (Pecht *et al.* 1997: 42).

These studies and lobbies highlighted the benefits of the greater use of commercial products and practices in military systems, compared to the costs of conventional Pentagon procurement policies. A DSB study, echoing the famous Packard Commission study, estimated that the Pentagon, by using growing numbers of commercial components and systems, could reduce costs from 10 to one at the component level and from between four and eight to one at the system level (Defense Science Board 1987a; Gansler 1987: 58). A 1989 follow-up DSB study further identified the chronic weakness of the relatively rigid Pentagon procurement and acquisition policies on semiconductor supplies. Originating in the 1960s, these policies included the "MIL SPEC" (i.e. military specification or mil-spec) requirements and procurement rules.[80] They were designed to control the quality and reliability of ICs used in US military systems which would operate in harsh environments. Some of them also aimed at ensuring onshore manufacturing capability of these components and at tracing these parts to sources (Pecht *et al.* 1997).[81] However, the study identified chronic problems inherent in the MIL SPEC policy package as follows: higher costs, constrained supply, unrealistic device qualification rules, and onshore manufacture and assembly (Defense Science Board 1989: A-15, A-16).

While mil-spec semiconductors experienced the aforementioned problems, commercial chips had been equal to or better than mil-spec ones in quality and reliability (Karatsu 1987; Kanz 1991; Slomovic 1991; Pecht *et al.* 1997). This trend "spurred a renaissance in US semiconductor quality driven by commercial and industrial customers and well documented by supplier, trade association, and customer records. Military parts directly benefited from this improvement, but were not the driving force behind it" (Kanz 1991: 338). Slomovic's (1991) study reached a similar conclusion. She thus urged Washington to institutionalize the process of spin-on even though spin-on had already taken place informally. Gansler (1988: 71) envisaged additional benefits that the spin-on policy might engender aside from gains in cost and quality:

> By using commercial parts, the Pentagon will have the advantage of buying equipment that has already met the market test for quality and price. This approach also makes it more possible to rapidly increase production during an emergency by switching production lines from commercial to military.

In response to the aforementioned demands, the Pentagon began a series of reforms in the 1980s which reflected its acceptance of the gradual impact of commercial standards, practices, and components on the military.[82] In particular, the Pentagon introduced a new grade of military chip products designated Qualified Manufacturers List (QML). The QML program was documented as MIL-I-38535. One advantage of the QML over the Qualified Parts List (QPL) is that the QML exercises concepts well accepted by the large commercial chip firms, which help standardize quality management in their mainstream businesses (Sheppard 1990). In November 1989, the Pentagon merged the QPL into the QML program in order to remedy the problems embedded in the QPL system. These problems included a limited number of listed parts, high costs, and a prolonged review process (i.e. two years) for new suppliers (ibid.; Kanz 1991; Pecht *et al.* 1997: 39–41; Winokur *et al.* 1999: 1494; Defense Supply Center – Columbus 2000). Additionally, MIL-I-38535 was revised to eliminate the onshore provision, thereby paving the way for devices assembled offshore to be acceptable for qualification (Pecht *et al.* 1997).

These reforms culminated in the 1994 Perry COTS Initiative. The Initiative represented the key top-down drive within the Pentagon to push for the greater use of COTS items in military systems, the further elimination of the restrictions imposed by military specifications and standards, and the greater use of commercial specifications and standards to assure implementation of Best Commercial Practices (Perry 1994; Pecht *et al.* 1997: 42; Winokur *et al.* 1999: 1494; Lorell 2001: 3).

To institutionalize further the spin-on process was on the Initiative's agenda. According to William A. Reinsch, who was involved in the Initiative in his capacity as the Under-Secretary of Commerce in the Clinton administration, Washington came to realize at the time that "the US military establishment needs high-end electronics and high-end chips, and wants to rely on US companies to produce them."[83] After all, state-of-the-art COTS chips would provide for improved performance and functionality as well as lower cost per function,[84] all of which would contribute to a high-tech military force.

Reinsch identified another consideration behind the reform, namely, the mismatch between the relatively short commercial chip cycle time and the long defense system development cycle time, a headache for defense policy-makers. In explaining the Pentagon's move in the direction of COTS for end-uses in US military systems over the past 15 years, he said:

> When I was in the government, I had a Pentagon official tell me seriously that the procurement cycle of the Pentagon is longer than the life cycle of some of the stuff they want to buy. So if you take the time to develop some specific specifications and then you wait for them to make that, by the time they produce the product, it is obsolete because the civilian market has moved on to even more powerful and even more competent chips. So the Pentagon moved in the direction of off-the-shelf items.[85]

The impact of the Initiative has been far-reaching. The actual cost, efficiency and performance benefits of spin-on spurred by the initiative were obvious. Michael C. Maher (2003), a principal engineer at National Semiconductor, estimated that COTS products had replaced mil-spec parts at a rate of 15 per cent per year. One plausible explanation for such a high replacement rate would be the apparent benefits the military had gained from the insertion of COTS into military systems. In the case of the USA, the Pentagon-related Institute for Defense Analyses (IDA) revealed, in

2007, that the Pentagon, through its suppliers, would buy roughly US$1 billion annually in military and aerospace chips, and about US$2–4 billion annually of additional COTS. In other words, the Pentagon's expenditure on COTS would be two to four times more than that for military and aerospace chips. Arguably, this development can be attributed to the actual benefits of spin-on (Cohen 2007).

Aside from benefiting the US military, the Initiative has had an overarching impact beyond US soil. Other countries (e.g. NATO member states, China, and Taiwan) have emulated the initiative by formulating their policies to push for greater use of COTS devices in their military systems. These reforms have been driven by the potential spin-on benefits expected from the new procurement programs (Edmonds *et al.* 1990; Defense & Aerospace Electronics 1991; Xu 2004; Zhang 2004).[86] We shall return to this theme in Chapter 4 and Chapter 6 when the case of China and Taiwan is examined.

The fourth initiative pertaining to semiconductor spin-on and dual-use refers to the Trusted Foundry program and complementary Accredited Trusted IC Suppliers scheme (Table 2.3). Both have been introduced since 2004 to implement the 2003 Wolfowitz memo on DTICS, another government initiative designed to further institutionalize the process of spin-on. An analysis of the origins, causes and impact of the Wolfowitz policy package is to follow.

The memo, issued by the US Deputy Secretary of Defense, Paul Wolfowitz, on 10 October 2003, envisaged a five-part plan to implement a defense trusted IC strategy. It stated:

> The country needs a defense industrial base that includes leading edge, trusted commercial suppliers for critical integrated circuits used in sensitive defense weapons, intelligence, and communication systems. The purpose of this memo is to establish a strategy to ensure that such suppliers exist.
>
> (Office of the Deputy Under-Secretary of Defense
> (Industrial Policy) 2004: 29)[87]

The strategy contained five objectives, and those relevant to the discussion here include facilities identification, product identification, and a healthy commercial industry. The Pentagon planned to identify facilities within the US semiconductor defense industrial base that could qualify as trusted sources for ASICs, based upon "special facility clearances or other government agency technical certification." Moreover, the Pentagon set out to identify products available from the trusted sources. To pursue a healthy commercial IC industry, the memo stressed that the Pentagon should "ensure the economic viability of domestic IC sources" because the health of the defense IC supplier community depends on that of the larger commercial IC base. This manifesto reflected the Pentagon's continued recognition of the almost irreversible trend of spin-on, and the need to exploit the larger commercial IC sub-sector in order to benefit the tangential defense IC industrial base.

So far, the objectives of facilities and product identification have been implemented through the Trusted Foundry program and the complementary trusted IC supplier accreditation scheme. The former scheme, initiated in 2004 and

Table 2.3 Accredited Trusted IC suppliers

Company	Location	Services
IBM Systems Technology Group	Vermont, USA	• Comprehensive suite of trusted technology manufacturing and technology services • 0.25 μm–65 nm digital CMOS technologies • 0.35–0.13 μm analog and mixed signal/ SiGe BiCMOS technologies • Basic foundry manufacturing
The Trusted Solutions Business Unit, National Semiconductor Corporation	Maine, USA	• CMOS and mixed signal processes at 0.65 μm, 0.35 μm, 0.25 μm, and 0.18 μm • BiCMOS at 0.35 μm and 0.25 μm • CMOS at 0.25 μm and 0.18 μm based on and compatible with TSMC processes and design rules
BAE Systems Information and Electronic Systems Integration, Inc.	Virginia, USA	• A complete design, fabrication, packaging, and test facility for VLSI wafer and module development and production • Rad-hard CMOS at 0.5 μm, 0.25 μm, and 0.15 μm
Northrop Grumman Space Technology	California, USA	Advanced InP and GaAs chip manufacturing processes
Honeywell International, Inc. Solid State Electronics Center	Minnesota, USA	• Rad-hard digital SOI CMOS, running at 0.15 μm, 0.35 μm, and 0.8 μm • High temperature and mixed signal digital SOI CMOS at 0.35 μm, and 0.8 μm • Multi-chip module package design
HRL Laboratories, LLC	California, USA	Services for compound semiconductor devices and ICs including InP HBT, InP HEMT, and GaN HFET
Intersil Corporation	Florida, USA	A variety of technologies including CMOS, BiCMOS, and SOI

Source: Trusted Access Program Office n.d.

administered by the National Security Agency (NSA), leveraged a 10-year, US$600 million contract with IBM to aggregate purchases of leading-edge IC fabrication technologies[88] for use in defense applications. The contract between the IBM and Washington obliged IBM to upgrade its facilities and implement enhanced security procedures at its 8-inch wafer fab in Vermont.[89] IBM thus became the Pentagon's first Accredited Trusted IC Supplier.[90]

The latter program, also administered by the NSA, aims to ensure that custom-designed chips for high Mission Assurance Category (MAC) (Department of Defense 2002)[91] and confidential environments would be obtained from an

Accredited Trusted IC Supplier. It excludes the participation of any offshore firms, representing a case of non-globalization. The criteria for a trusted domestic supplier would be more achievable and affordable than that for a foreign-based one. NSA has reached out to other commercial companies in order to include them in the scheme. As a senior executive at AMD recalled, an official came to his hotel room in Washington, DC, in the autumn of 2004 to persuade AMD to take part in the program.[92] By early 2007, six additional firms had become accredited trusted IC suppliers with eight more accreditations in progress (Cohen 2007). By November 2007, seven chip firms had become accredited trusted IC suppliers. Semiconductor technologies available at these foundries include rad-hard CMOS and compound semiconductor technologies.

A confluence of four factors accounted for the Wolfowitz memo and the subsequent implementation schemes. The first factor was related to the impact of globalization. The policy change intended to mitigate security and access concerns which could be exacerbated by the increasing globalization of the US industry (McCormack 2004; Defense Science Board Task Force 2005).[93] The second factor involved the Pentagon's recognition that the commercial IC firms had been technology drivers and that it would be in the government's interest to exploit the larger commercial industrial base in order to satisfy its military needs. This rationale was documented in the Wolfowitz memo, as mentioned above. The project can be seen as another attempt to institutionalize the process of spin-on. As Donofrio of IBM put it, "It just reflects the pragmatic nature of the industry, the fact that . . . the industrial technologies are becoming capable enough to do those kinds of things the government wants now."[94] Similarly, a 2002 study conducted by IDA concluded that it would not be economical to run a captive domestic semiconductor unit to mitigate security and access concerns engendered by the increasing globalization of the US chip industry. A viable solution, it added, would be to exploit resources available from the commercial chip industry at home to assure a trusted supply of custom chips for military end-uses (Cohen 2007).

The third drive concerned the depleting US domestic defense supply base of classified ICs for critical military systems due to two reasons (Defense Science Board Task Force 2005: 9). One was that giant American commercial chip firms withdrew from the military market in the mid-1990s.[95] Second, US government-owned, government- or contractor-operated or dedicated facilities at home, such as those at NSA and Sandia, were becoming increasingly non-viable suppliers of classified military ICs. This was in spite of the fact that they had primarily satisfied the US need for classified military chips throughout the past ten years.

The fourth motivation concerned the inadequacies of COTS ICs for military end-uses. A Sandia study argued that the decreasing robustness of COTS ICs and the focus of the mainstream commercial industry on large volumes of standardized parts would render COTS ICs insufficient to meet US defense needs (Dellin *et al.* 1998). Although the Perry Initiative further institutionalized greater use of COTS in military systems, COTS chips alone still cannot satisfy all defense needs (McCormack 2004). The inadequacies of the Initiative thus became another reason why the Wolfowitz memo came into being.

So far, the benefits of spin-on as a result of implementing the memo have been obvious. According to IDA, the IBM scheme has been successful for two reasons: cost reduction and mission accomplishment. Multi-project wafers (MPWs) and sharing both the access fees and infrastructure have allowed the trusted foundry customers to achieve substantial savings. As for mission accomplishment, 26 program requirements were met by early 2007 and 13 MPWs were assembled and fabricated with an average of 15 IC designs on each in 2006 (Cohen 2007).[96]

In short, the memo has facilitated successful transfers of resources from the commercial side of the US chip industry to the defense side through the trusted IC supplier schemes. Although it is premature to assess the effectiveness of the relatively recent initiative, data currently available seem to indicate that the resultant actual benefits of spin-on have been obvious.[97]

The four US initiatives analyzed so far have been undertaken with the recognition of spin-on and dual-use in the semiconductor arena, although the resultant benefits of spin-on for the US defense microelectronics community vary. In view of dual-use and spin-on in the chip sector, how would we assess the defense significance of the semiconductor sector in a country, characterized by one shared technology and two markets (i.e. commercial and military)?

Implications for the defense significance of the industry

The answer is as follows: the consideration of such dominant technological trends has enabled us to assess the defense relevance of the chip industry in a country in three major ways. First, the mainstream commercial sub-field of the overall domestic industrial base can contribute to national defense and power if it transfers resources, hardware and expertise alike, to satisfy the needs of the domestic defense microelectronics community. Figure 2.4 summarizes possible resource transfers from the mainstream commercial sub-field, which is relatively open and technologically strong, to the tangential military sub-field, which is relatively closed and technologically weak.[98]

Hardware transfers may involve using the commercial chip sub-sector's supply of high-performance and cost-effective COTS items in national military systems,

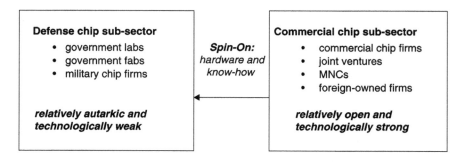

Figure 2.4 Resource transfers through spin-on in the semiconductor industry.

especially those which can operate in a relatively benign environment. This possible transfer is related to the opportunities that the evolution of the commercial chip industry has created for defense microelectronics (Dellin *et al.* 1998). The commercial sector has evolved to target a mainstream high-volume market, and commercial firms, in order to survive in this marketplace, have churned out large quantities of standard parts to be used in systems which could operate in a benign environment and which would normally become obsolete in five years or less. Thus, the commercial chip sector has often emphasized the initial quality of components and one-year part warranties. This evolution of the commercial sector has created opportunities for defense applications which could operate in a similarly benign environment. The bond between the commercial sector and defense microelectronics has been strongest in the case of COTS parts.[99] Hence, the military would benefit from the aforementioned hardware transfer: it gains access to home-grown COTS semiconductors with improved performance, increasing functionality and a decreasing cost per function.[100] Such spin-on at the hardware level may further become institutionalized because of related government defense procurement policies, such as the Perry Initiative and its counterparts elsewhere.

Expertise transfers may involve the flows of talent and technologies from the commercial sub-sector to the defense one.[101] Alternatively, they may refer to the formulation of joint ventures between civilian firms and members of the defense establishment. Another possibility may result from the decision by the conventional members of the defense chip industrial base (e.g. government-run agencies) to outsource some of its defense ASIC design and fabrication projects to civilian firms. The transfer of technologies may involve adjusting certain commercial process technologies in order to make military chips. If a country's chip industry has possessed cutting-edge manufacturing technology to fabricate chips chiefly for commercial end-uses, related technology can be used to make chips for military applications, depending on the design.[102] This has to do with the dual-use nature of the process technology for military and non-military chips, as discussed earlier. More importantly, such an adaptation is not necessarily challenging in technological terms. According to a R&D chief at the world's leading pure-play foundry, it would take his team only a year at the maximum to modify, by 10 per cent, its state-of-the-art process technology for the fabrication of civilian chips in order to make mil-spec chips.[103]

If the above analysis of the potential defense significance of a solid commercial sub-field of the overall domestic chip industrial base is substantiated, it is logical to refute the claim that many commercial chip industry interviewees have made regarding what they describe as the absolute irrelevance of their business to defense power and security. After all, resources in the commercial chip sector, which has been the technology driver, can be transferred to benefit the military through the process of spin-on in ways depicted above.

Second, the military sub-field of the overall national chip sector is still important to national defense power and security if it supplies home-grown trustworthy ICs to the national military, especially those unavailable in either the domestic or international mainstream market. These ICs are often different from mainstream commercial parts due to longer lifetimes, harsher environments, and higher consequences of parts failure. (For a comparison of military and commercial ICs, see Table 2.4.)

It is strategically important for a country to retain a tangential and yet viable military chip sub-sector at home for two reasons. First, the availability of such a domestic military IC supply helps to mitigate security risks due to dependency on offshore supplies of ICs. These security misgivings have deepened in recent years in several countries, such as the USA, as the globalization of semiconductor production activities intensifies (General Accounting Office 2002; Lieberman 2003; Defense Science Board Task Force 2005). Second, the defense significance of retaining such an indigenous military chip sub-sector has been heightened by challenges due to the evolution of the commercial chip industry. One of these challenges is the radiation hardness of COTS ICs, which decreases as technology scales into the deep sub-micron level.[104] As military ICs are often required to be radiation-hardened in order to survive in harsh environments, COTS ICs with minimal radiation hardness will not be suitable. To meet this challenge, a viable solution lies in a vibrant defense chip sub-field at home that can churn out high-value, mil-spec ICs that would differ from COTS items.

This, however, does not exclude the possibility that the revamping of the tangential military industrial base can benefit from the mainstream commercial chip sub-field. A telling example involves the implementation of the Trusted Foundry program and the complementary accredited IC supplier scheme. As

Table 2.4 Military and commercial semiconductors: one technology, two markets

	Mil-spec chips	*Commercial chips*
Volume	Low volume of speciality parts	High volume of standardized parts
Global market share	Tangential	Mainstream
Operating environment	Relatively stringent and harsh	Relatively benign
Parts requirements	Ruggedized and radiation-hardened	Less so
Reliability requirements	Higher	Lower
Acquisition cycle time	Longer (e.g. 8–12 years)	Shorter
Parts selection bias	• Older technology preferred • Bias against offshore manufacturing for classified ICs	• Newer technology preferred • Less bias against offshore manufacturing
Cost per unit	Higher	Lower
Quality and reliability control mechanism	Specific rules and standards for design, inspection, test, and certification	Rules and standards for design, inspection, test and certification

Sources: Interviews and secondary data.

discussed above, the Pentagon launched these initiatives partly because COTS ICs alone could not satisfy all the US military needs and because there is a need to infuse new blood into a depleting domestic defense industrial base. So far, the production base has been revamped through the process of spin-on: several eligible domestic commercial chip firms, through a stringent security control mechanism, have been absorbed to nurture the military's ambition to gain access to trustworthy and reliable classified ICs in its march towards defense modernization. Although some of the participating commercial firms have operated in both the commercial and defense chip sub-fields of the overall sector, they, or at least their operations which have been dedicated to the trusted IC supplier schemes, have become new members of the tangential defense production base. The case above reflects the US policy-makers' recognition of the need to retain a viable, albeit small-scale, defense chip sub-field at home in order to advance its national defense and security. Such recognition, as will be discussed in Chapter 4, has been shared by policy-makers in China.

If the defense significance of the tangential defense chip sub-sector holds true in an era dominated by commercial chip end-uses, this will challenge the claim that the industry has little relevance to defense because the defense chip sub-field has been marginalized. After all, marginal domestic defense chip production activities can contribute to national defense and security in ways that the commercial field cannot, although the latter may benefit the former through the process of spin-on.

Third, a country has to adopt a dual-track IC acquisition strategy in order to exploit both commercial and defense chip sub-fields of the domestic chip industrial base if it aspires to ensure the availability and reliability of COTS ICs and custom ICs for end-uses in its defense, intelligence, and critical infrastructure systems.

On the one hand, the greater use of COTS ICs nurtures national defense and security for cost, performance, and efficiency considerations, especially when these ICs are to be inserted in systems which operate in a relatively benign environment. On the other hand, the design and production of custom ICs by the tangential defense sub-field of the industrial base advances national defense and security in a different way. The trusted domestic production environment reduces security risks associated with foreign dependency, and the ICs involved are often not readily available on either the national or the international commercial market since they are to be inserted in systems which operate in harsh environments. To best maximize national defense and security, a country needs to gain access to both sub-fields.

Although the discussion so far has focused on the defense relevance of a national chip industrial base, some parts of the defense chip sub-field have become globalized and ICs, both COTS and ASICs, for military end-uses have not been wholly produced at home. Washington has adopted the aforementioned dual-track IC acquisition policy and has obtained commodity ICs and certain types of ASICs for its military end-uses from the commercial chip sub-field, both home and abroad. Suppliers of these ICs have not been confined to onshore ones. Certified offshore commercial firms, US-owned or foreign-owned, have become part of the globalized US defense chip production base, as will be analyzed in detail in Chapter 3. For this segment of military IC supply, location does not matter. However, Washington has still acquired classified ICs for mission-critical military systems from a limited number of onshore suppliers, which form a closed and small-scale defense chip production base. These suppliers include government-owned or -run facilities and certified commercial firms. For this segment of military IC supply, location does matter. As an American industry player put it, "A country has to maintain a strategic supply to make sure that they have that piece [mil-spec chip] covered, but can get a lot of commodity that is not entirely tied to the location."[105]

To conclude, the dual-use nature of semiconductors and the dominant spin-on practices have complicated the defense significance of a national chip industry. The commercial and defense sub-fields of the national semiconductor sector are of varying security importance to the country in question, and the country should adopt a dual-track acquisition strategy in order to exploit both sub-fields.

Conclusion

In a nutshell, the significance of the chip industry in national power and security terms has been profound and yet complex. The industry has strong linkages to economic power and security for four reasons: (1) the role of the semiconductor as an enabling technology; (2) the role of the industry as an enabler, spurring innovation and growth in other industries; (3) the sector's contribution to growth and productivity, economic competitiveness, and job creation; and (4) the role of the industry as the cornerstone of economic security.

Moreover, the sector is important to defense power and security in four major ways. First, the industry supplies chip components – both mil-spec ICs and COTS devices– which are the building blocks of modern information-dependent military systems. Semiconductors have not only enabled the Apollo Program and the Minuteman II ICBM in the 1960s, but also underpinned precision forces, C4ISR and NCW as major elements of RMA in recent years. The defense importance of advanced chips is expected to continue if the move towards RMA progresses unabated. Improvements at the chip component level will continue to drive innovation at the system level. The relevance of a strong indigenous chip capability to national defense and security will grow accordingly. Second, even though the strength of national defense goes beyond semiconductors, a strong indigenous chip industry remains significant in defense terms – especially for countries with difficulties accessing offshore supplies of military chips, such as China. Third, since chip-based electronics systems are susceptible to semiconductor-focused IW attacks, strong domestic chip capability contributes to national defense security by advancing any given country's defensive and offensive IW capabilities. Fourth, technological changes in the chip sector (i.e. dual-use and spin-on) have complicated the ties between the industry and defense power and security in three major ways:

1   The tangential defense chip sub-sector in a country matters in national security terms if it provides classified chips for critical military systems which are not available on the commercial market.
2   The mainstream commercial chip sub-sector can be of defense importance if it manages to transfer resources to satisfy national defense needs.
3   A dual-track IC acquisition strategy which aims to exploit both commercial and defense sub-fields of the national chip sector is important to ensure the availability and reliability of chips for critical national defense systems.

Hence, the chip technology evolution over the past six decades has created a strategic industry whose importance cannot be underestimated in economic, technological, and defense terms. The sector has occupied a unique position in the discussion of economics, security, and international relations today. Nevertheless, as chip production activities are becoming increasingly globalized, the outcome of these globalization forces has far-reaching security implications. To comprehend the chip–security linkages in the age of globalization, it is key to understand the nature of the sectoral globalization. Chapter 3 will explore this theme.

Notes

1   Interview, 31 January 2005, Washington, DC, USA.
2   Interview, 2 February 2005, Alexandria, VA, USA. The official was from the DTSA.
3   Interview, 24 January 2005, Washington, DC, USA. At the time of the interview, he was the vice-chairman of National Research Council's Board on Science, Technology, and Economic Policy.
4   Also see interview with Nicholas M. Donofrio, Senior Vice-President of Technology and Manufacturing at IBM, 8 February 2005, Armonk, New York, USA.
5   The average growth rate from 1978 to 2005 reached 15 per cent per annum. See Semiconductor Industry Association 2003: 7; Semiconductor Industry Association 2006.
6   However, transistor density improvement rates differ among a wide range of chips. See Barbe 1980; Hutcheson and Hutcheson 1996.
7   The minimum feature size of the transistors on the IC is one of the defining aspects of advanced semiconductor technology, and is often used to define the current level of semiconductor technology.
8   Nanotechnology refers to a broad field of science in which materials are manipulated at dimensions that approach the size of individual atoms or molecules. See Lieber 2001.
9   ITRS is a global effort by the semiconductor industry and research community to define the future development and requirements of the chip technology for the next 15 years.
10   Ruttan (2006: 191–4) regarded semiconductors as one of the seven military general-purpose technologies.
11   Interview with a senior American engineer with more than 40 years' experience in military avionics system design, 22 February 2005, St Louis, MI, USA.
12   Interview with the CEO of an IC design house who has designed military chips in the US, 7 September 2005, Beijing, China.
13   Interview, 4 January 2005, Austin, TX, USA. As will be seen in Chapter 4, Chinese policy-makers have held a similar view of the semiconductor technology.
14   Also see interview with a senior industry player in the Chinese industry, 17 September 2005, Shanghai, China.
15   Interview, 12 January 2005, Dallas, TX, USA.

16 The industry defined in the government statistics includes the semiconductor and electronic components segment, as well as the semiconductor machinery manufacturing segment. See Office of the Governor 2007: 3–4.

17 The Soviet military did the original theorizing of RMA in the 1970s and 1980s.

18 Interview, 7 September 2005, Beijing, China.

19 Another development concerned the Pentagon's sponsorship of semiconductor R&D. See Frost & Sullivan 1981: 23; Braun and Macdonald 1982: 71; Flamm 1985: 40. For the cases of the United Kingdom and Israel, see Dickson 1983; Morris 1994; De Fontenay and Carmel 2004.

20 In 1965, producers of the Minuteman missile accounted for 20 per cent of all sales by the US chip sector. See Flamm 1985: 41.

21 Also see Mackenzie 1990: 207, 326.

22 In 1963, NASA launched the Interplanetary Monitoring Platform (IMP) satellite, the first space vehicle to use ICs. See Mackenzie 1990: 138–9, 140–9.

23 For the US Navy's vision on information in warfare, see Panel on Information in Warfare 1997.

24 For critical reviews of RMA, see Nye and Owens 1996: 23–4; Hayward 2000: 129; O'Hanlon 2000; Williams 2001; Betz 2006; Dombrowski and Gholz 2006: 137.

25 Interview with president and CEO of SEMI, 10 December 2004, San Jose, CA, USA.

26 Interview with a manager at Freescale, 7 January 2005, Austin, TX, USA.

27 Interview, 24 August 2005, Beijing, China.

28 Interview, 6 January 2005, Austin, TX, USA.

29 According to Duncan Lennox, editor of *Jane's Strategic Weapon Systems*, the ability to assess the target over the last few seconds of weapon flight has only recently been achieved. See Lennox 2008.

30 Interview, 31 January 2005, Washington, DC, USA.

31 ASIC is an IC customized for a particular use. The nature of VHSIC will be reviewed later. See Sheppard 1990; Fong 2000: 153, 166–8.

32 Interview with a former head of a military semiconductor fab at TI, 12 January 2005, Dallas, TX, USA. In 1964, TI began to develop the Paveway series of laser-guided systems for bombs.

33 Interview, 24 January 2005, Washington, DC, USA.

34 For an industry's view on the digitalization of battlefield, see Finley 1998.

35 Also see Alberts *et al.* 1999: 88; Dombrowski and Gholz 2006: 9.

36 Interview, 21 January 2005, Alexandria, VA, USA.

37 Interview, 9 August 2005, Taipei, Taiwan.

38 Interview, 9 August 2005, Taipei, Taiwan. CSIST is Taiwan's leading institution for the research, development and design of defense technology.

39 Interview, 31 January 2005, Washington, DC, USA.

40 Examples include radar operations and night vision goggles. See Frost and Sullivan 1981: 124–45; Zolper 2005: 119–20; interview with the president of a Taiwanese IC design house, who was involved in analog IC design for a US-military commissioned project on night vision goggles, 12 August 2005, Hsinchu, Taiwan.

41 A country's strong IC design capability is most relevant to its security because it is the most skill-intensive sub-sector of the industry. However, in view of the increasing fab-lit trend, a country's manufacturing capability, which includes its foundry capability, is also important to national security – albeit to a lesser extent. This is especially so when overt dependence on foreign-located foundry services engenders risks. See Chapter 6 for related discussions.

42 Interview, 22 February 2005, St Louis, MI, USA.

43 Interview, 22 February 2005, St Louis, MI, USA.

44 Interview, 8 December 2004, San Jose, CA, USA.

45 Interview, 28 January 2005, Washington, DC, USA. In 2000, he became a commissioner of the US–China Economic and Security Review Commission.

46 Interview, 6 January 2005, Austin, TX, USA. Ross J. Anderson, Professor of Security Engineering at the University of Cambridge, partly echoed this view. See Anderson 2008a.
47 Interview, 5 August 2005, Taipei, Taiwan.
48 Interview, 26 January 2005, Washington, DC, USA. Also see interview with a former Pentagon official, 26 January 2005, Washington, DC, USA; interview with James Mulvenon of DGI, 21 January 2005, Alexandria, VA, USA.
49 Interview, 2 February 2005, Alexandria, VA, USA.
50 Interview with Roger Cliff, Senior Political Scientist at Rand Corporation, 31 January 2005, Arlington, VA, USA.
51 For studies of IW, see Cobb 1999; Denning 1999; Bishop and Goldman 2003; Bolt and Brenner 2004; Barrett 2005; Knapp and Boulton 2006; Anderson 2008b.
52 Also see interview with Joe Yu-wu Chen, former president of CSIST, 5 August 2005, Taipei, Taiwan.
53 The Office of the Secretary of Defense's definition of IW captures the binary dimension of IW. See Panel on Information in Warfare 1997: 8; Denning 1999: 10. For alternative definitions, see Lorber 2002; Barrett 2005.
54 These weapons emulate the kind of damage that nuclear-generated EMP can inflict upon electronics but at far less range. See Sample 2000; Walling 2001: 92; Bishop and Goldman 2003: 119; Schiesel 2003; Anderson 2008b: 584–6.
55 This assumes that the COTS ICs in question are not radiation-hardened or are not protected by a sufficient amount of shielding at the unit or subsystem level.
56 Also see interview with a former IC design house chief with military microelectronics design experiences in the USA, 20 July 2005, Taipei, Taiwan.
57 These countries include the USA, the United Kingdom, Russia, France, South Korea, China and Taiwan. See Chien 2000; Pillsbury 2001: 17–18; Walling 2001: 97; Abrams 2003; Schiesel 2003; O'Rourke 2007.
58 Cohen, a principal member of technical staff at Sandia National Laboratories, confined chipping to tactics conducted at the hardware level. See Cohen 2000: 16. Also see Critchlow 2000.
59 The supply of counterfeit chips can be driven by economic or defense considerations.
60 An urban legend concerning the First Gulf War involved chipping. According to the original news report, US agents planted a destructive chip in a French printer that was then smuggled into Iraq. Software written on the chip bypassed Iraqi electronic security measures, and devastated the computer network for the Iraqi air defense system, erasing information on display screens. This contributed to the US-led victory. However, the report was credibly debunked. Still, the urban myth continued to be recounted as true and was cited as a "successful" example of the "chip war." See Slade 1992; Schwartau 1994b: 249–52; Dornheim 1998; Fu and Li 2004; interview with Abe Lin, director general of Integrated Assessment Office at Ministry of National Defense, 27 June 2005, Taipei, Taiwan; Qinsheng Wang, Chair of Huada Electronic Design (HED) and Chair of China Integrated Circuit Design Center (CIDC), 30 August 2005, Beijing, China. Wang was formerly involved in R&D for state projects on computer networks and satellite control systems at China's Ministry of Machinery and Electronics Industry. At the time of the interview, she also served as director chief of the IC design branch of China Semiconductor Industry Association (CSIA). CIDC was established in 1986 as China's first independent and mainly commercially oriented design house and was the major undertaker of the 908 Project and the 909 Project in the area of IC design and EDA tool development. See Huada Electronic Design.
61 For other research projects, see Roy *et al.* 2008.
62 For Vogel's work, see 1992.
63 Interview, 5 January 2007, Austin, TX, USA.
64 For instance, dual-use technologies accounted for 14 of 22 "critical technologies" list issued by the Pentagon and the US Department of Energy in 1989. See Samuels 1994: 27.

65 Also see interview with C. Mark Melliar-Smith, former president and CEO of International SEMATECH, 4 January 2005, Austin, TX, USA; interview with a senior manager at AMD, 6 January 2005, Austin, TX, USA; interview with a manager at Freescale, 7 January 2005, Austin, TX, USA.

66 Interview, 24 January 2005, Washington, DC, USA.

67 Also see interview with a manager at Freescale, 7 January 2005, Austin, TX, USA. In a similar vein, Okimoto *et al.* (1987) argued "The know-how for the creation of leading-edge military systems lies primarily in the dual-use electronics infrastructure created for commercial production."

68 8 February 2005, Armonk, New York, USA.

69 Also see interview with Yoshio Nishi, director of Center for Integrated Systems at Stanford University and former senior fellow at TI, 6 December 2004, Palo Alto, CA, USA. Nishi was also a former vice-president of R&D for the semiconductor group at TI.

70 Interview with C. Mark Melliar-Smith, former president and CEO of International SEMATECH, 4 January 2005, Austin, TX, USA.

71 Interview, 12 January 2005, Dallas, TX, USA. Also see interview with a manager at Freescale, 7 January 2005, Austin, TX, USA.

72 Field-programmable gate arrays (FPGAs), DSPs, CPUs and memory chips are also examples of dual-use IC goods. See Lieberman 2003; Defense Science Board Task Force 2005; Office of the Under-Secretary of Defense Acquisition 2005.

73 Interview, 7 January 2005, Austin, TX, USA.

74 As to why higher volume production in the commercial sector helps improve quality and reliability, see Kanz 1991: 303–4.

75 Intel had spent more than US$1 billion to develop the processor. The redesign effort involved NASA, the Air Force Research Laboratory and the National Reconnaissance Office.

76 Also see interview with an American defense industry player, 13 December 2004, San Francisco, CA, USA. For other dual-use chip technology development schemes, see Gansler 1988: 71; Sheppard 1990; Grindley *et al.* 1994; Randazzese 1996; Browning and Shetler 2000; Carayannis and Gover 2002; Pittman 2003; Berlin 2005: 282–3, 295–6, 301–2; Zolper 2005.

77 Also see interview with a former head of a military semiconductor fab at TI, 12 January 2005, Dallas, TX, USA.

78 These firms make chips for the open market.

79 Throughput refers to a parameter used to measure the processing capacity of a chip. The throughput measurement units are gate Hertz, which are the product of the number of gates on a chip times the clock frequency in Hz. See Barbe 1980: 20.

80 In November 1969, the Rome Air Development Center issued MIL-M-38510 as part of the procedure for line certification and parts approval. Suppliers who met the mil-spec would have their products listed on the QPL. See Defense Science Board 1989.

81 The program was administered by the Defense Electronics Supply Center, which subsequently became the Defense Supply Center – Columbus (DSCC).

82 These unfolded amid debates between those in favor of autarkic approach to protecting defense critical industries and those in support of dual-use products for commercial and military usage. See Kanz 1991: 327.

83 Interview, 28 January 2005, Washington, DC, USA.

84 Pecht *et al.* (1997) compared the costs of COTS and QML parts by using a class of TI DSP as an example.

85 Interview, 28 January 2005, Washington, DC, USA.

86 Also see interview with Abe C. Lin, director general of Integrated Assessment Office at Ministry of National Defense, 27 June 2005, Taipei, Taiwan.

87 Also see Defense Science Board Task Force 2005: 7, 85–6.

88 By November 2007, these included CMOS/SiGe Bipolar Complementary Metal Oxide Semiconductor (BiCMOS) 65 nm–130 nm.

89 These procedures include cleared facility and personnel. Only US citizens and Green Card holders can work in the foundry. See McCormack 2004; Carlson 2005; Cohen 2007; Trusted Access Program Office n.d.; interview with an experienced American military IC designer, 3 October 2008, Cambridge, England.
90 Trusted Access Program Office (TAPO) under the NSA manages the contract.
91 According to the Department of Defense (DOD) directive, MAC is applicable to DOD information systems, and it reflects "the importance of information relative to the achievement of DOD goals and objectives, particularly the warfighter's combat mission." The Pentagon has three defined MAC and each requires different levels of safeguard measures. They range from the most stringent protective measures, additional safeguards beyond best commercial practices, to protective measures, techniques or procedures commensurate with best commercial practices.
92 Interview, 15 December 2004, San Francisco, CA, USA.
93 Also see interview with a DTSA official, 2 February 2005, Alexandria, VA, USA; interview with a US defense industry player, 13 December 2004, San Francisco, CA, USA.
94 8 February 2005, Armonk, New York, USA.
95 Examples include Intel and Motorola. Also see McHale 1996; Pecht *et al.* 1997.
96 The MPW program has been developed in the foundry industry in response to today's SOC development methodologies, which require the independent development, prototyping, and validation of several modules before they can be integrated into a single device. MPWs take several IC designs and combine them onto a single manufacturing run, distributing high non-recurring expenses associated with mask development. By participating in an MPW service, a chip designer can enjoy reduced prototyping costs and greater confidence that the design will be successful. The service thus accelerates time-to-market for device designers.
97 For skeptical views on the initiative, see Bryen 2008; interview with an experienced American military IC designer, 3 October 2008, Cambridge, England.
98 The figure is a simplified one because commercial firms may operate in both spheres and members of the defense chip sub-field may change over time.
99 The bond is also strongest for modified-off-the-shelf (MOTS) parts.
100 These supplies can also come from overseas because of the global diffusion of semiconductor technology and free trade in COTS ICs.
101 Examples include the Intel–Sandia case and the VHSIC program.
102 Interview, 7 January 2005, Austin, TX, USA.
103 Interview, 30 June 2005, Hsinchu, Taiwan.
104 Also see interview with a former IC design house chief with military IC design experience in the USA, 20 July 2005, Taipei, Taiwan.
105 Interview with a manager at Freescale, 7 January 2005, Austin, TX, USA.

3 The globalization of the semiconductor industry

> Significantly, the semiconductor industry was the first to which the label "global factory" could be applied. It was here that a spatial hierarchy of production at the global scale first became apparent, with clear geographical separation between different stages of the production process.
>
> (Peter Dicken 2007: 317)

Introduction

This chapter examines the globalization of production in the semiconductor industry for both commercial and military end-uses. Following a brief introduction, the rest of the chapter comprises four sections. The second section examines the economic characteristics of the three separable and distinctive stages of semiconductor production operations, namely, design, fabrication, packaging, and testing (Figure 3.1).[1] The third section explores the different degrees of economic globalization during each of the major production processes, as measured by cross-border business activities such as in-house offshore operations and foreign outsourcing. The section further analyzes the driving forces behind the sectoral globalization, namely, the enabling, push and pull factors. The fourth part of the chapter examines the globalization of semiconductor production activities for military end-uses in the US

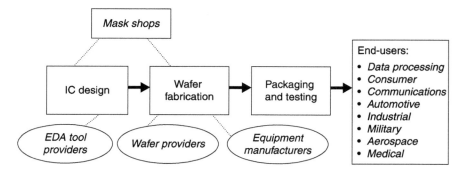

Figure 3.1 Basic stages of the semiconductor industry supply chain.

context. Whereas the production of some military ICs has increasingly globalized, that of the most sensitive chips for mission-critical US weaponry systems has not. The final section summarizes the key arguments of the chapter.

Semiconductor economics

This section summarizes major semiconductor production processes before examining the economic characteristics of IC design, fabrication, packaging, and testing in terms of skill requirements, capital intensity, and specialized input. This examination sets the scene for the analysis that follows because the different economic features of each production stage in the chip supply chain have often affected firms' decision to globalize production activities.

Originally, semiconductor design involved laying out the blueprints of the desired electronic circuits design by hand on paper. In the 1970s, the later stages of the IC design were computerized because semiconductors became increasingly complex. In the 1980s, these stages of design became automated due to the availability of EDA software tools (Brown and Linden 2005: 301).

What bridges IC design and fabrication is mask making, which involves the production of celluloid filaments that contain the microscopic electronic circuits (Henderson 1989: 31). By turning the abstract representations of electronic circuits developed during the design stage into physical presence, mask making thus enables IC fabrication to take place. "It is a process that transfers the virtual world into the real world," observed Parkson Chen, president of Taiwan Mask Corp.[2] Because of increases in IC integration densities, mask costs have become exorbitant. They have grown from US$800,000 at the 90 nm node, to US$1.2 million at the 65 nm node, to several million dollars at the 45 nm node (Hilkes 2007).

During the wafer fabrication stage, which is one of the most complex manufacturing processes in the world today, intricate miniature circuits on the mask are transferred to a thin silicon wafer and etched into its surface through complex manufacturing steps. A cycle of these manufacturing procedures simultaneously creates many identical chips on a wafer, which is coated with resist, a kind of photosensitive material. Each cycle starts with a different pattern, which is projected repeatedly onto the wafer. A chip is made in each place where the image falls. Subsequently, the photosensitive coating is removed. Gases etch the light-exposed areas. These areas are showered with ions (or "doped"), creating transistors. The transistors are connected as successive cycles add layers of metal and insulator (Slomovic 1991: 170–2; Hutcheson and Hutcheson 1996). What follows is a wafer probe test to check the individual dice (i.e. the separate chips-to-be) on a wafer for functionality. Hereafter, the wafers are scored and sectioned into separable dice with the defective dice discarded.

During the packaging and testing stage, electrical leads are connected to the individual dice, which are encapsulated in plastic or ceramic shells, called packages. These packages contain connections to other components. Packaged chips are then subjected to a final functional testing and burn-in before shipment.

Because each of the major semiconductor production stages differs from the others and evolves over time, its respective economic characteristics change accordingly. Hence, it is essential to examine semiconductor economics segment by segment.

The economics of IC design

The economic features of IC design comprise four components: (1) IC design as the most value-added and skill-intensive segment of the industry; (2) increasing automation in design; (3) increasing costs in design; and (4) changes in design composition and cost trend (i.e. rising importance of specification and software stages of design). First, IC design is the most value-added and skill-intensive part of the supply chain. The IC design flow can be understood in two ways. It can be viewed as a hierarchical procedure involving three stages of design (i.e. specification, logic design and physical design), and two additional tasks of validation and software. Alternatively, it comprises hardware (i.e. architecture, verification, physical design and validation) and software tasks (Brown and Linden 2005: 299–300; International Business Strategies 2007).

The highest-level design stage involves the specification (i.e. architecture) of how a chip will function in the system of which it is a part. A high-value added function in the supply chain, the specification stage, requires a combination of inputs (e.g. the chip company's market knowledge and intellectual property) in order to identify which feature set of the chip will maximize profits. During the next stage of logic design, symbolic abstractions are developed to describe how signals will be processed within the chip. During the phase of physical design, engineers translate the abstract version into a map of actual devices and wires interconnecting across multiple layers on the silicon surface. The later stage of physical design, called "place and route," is less skilled than the earlier one (Brown and Linden 2005: 309).[3] Once the chip reaches the prototype stage, it needs to be validated in a hardware simulation of a complete system. Repeated verifications of the design follow. Engineers need to write the software that will become part of the chip and will run on it.

The success of IC design relies on access to a supply of highly qualified IC designers. It also depends on the availability of expensive EDA tools as engineers use these software systems for the realization of their designs or for the automation of the specific parts of design-flow with less engineering input.

The second economic feature of IC design is its increasing automation. This feature results in two major developments. To begin with, the move towards automation in IC design has enabled engineers to specify the parameters of standardized fabrication process technology in the design software, thereby uncoupling design from manufacturing in many IC products. This, in turn, has fostered the emergence of the fabless-foundry business model, which will be explored in detail later. As Leachman and Leachman (2004: 206) argue, the detachability of design from manufacturing "created the potential for the emergence of the fabless semiconductor firm and its binary, the semiconductor foundry." More importantly, increasing automation in design has partly facilitated the globalization of design activities. Computerization allows "the easy transference of designs through computer networks to anywhere in the world." This, along with the debut of high-bandwidth telecommunications, has enabled firms to spread the design process across multiple locations.

The third economic feature of IC design refers to increasing costs in design. Costs at each design stage, which are measured by engineering hours and loaded overhead costs,[4] have increased over succeeding generations of process technology from 0.35 µm to 90 nm line widths, according to a study by a US market research firm (Table 3.1) (International Business Strategies 2007: 1, 16).[5] Over the past five technology generations, for instance, design hours and loaded costs for the representative chip, a logic chip with one million transistors at each generation, have increased respectively by 239 per cent. Engineering hours have increased from 1530 hours for 0.35 µm line widths to 5185 hours for 90 nm line widths.[6]

The fourth feature is the rising importance of the specification and software stages of design, which marks a significant change in design composition and cost trend. The aforementioned study by a US market research firm concludes that design composition over the past five succeeding generations of process technology for the representative project has changed. The number of verification and physical design engineering hours for each million transistors has grown by a modest 29 per cent and 73 per cent respectively. This modest increase results from the growing automation of these parts of the design flow. By contrast, the number of validation engineering hours has increased by 148 per cent for each million transistors. This relatively huge growth is due to the greater complexity of chips and the pertinent software tasks (Brown and Linden 2005: 301).

The most dramatic changes concern specification/architecture and software tasks. Over the succeeding generations, the engineering hours required for the specification/architecture tasks have exploded by 2471 per cent, and the loaded costs for the same operations have increased by 2467 per cent. The software effort has grown by 660 per cent measured by engineering hours and by 665 per cent measured by loaded costs. Another striking change in design composition is the fact that the importance of software relative to hardware during the design process has grown dramatically. At 0.35 µm technology node, software costs in terms of engineering hours are minimal compared to those of hardware. At 90 nm technology node, however, it accounts for more than half of the total design cost. In terms of loaded costs, the significance of software relative to hardware has also increased. This has important implications for the competitiveness of firms in the sub-sector. That is, although software work is not conventionally regarded as part of chip design as such, software expertise has become increasingly instrumental to the competitiveness of IC design firms. Various industry interviewees have confirmed such a development. As the former CEO of a fabless house observed in 2005, the ratio between hardware and software expertise in a typical IC design company that focuses on SOC has been changed from 3:1, to 1:1, to 1:2 or 1:3 in order to ensure firm competitiveness.[7]

Table 3.1 Engineering hours and loaded costs required to design one million logic transistors

Feature dimension	0.35 μm		0.25 μm		0.18 μm		0.13 μm		90 nm	
Logic transistor count	1M		1M		1M		1M		1M	
	Engr. hours	Loaded costs ($M)	Engr. hours	Loaded costs ($M)	Engr. hours	Loaded costs ($M)	Engr. hours	Loaded costs ($M)	Engr. hours	Loaded costs ($M)
Hardware										
•Architecture	23.0	0.003	29.8	0.004	91.4	0.012	271.6	0.035	591.4	0.077
•Verification	714.2	0.093	738.4	0.096	756.4	0.098	837.7	0.109	921.4	0.120
•Physical design	311.0	0.041	357.2	0.047	391.7	0.051	473.5	0.062	538.6	0.070
•Validation	103.7	0.014	127.6	0.017	164.5	0.021	197.4	0.026	257.4	0.034
Subtotal (design engineering resources)	1,152	0.150	1,253	0.163	1,404	0.183	1,780.2	0.232	2308.8	0.301
Software	378.4	0.049	672.4	0.088	985.7	0.128	1,798.3	0.234	2876.5	0.375
Total	1,530.4	0.199	1,925.4	0.251	2,389.7	0.311	3,578.5	0.466	5,185.3	0.675

Source: Data from Table 1.3, International Business Strategies 2007.

What accounts for these dramatic changes in design composition? The answer is the increase in IC design technological complexity as increased SOC has emerged as the dominant trend in the industry today.

> Design task composition is changing with the reduction in feature dimensions, and there is increased emphasis on implementing the architectural concepts in silicon. Thus, the costs of developing the architectural concepts are migrating from the system to the IC design phase, and software development costs are migrating from the system development environment to IC development.
>
> (International Business Strategies 2007)

As system software needs to be developed in parallel with the system-level chip hardware for reasons of coherence and time-to-market, the need for scrupulous hardware-software co-design results in the explosive design cost for the specification/architecture segment of the design flow.

To conclude, chip design is the most value-added and skill-intensive segment of the industry. Moreover, the move towards automation in IC design has enabled firms to globalize part of its design operations by transferring information across national borders via high-speed communications interfaces. Design automation has also helped to decouple design from manufacturing, thus facilitating the rise of the fabless-foundry business model. In addition, design costs have increased over time as a function of reduction in feature dimensions. Finally, the growing cost for specialization and software tasks has been striking because of the increase in design complexity amid the move towards SOC. Because of the aforementioned features of IC design, the competitiveness of firms in this sub-sector thus relies on three inter-related elements: (1) access to a limited number of qualified and experienced IC engineers with hardware and software expertise; (2) access to EDA tools; and (3) the ability to tackle the challenge of the longer time-to-market for design completion (ibid.).

The economics of IC fabrication

The economic features of IC fabrication include the following four dimensions: (1) IC fabrication as a skill-intensive and capital-intensive process; (2) manufacturing equipment cost as a substantial portion of fab expenditure; (3) process yield as the single most important determinant of economic productivity for an IC manufacturer; and (4) increase in wafer size as another determinant of IC fabrication productivity.

First, IC fabrication is skill-intensive and capital-intensive. Highly skilled engineers and technicians, aside from operators, are required during the laborious manufacturing process in spite of the heavy use of computer-controlled processing and tracking systems. Moreover, this sub-sector is highly capital-intensive because the build-up of a modern semiconductor fabrication facility (hereafter, fab) requires a large amount of fixed investment for its costly manufacturing equipment, its ultra-clean rooms built to stringent specifications to maintain a low density of airborne particles and the availability of suitable utilities. In 1965, the cost of a state-of-the-art fab was about US$1 million. By 1981, it had reached US$50 million (Henderson 1989: 43). Today, a leading-edge 12-inch wafer fab typically costs in excess of US$2 billion, making it one of the most expensive types of factories in the world. For instance, Intel's 12-inch wafer fab in northern China involves a US$2.5 billion investment (Clendenin 2007; Intel 2007b). According to one account, the capital investment for chip manufacturing was about ten times that for packaging and testing on a per-chip basis (Leachman and Leachman 2004: 205).

Second, a substantial proportion of fab cost arises from the expensive chip manufacturing equipment rather than labor. The manufacturing equipment accounts for 16 per cent of costs (including depreciation) in a typical 8-inch wafer fab and less than 10 per cent in a 12-inch fab in the USA (Howell *et al.* 2003; Leachman and Leachman 2004: 288). Among the exorbitantly expensive equipment, lithography deserves special attention. Its advances are important to the success of a fab because they determine the smallest possible features that can be created on ICs. In 1996, the price of lithography equipment had risen at 28 per cent a year although the size of these smallest features had decreased about 14 per cent annually since the inception of the industry. In the early days of the sector, each new generation of lithography equipment cost ten times as much as the previous one did. Nevertheless, these steep price increases were reduced to a doubling of price with each new major lithography development. The price of other types of chip fabrication equipment had also augmented over the succeeding generations of technology development.[8]

Third, the total process yield is one of the single most important determinants of economic productivity for an IC manufacturer. It refers to the fraction of individual dice that survives all stages of production and testing to emerge as marketable packaged chips. Total yield varies. Whereas it can be 80 per cent or higher for relatively simple chip manufacturing with mature process technologies, it can be as low as 10 per cent for highly complex chips in the early stage of production.

Fourth, the increase in wafer size also determines IC fabrication productivity. For instance, a 12-inch wafer is more valuable than before because its area is 2.25 times that of an 8-inch one. However, wafer size increase contributes to changes in the workforce composition in a fab. Because a 12-inch wafer is heavier and more awkward to handle than an 8-inch one, the risk of the wafer being dropped by human handlers increases. Moreover, an increase in wafer size often requires major re-engineering of the equipment and process technology. Hence, both information systems and materials handling become more automated in order to ensure the safe handling of the increased weight and value of each wafer and to minimize human error. Today, a 12-inch wafer fab has total automation of materials handling and wafer processing. The increasing automation means that the need declines for operators and increases for engineers. Because a 12-inch wafer fab processes advanced chips often by using high-end process technology, the amount of inspection and in-line engineering-related activities required in the fab is significantly higher than that in an 8-inch plant. Thus, the increasing auto-mation in a 12-inch wafer fab does not reduce the number of workers involved; instead, it changes task composition in the factory.

Brown and Campbell's study (2001) supports the analysis above. They compared workforce composition in matched 6-inch and 8-inch fabs based on a sample size of 14 fabs in the 1990s. They concluded that engineers increased from 15 per cent to 24 per cent of the total workforce between 6-inch and 8-inch generation fabs, with a corresponding decline in operators from 73 per cent to 62 per cent. However, the total number of workers in each fab stayed the same at about 750.[9]

To conclude, chip manufacturing is skill-intensive and capital-intensive, with the expensive fabrication equipment accounting for a large proportion of fixed capital investment in a modern fab. The determinants of economic productivity for an IC manufacturer include the pursuit of high process yield as well as wafer size increase. However, wafer size increase results in changes in the workforce composition in a fab. As noted, while advanced automation in a 12-inch wafer fab reduces the number of operators needed, the requirement in the same fab for highly skilled engineers increases.

The economics of IC packaging and testing

The packaging and testing sub-sector is the most labor-intensive segment of the semiconductor supply chain in spite of its move towards automation since the 1980s. It is less skill-intensive and technology-intensive than design and fabrica-tion. Although packaging and testing equipment is also expensive, a plant in this sub-sector is not as expensive as a fab. In spite of the need for a clean production environment in a packaging and testing site, this is less critical than for fabrication (Dicken 2007: 318).

Capital equipment costs for standard digital testers are low. Moreover, test costs per hour for digital devices account for less than 5 per cent of the overall device cost, although they are dramatically more expensive as ICs become more integrated. For instance, a high-speed analog tester can cost hundreds of dollars per hour, so the cost of the final test can reach as high as 30 per cent of the device cost (Hilkes 2007).

To what extent has semiconductor economics enabled firms to pursue the inter-national division of labor? As will be examined later, the separation of the produc-tion processes spatially is far from unified for IC design, fabrication, and packaging and testing.

The economic globalization of the semiconductor industry

Although the globalization of production in the semiconductor industry varies in degree across and within the three stages of the production process, it has consist-ently increased over time. According to the *2005 World Investment Report* (United Nations Conference on Trade and Development 2005: 173), "The chip industry was one of the earliest to globalize production." The sector has successively offshored packaging and testing, fabrication, and then design. The most globalized segment is the relatively low-end and labor-intensive areas of packaging and testing. Firms in this sub-sector began their offshore operations as early as the 1960s, giving the industry the nickname of the "global factory" (Dicken 2007: 317). The latest wave of semiconductor globalization encompasses all three main segments of the industry's food chain, contributing to what George Scalise, president of SIA, described as a "new outsourcing" trend.[10] In this new model, production activities in the three sub-sectors have become concurrently internationalized. Excluding Japan, the Asia-Pacific countries, particularly China, have become prominent spots in the investment profiles of global firms (Semiconductor Industry Association 2009: 14, 17). This section first of all explores the evolving patterns of production globalization for each of these production stages, and second, analyzes the driving forces behind the pertinent globalization.

The globalization of packaging and testing segment

The most globalized segment of the industry is packaging and testing. Since the 1960s, firms in this sub-sector have led the locational shifts from the developed countries to the lower-labor-cost developing ones, resulting in the emergence of Asia-based suppliers over time as the dominant players in the sub-sector. Five points follow.

First, Flamm's study (1985) concludes that semiconductor firms headquartered in different national systems began to pursue the varied degrees of internationalization of production in the 1960s. Since then, US chip firms have been the most active in spearheading the geographically diversified spread of assembly operations of semiconductors for export to the USA. Fairchild pioneered these offshore operations by setting up a manufacturing affiliate in Hong Kong exporting to the US market. Other US companies followed suit, spreading overseas assembly activities to countries including Korea, Taiwan, Mexico, Malaysia, Indonesia, Thailand and the Philippines in the next 20 years.

Second, chip supply within Asia in this sub-sector has become diversified over time. According to Flamm (ibid.: 68–85), exports of assembled chips to the US from Hong Kong continued to decline from the 1960s to the 1980s, whereas those from Singapore and Malaysia increased. During the mid to late 1970s, Indonesia, the Philippines and Thailand were major hubs for chip assembly production activities in Asia. By the early 1980s, Malaysia became the favorite Asian destination for the assembly of US semiconductors, followed by the Philippines, Korea, and Taiwan.[11]

Third, while the US firms initially engaged in in-house off-shoring assembly activities by setting up overseas subsidiaries, off-shored outsourcing has gradually become the dominant trend. Today, most US giant chip firms still own and operate assembly plants in Asia, although many of them have outsourced a large portion of their operations to Asia-based foreign suppliers.

This has resulted in the emergence of Asian suppliers over time, the fourth feature of globalization in the subsector. Since the early 1980s, about 85 per cent of worldwide packaging and testing capacity has been concentrated in Southeast Asia. In terms of location and business boundaries, this segment of the supply chain has remained relatively stable, with a few exceptions such as Intel's facility in Costa Rica (Leachman and Leachman 2004: 205). As the revenue of the worldwide contract semiconductor assembly and test services (SATS) market reached $20.2 billion in 2008, nine of the top ten SATS firms are headquartered in the Asia-Pacific region, accounting for 49 per cent of global revenue. Hence, "the initial moves offshore can have unforeseen dynamic consequences such as the emergence of foreign suppliers who dominate the industry segment" (Brown and Linden 2005: 282–3).

Finally, the outsourcing trend in the sub-sector is expected to continue. While the outsourcing segment's share accounted for 43.6 per cent of the worldwide packaging market in 2006 (Semiconductor International 2007), it is expected to reach US$33.12 billion in 2011 from US$15.17 billion at a compound annual growth rate (CAGR) of 11.7 per cent. While IDMs outsourced only 28.7 per cent of their packaging and testing needs in 2001, the percentage is expected to reach 48.1 per cent by 2011.

The globalization of the IC fabrication segment

Two distinctive features characterize the globalization of semiconductor fabrication. First, since fabrication offshore investment began in the 1960s, the global chip fabrication capability has shifted geographically, resulting in the relative decline of the US and Japan and the relative rise of the Asia-Pacific region, excluding Japan. Second, the development of the fabless-foundry business model in the late 1980s has spurred the emergence of outsourced offshore fabrication operations in the Asia-Pacific region, and the impact of these Asia-based contract manufacturers on the global semiconductor landscape continues to the present day. These two features will be detailed as follows.

Offshore investment

In the 1960s, US chip firms began to invest in offshore fab facilities primarily to gain market access in Western Europe and Japan rather than to lower production costs (Henderson 1989: 45, 118–38; Brown and Linden 2005: 288). This development, however, has been partially offset by reciprocal investment from leading European and Japanese producers.

US chip companies began to invest in wafer fabs in Scotland in the 1960s because of the existence of a growing and substantial market in Western Europe, which was protected by the 17 per cent high European Economic Community (EEC) tariff barrier on semiconductors. Because the high tariff was levied on the value added during the fabrication process, these companies could produce most of the value-added in their European fabs, cheaply assemble and test the chips in their Asian facilities, and then import the completed devices back into the EEC market. For these companies (with the exceptions being Motorola, and to a lesser extent, National Semiconductor), a major portion of their European customers was the British and European defense end-users. Because defense end-users of semiconductors often imposed stringent quality control over the supplies of these components, the need for systemic and regular customer–supplier liaison ensued. Hence, "it has become increasingly difficult for a component manufacturer to break into a national defense market without producing at least the technological core of the component within the particular national or regional boundaries themselves" (Henderson 1989: 126). This explains why Hughes Aircraft, a major supplier of military chips then and in the 1980s, pioneered the investment of semiconductor fabs in Scotland in 1960.[12]

Since offshore investment in semiconductor fabrication started in the 1960s, the geographical distribution of worldwide fabrication capacity has changed accordingly. This change is characterized by the relative decline of Japan and North America, the stagnant performance of Europe, and the explosive rise of the Asia-Pacific region (Figure 3.2).

In the 1950s, the commercial production of semiconductors began in the USA, and since then the USA has dominated global chip production for more than two decades. Japan caught up in the 1980s surpassing the USA by 1985. However, the share of the global manufacturing capacity by Japan and North America decreased from 80 per cent in 1980 to 49 per cent in 2001. Europe has had stagnant

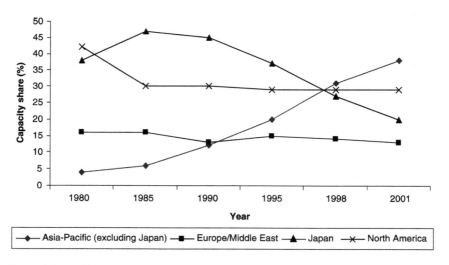

Figure 3.2 Regional percentage share of the worldwide fabrication capacity, 1980–2001*.

Source: Table 8.2 in Leachman and Leachman 2004.

Note: *All figures are fractions of total global fabrication capacity and may not add to 1.0 due to rounding. The Europe/Middle East region includes Russia, Turkey, and Israel. Fabrication capacity located outside the four regions was negligible in all years.

performance in the sub-sector because of the continuous shift of related investments away from European economies. The UK, in particular, has become a marginal player in semiconductor fabrication today. In October 2007, TSMC decided to purchase Atmel's 8-inch wafer fab equipment in the UK in order to expand its fabrication capacity in China (Walko 2007; TSMC 2007d). The fate of two other strategic fabs in the country is mixed. By early 2009, Freescale was in the process of winding down manufacturing at its Scottish fab because it planned to close the plant that had operated since the 1960s (Lammers 2009). In March 2008, Filtronic sold its GaAs fab near Country Durham, which is Europe's largest, to chip-maker RF Micro Devices (RFMD). In view of the erosion of the chip fabrication base in the UK, an industry observer lamented that the majority of the fabs "built by big foreign players in the heady 1990's could be mothballed soon, leaving the U.K. with less chip manufacturing capacity than all other major European economies" (Walko 2007).

By contrast, over time the Asia-Pacific region has steadily augmented its share of the worldwide fabrication capacity. Its share rose from 4 per cent in 1980 to 38 per cent in 2001. In 2006, the region accounted for 43 per cent of the worldwide installed wafer fabrication capacity, followed by Japan (25 per cent), North America (20 per cent) and Europe (12 per cent) (Solid State Technology 2007a). Dicken (2007: 319) described this development as a "major global shift" in the semiconductor arena.

Within the Asia-Pacific region, Taiwan, and South Korea have emerged as the leading chip-makers. Both countries accounted for 83 per cent of the fabrication capacity located in the region in 1990, 89 per cent in 1995, 88 per cent in 1998 and 86 per cent in 2001. Whereas Korea was the region's top chip-maker in 1990 and 1995, Taiwan took the lead in 1998 and 2001. Singapore had the next largest share during the period studied, about 9–11 per cent. In 2001, Malaysia had a 3 per cent share of the regional total (Leachman and Leachman 2004: 209–10). By 2006, Taiwan and Korea accounted for 89 per cent of the installed wafer capacity in the region; Taiwan alone had 56 per cent of the total capacity, followed by Korea (33 per cent), China (7 per cent), and Singapore (5 per cent) (Solid State Technology 2007a).

In recent years, China has emerged as a favorite destination for foreign investments in chip fabs. In March 2007, Intel announced that it would build a 12-inch wafer fab in China, marking the first time in 15 years that the global chip giant had decided to build a new manufacturing plant overseas (Clendenin 2007; Intel 2007b). Four foreign IDMs had some form of invested IC wafer fabrication capacity in China by the end of 2009; they included Hynix, Intel, Nippon Electric Company, Ltd. (NEC), and ProMOS (PricewaterhouseCoopers 2010: 42). What have been emerging are "geographical shifts" in the global chip sector, whereby semiconductor firms are moving their production activities to China to take advantage of the world's fastest-growing economy.[13]

However, the fall in capacity owned by US firms is less severe than the decline in capacity located in the US though the rise of capacity owned by companies in the Asia-Pacific region mirrors that of regional capacity (Figure 3.3) (Leachman and Leachman 2004).

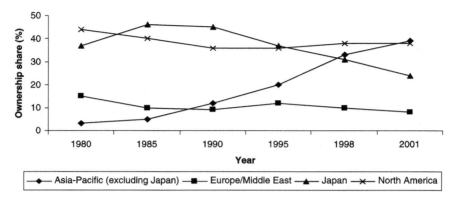

Figure 3.3 Regional ownership percentage share of the worldwide fabrication capacity, 1980–2001*.

Source: Table 8.4 in Leachman and Leachman 2004.

Note: *The ownership total for 2001 adds to more than 100 because the capacity of jointly owned fabs is credited to all owners.

In 2001, chip firms headquartered in the Asia-Pacific region accounted for 39 per cent of the share of ownership of fabrication capacity, followed by North America (38 per cent), Japan (24 per cent), and Europe (8 per cent). In addition, the shares of global capacity wholly or jointly owned by Japanese and North American firms did not decline as much as did the shares of fabrication capacity located in those regions. Nevertheless, the decline in the Japanese ownership share was steep.

The concentration of both fabrication activity and ownership of fabrication capacity in the Asia-Pacific region has resulted from two changes that occurred during the second half of the 1990s. The first change refers to the increase in investments in foreign-located capacity by North American, Japanese, and European firms; these investments mostly targeted in-process technologies or contract manufacturers based in the Asia-Pacific region. The second change involves the relatively few investments by Asian-Pacific firms in fabrication capacity located outside their region (ibid.: 211; Brown and Linden 2005: 294).

Offshore outsourcing

The second feature of the globalization of IC fabrication refers to the rise of outsourced offshore fabrication operations, which have been geographically concentrated in the Asia-Pacific region since the late 1980s following the debut of the fabless-foundry business model.

In the mid-1980s, the birth of the pure-play foundries helped some of the industrializing countries, especially Taiwan, successfully build up local fabrication capability. This occurred in spite of the relatively few cases of US-invested fabs in these countries even with the rich subsidies Singapore and others had offered. Originated in Taiwan in 1987, pure-play foundries are contract manufacturers: they do not sell chips of their design but concentrate on fabricating chips designed by other firms. "It's a company that does not sell its own products; what it basically sells is its manufacturing service," said Klaus C. Wiemer, former president of TSMC, the world's first pure-play foundry.[14]

Today, the Asia-Pacific region dominates the global pure-play foundry capacity. Its share of the global total increased from 70 per cent in 1995 and 76 per cent in 1998 to 89 per cent in 2001. In 2007, it was home to nine of the world's top ten contract manufacturers, with the only exception being Europe-headquartered X-Fab (Table 3.2). Measured by monthly 8-inch wafer production, Taiwan's foundry capacity remained the strongest worldwide in July 2009 (IC Insights 2009).

The ascendancy of the predominantly Asia-based pure-play foundries has changed the global semiconductor industry in four ways. First, the birth of the world's first pure-play foundry in Taiwan has proven to be a successful business model, thereby fostering fabless-foundry operations and the vertical specialization of the industry. According to F. C. Tseng, vice-chairman of TSMC, pure-play foundries allow IC design houses to focus on design, and the success of this model "enables the semiconductor industry to become disintegrated,"[15] contributing to the subsequent restructuring of the sector. Over time, the rise of these Asia-based pure-play foundries has enabled chips designed by the US, European or Asian fabless houses to be fabricated in these operations.

Although some IDMs in the USA and Japan could operate as contract manufacturers just as pure-play foundries did,[16] pure-play foundries outperformed IDMs as the most important service providers of outsourced fabrication to the rapidly

Table 3.2 Top ten pure-play foundry companies in 2007

Rank	Company (headquarters)	2004 sales ($M)	2005 sales ($M)	2006 sales ($M)	2007 sales ($M)
1	TSMC (Taiwan)	7,648	8,217	9,748	9,813
2	UMC Group (Taiwan)*	3,900	3,259	3,670	3,755
3	SMIC (China)	975	1,171	1,465	1,550
4	Chartered (Singapore)	1,103	1,132	1,527	1,458
5	Dongbu (S. Korea)	228	347	456	510
6	Vanguard (Taiwan)	474	353	398	486
7	X-Fab (Europe)	177	202	290	410
8	SSMC (Singapore)	260	280	325	350
9	HHNEC (China)	324	313	315	335
10	HeJian (China)	230	250	290	330

Sources: IC Insights 2006; 2008; company reports.

Note: *This includes UMC Japan sales.

growing fabless sector in the USA and subsequently in other countries (Brown and Linden 2005: 289–90).[17] Because IDMs typically run a design business, fabless firms fear that when they outsource fabrication business to IDMs, IDMs may steal their proprietary intellectual property (IP) pertaining to their products. By contrast, foundries do not design or market products under their brand name, and thus are not viewed as competitors by their customers (Hurtarte *et al.* 2007: 7).[18] Thanks to the rise of pure-play foundries, "a new breed of fabless companies began to emerge that were able to take an outsourced manufacturing strategy as a given" (ibid.: 5).

Second, the rise of the Asian foundries, by legitimizing the fabless business model and by facilitating the blossoming of US fabless design houses, contributed to the recovery of the US semiconductor industry in the 1990s (Macher *et al.* 1998; Hurtarte *et al.* 2007: 3–5). Because of the leading role played by the USA in fabless design, the recovery and subsequent blossoming of US fabless design houses have reinforced the formidable role that fabless firms have played in the global industry since the 1990s. In 1995, Cirrus Logic became the first fabless firm to surpass the billion-dollar annual revenue barrier in 45 quarters. Broadcom followed in 2000 – all in 36 quarters. Nvidia soon followed in 2001 – all in 32 quarters. By 2006, all of the world's top ten fabless companies had achieved billion-dollar annual revenue. The performance of the global IC fabless sector is equally striking. From 1998 through 2006, fabless design firms increased at almost three times of the 9 per cent average annual growth rate of the worldwide IC market. During the same period, fabless IC company sales grew almost six-fold, whereas the total IC market increased by only 93 per cent. In 2010, their sales reached US$73.6 billion, accounting for 24.5 per cent of the US$299.9 billion semiconductor sales total (Global Semiconductor Alliance n.d.).

Third, pure-play foundries have further facilitated the restructuring of the global chip sector by allowing IDMs to hedge the enormous risk of building expensive new fabs through utilizing foundries for buffer capacity. IDMs often provide pure-play foundries with advanced process technologies in exchange for a guaranteed supply of chips from these fabrication service providers. This technology-in-exchange-for-capacity model can be traced back to the early days of TSMC operations when some of the US IDMs, such as TI, provided the Taiwanese start-up with much needed technology. "We were trading capacity or capital investment for technology," recalled Wiemer.[19] Citing the case of TSMC-US IDMs, Wiemer argued that it would make business sense for IDMs to shift businesses to foundries in order to minimize the risk of having unused fabrication capacity with a huge fixed depreciation expense. The increasing capital requirements of semiconductor manufacturing thus provide another impetus for vertical specialization (Macher *et al.* 2007). As IDMs began to shift business to contract manufacturers in the mid-1990s, they accounted for about 45 per cent of foundry revenue around 2004 (Brown and Linden 2005: 292). In 2007, pure-play foundries accounted for 84 per cent of the worldwide foundry market (IC Insights 2006). These contract chip-makers are expected to become increasingly important as outsourcing from IDMs continues to be on the upswing.

Fourth, and finally, the world's leading pure-play foundries have shaken the global industry by demonstrating technological capabilities close to the sector's leading giants. The technology level of TSMC today is close to the sector's leaders such as Intel and IBM, which are first to put a new technology generation into production in order to enjoy first-mover advantages (Brown and Linden 2005: 289–90).[20]

The globalization of the IC design segment

So far, we have discussed the globalization of packaging, testing, and IC fabrication. The following section examines the four features of the globalization of IC design sub-sector. These features are summarized as follows. First, the US chip firms have remained a dominant force in the IC design segment. Although they began their in-house offshore designs in the 1970s and have expanded their offshore investments in IC design since the 1990s, a substantial geographical shift of worldwide IC design capability has not ensued. Second, IC design firms in the US, Europe, and Asia tend to outsource their design work domestically or in their region rather than internationally. Third, even when international outsourcing of design occurs, most of the vital IP critical to the core competency of a company is often developed and retained within or close to firm headquarters. In contrast, the lower-skilled and less sensitive part of design steps is frequently outsourced internationally. Hence, IC design is the least globalized compared to fabrication, packaging, and testing. Macher *et al.* (2007) thus labeled the internationalization of innovation-related activities of semiconductor companies, including IC design houses, as a case of "non-globalization." Fourth, the Asia-Pacific region has gradually emerged as a favorable destination for offshore design investments and third-party outsourcing because of the availability of local design engineers in spite of the uncertainty over the future course of this trend.

Offshore investment

First, the US chip firms started to invest offshore in in-house design activities in the 1970s, and offshore investments in IC design have expanded since the 1990s. However, the globalization of IC design activities so far has not resulted in an obvious geographical shift in design capability.

Through the 1970s, US chip firms engaged in most early offshore design investments in Western Europe and Japan. These design centers adapted standard designs to customer requirements instead of engaging in innovative work (Henderson 1989: 45–8). In 1978, TI set up a design center in Northampton, England, as a backroom next to its IC fab. After TI sold the plant in 1995, the hub nevertheless continued to be a TI subsidiary engaged in the design of DSP, TI's flagship product (Daly 1996). By the mid-1980s, some of the advanced economies in East Asia outside Japan, including Hong Kong, Taiwan, and Singapore, also became destinations for US design investments. These Asian design hubs adapted existing chips to local market needs, performing tasks similar to those carried out at their counterparts in Western Europe and Japan. By contrast, the headquarters in the USA carried out all the major innovative design work (Henderson 1989: 56–8).

In 1985, TI became the first US IDM to invest in in-house design operations in India by setting up a software center in Bangalore, focusing on developing design automation software for internal use (United Nations Conference on Trade and Development 2005: 167). In 1988, TI-India began to design mixed-signal chips. In 1995, the subsidiary became involved in the design of DSP. In 1998, the subsidiary announced that it had taken its first DSP core from specialization to working silicon over the preceding two years and that it had integrated a controller with DSP function for the first time. By 2003, the technological capability of TI-India was such that it had had 225 US patents. In 2004, TI introduced a high performance analog-to-digital (A/D) converter that was designed primarily in Bangalore (Lammers 2004; Brown and Linden 2005: 314–15).

Since the 1990s, the amount of offshore IC design in industrializing countries has expanded. Some firms have followed in the footsteps of TI by setting up subsidiaries in India. STMicroelectronics set up its design center in India in the 1990s. Freescale India was founded in 1998. Intel India Development Center, the largest non-manufacturing international site for the chip giant, started in July 1999. Today, major semiconductor players run their design centers in India (Giridhar 2006; Freescale 2007). Since 2001, semiconductor and systems companies have established their presence in India at an accelerating pace. For instance, Freescale expanded its operations in India in March 2007 by inaugurating a new chip design campus in Noida. These subsidiaries perform various functions, ranging from sales, outsourcing part of their design work to local design service companies, or shouldering in-house design tasks by taking advantage of lower costs in India (Nadamuni 2004: 3). According to *Gartner Dataquest*, electronic equipment manufacturing typically outsources to China, while product design and software engineering support often go to India (Koh 2004: 2).

In-house design off-shoring, however, works both ways. Many foreign IC design firms also invest in the USA and maintain their design centers on US soil, chiefly to take advantage of the highly-skilled engineering talent there and to gain access to their US customers. The case of Spreadtrum Communications (Shanghai) Co., Ltd., one of China's top ten fabless firms, is telling. Wu Ping, the CEO of the company, observed in 2005 that his company's design center in Silicon Valley enabled the firm to design on a 24-hour cycle amid the enormous pressure to reach the market ahead of its competitors. In his words:

> We utilize global resources. Our firm operates 24 hours every day. Our research and development is done in Silicon Valley during the day. In the evening, the result achieved there is submitted to China enabling China-based engineers to continue the pertinent task. Subsequently, the result accomplished in Shanghai is sent back to the US.[21]

The firm's bi-national arrangement is common among IC design houses I interviewed. However, other firms manage multiple design centers in more than two countries. For instance, one start-up in Beijing adopted a rolling cycle between design hubs in the USA, China, and France,[22] whereas the other in Taipei did so between operations in the USA, China, and Taiwan.[23]

In spite of the accelerating pace of in-house design outsourcing since the 1990s, there has not been a substantial geographical shift of worldwide IC design capability. This differs from the development in the IC fabrication sub-sector as the pertinent production globalization has resulted in the wide distribution of semiconductor manufacturing capability among developing and developed countries of the global economy. So far, the USA has continued to dominate global chip design activities, especially those associated with fabless design functions (Macher *et al.* 2007). It has remained the world's strongest IC fabless design powerhouse, accounting for over 70 per cent of global fabless revenue. Taiwan has maintained its second place in the global chip design race, generating more than 20 per cent of the worldwide total (Jen 2004). In 2006, North America (which was dominated by US companies) represented 75 per cent of global fabless revenue, followed by Asia with 21 per cent (which concentrated in Taiwan), Europe with 3 per cent and India with less than 1 per cent (Taylor 2006). In 2010, out of the top ten fabless design houses by revenue, eight were headquartered in the USA, one in Taiwan, and one in Europe (Table 3.3). In the same year, North America housed 500 fabless companies. It was followed by Asia with 500, Europe, the Middle East, and Africa with 200 (Global Semiconductor Alliance n.d.).

Domestic and international outsourcing of design activities

The second characteristic of the globalization of IC design activities is that IC design firms in the USA, Europe, and Asia tend to outsource their design work domestically or in their region rather than internationally.

According to Brown and Linden (2005: 305–6), most IC design outsourcing by US companies had taken place domestically rather than internationally: "Many

Table 3.3 Top ten fabless design companies by revenue in 2010

Rank	Company	Company headquarters	2010 revenue (US$)
1	QUALCOMM – QCT Division	USA	7,204,000
2	Broadcom – Product Division	USA	6,589,270
3	AMD	USA	6,494,000
4	MediaTek	Taiwan	3,909,158
5	Marvell Semiconductor	USA	3,611,893
6	NVIDIA	USA	3,543,309
7	SanDisk – OEM Division	USA	2,776,800
8	LSI	USA	2,570,047
9	Xilinx	USA	2,310,613
10	ST-Ericsson	Switzerland	2,293,000

Source: Global Semiconductor Alliance n.d.

interviewees reported that they outsourced physical design to small local companies on an as-needed basis." In an EDA user survey in 2006 (Electronic Engineering Times), IC designers in North America, Europe, and Asia said they most often outsource a portion of their designs in their region.[24] In the study, 68 per cent of North American respondents said their firms outsource to third parties in North America, a drop from 90 per cent in the 2005 survey. About 70 per cent of Western European respondents said their firms outsource to companies in the Continent. About 50 per cent of Asian respondents work with outsourcing partners in China and Taiwan, and 48 per cent in India. The *Electronic Engineering Times/Deutsche Bank* EDA user survey in 2006 (Goering 2006a) revealed a similar trend: 90 per cent of respondents said their companies use providers in North America, compared with 24 per cent for India and 21 per cent for China.

A case of non-globalization

The third aspect of the globalization of IC design activities is that IC design is the least globalized in the chip supply chain and that the internationalization of innovation-related activities of chip companies, including IC design houses, can be dubbed a case of "non-globalization."

To begin with, Macher *et al.* (2007) noted that the innovative activities of chip firms, many of which are related to IC design, are not as globalized as the other production stages of the supply chain. Based on an examination of four measures of the globalization of innovation-related activities in the semiconductor industry, they concluded that the "innovation-related activities of otherwise global firms in this industry remain remarkably 'nonglobalized'." In particular, their study of the patenting activities of US fabless chip firms during 1991 to 2003 indicated a modest growth in offshore inventive activities for the period studied. They argued that such a trend is not limited to US companies because the "homebound" character of the US firms' patenting is similar to that of their counterparts elsewhere. The results of their analysis of patents provide the strongest support for the seminal Patel–Pavitt study (1991) of the non-globalization of innovation-related functions of giant firms in the world economy.

The Patel–Pavitt (1991) study is based on an analysis of the US patenting by 686 of the world's largest manufacturing firms in different sectors and different countries. It concluded that the innovative activities of these companies are among the least internationalized of their functions and that "their technological activities remained far from globalized" (ibid.). Although Vernon (1979) and others suggested in their studies that large firms are increasingly footloose in their R&D activities, contributing to the trend towards "techno-globalism," the Patel–Pavitt findings showed the contrary: for the high-end innovative activities, giant firms tend to prefer a national bias rather than globalization.

Various scholarly studies and industry surveys (Brown and Linden 2005: 305–6; Electronic Engineering Times 2006; Macher *et al.* 2007) concluded that the least skill-intensive and the least value-added stages of the IC design flow are more easily outsourced than the most skill-intensive and most value-added ones.

In theory, all parts of an IC design flow can be outsourced domestically or internationally. In practice, the more strategic the service function is, and the closer it is to the core competency of a firm, the less likely it is to be outsourced to third parties, domestic and foreign alike. The higher-end architecture/specification task is the least likely object of outsourcing, compared to physical design and validation tasks during the hardware design phase.

According to Brown and Linden's study (2005: 305–6), the easiest part of IC design to outsource is physical design because it is a relatively standardized task and the least sensitive part of design in terms of revealing the customer's IP. Logic verification is another design function that is often outsourced. In contrast, "Architectural design and the design of key functional blocks containing proprietary algorithms are the least likely to be outsourced because of the risk of exposing proprietary knowledge."

An industry survey of 367 IC design houses (Electronic Engineering Times 2006) reached a similar conclusion. IC physical verification and IC place and route are the two hardware IC design functions that are most frequently outsourced to third parties. In North America, 15 per cent of respondents said their firms turn over IC physical verification to a third party, 14 per cent cited place and route, and 12 per cent cited design for manufacturability. Asians outsource the least, with 7 per cent, 9 per cent and 12 per cent in the three respective categories. European firms outsource more, with 28 per cent, 23 per cent and 19 per cent in these categories, respectively.

Two factors account for the relative homebound character of higher-end IC design activities. First, IC design firms seek comparative advantages from their proprietary technologies, so they prefer to retain the work essential to their core competency at firm headquarters instead of having it outsourced. Second, because access to highly skilled and experienced IC designers is critical to the competitiveness of an IC design house, the relative unavailability of skilled IC design professionals abroad constrains these firms from outsourcing key design functions overseas.

The emergence of the Asia-Pacific region

The fourth and final characteristic of the globalization of IC design refers to the ongoing shift of pertinent production activities to the Asia-Pacific region, though this trend has not dramatically augmented the design capability of the overall region, with the exception of Taiwan.

According to the *2005 World Investment Report* (United Nations Conference on Trade and Development 2005: 173), "design and development work in this industry is following on the heels of manufacturing by moving towards Asia," particularly leading electronics export countries in the region. Several scholarly studies have reached the same conclusion (Brown and Linden 2005; 2009; Ernst 2005; Macher *et al.* 2007). From practically nothing during the 1990s, the region's share of global IC design revenue reached around 30 per cent in 2002, demonstrating its growing share of worldwide chip design activities (United Nations

Conference on Trade and Development 2005: 173). This is in spite of the fact that IC design firms in Taiwan alone account for more than 20 per cent of the regional total. Moreover, Southeast Asia and East Asia are two of the fastest-growing markets for EDA tools, which are indispensable equipment for IC design companies today. In the first quarter of 2004, Asian market for EDA tools expanded by 36 per cent, compared to 5 per cent for North America, 4 per cent for Europe and minus 2 per cent for Japan. In addition, Asian and European IC design engineers were slightly behind their counterparts in North America in their use of 65-nm technologies in 2006. While 16 per cent of chip designs used 65-nm process technologies in North America, they were closely followed by 14 per cent in Asia and 12 per cent in Europe. This overturns the conventional view that Europe is ahead of Asia in IC design capability. However, a comparison of the number of gates used in IC designs by firms in different regions in 2006 (Electronic Engineering Times 2006; Goering 2006a) indicated that while the mean total equivalent gate count reached 14.1 million in North America, it amounted to 10.6 million in Europe, and 6.1 million in Asia. On this measure of IC design complexity, Asia still lags behind North America and Europe.

As IC design activities continue to shift to the Asia-Pacific region, China and India have received the most noticeable attention. Since the 1990s, offshore IC design in industrializing countries has expanded, with some setting up subsidiaries in India and others in China. By 2006, the aforementioned EDA user survey identified a growing offshore outsourcing trend for North American IC design firms, with 34 per cent saying they outsource to India and 16 per cent to China/ Taiwan. According to a study by the Indian Semiconductor Association and Frost & Sullivan (Giridhar 2006), the Indian chip design services industry grow at 30 per cent annually and the country is emerging as a key destination for IC, FPGA and SOC designs. As for China, Brown and Linden (2005: 312) argued that "China is yet an important destination for design offshoring" by US firms. As of June 2005, only a handful of the 15 US firms had opened research centers in China, whereas 13 had done so in India.

Despite the emergence of the Asia-Pacific region in the globalization of IC design activities, this development is yet to bring a comparable shift in design capabilities at the global level. Hence, the conventional wisdom of industrial catch-up through geographical production shift does not hold in the arena of IC design. This observation is nonetheless tentative because the globalization of IC design production activities is still an ongoing process.

Determinants of the economic globalization of the semiconductor industry

Following an analysis of the varied degrees of economic globalization of the chip industry, what accounts for the sectoral globalization? The answer includes a synergy of enabling factors, pull factors pertaining to host countries, and push factors related to countries of origin and changes in the industry (Flamm 1985; Henderson 1989; Leachman and Leachman 2004; Brown and Linden 2005;

Ernst 2005; United Nations Conference on Trade and Development 2005: 173–7; Hung *et al.* 2006; Dicken 2007).[25]

Enabling factors

The enabling factors comprise four aspects of the economics of the semiconductor industry. These include: (1) the split in chip production stages; (2) the physical features of chips, including their high value-to-weight ratio; (3) the increasing automation; and (4) the complexity of design that applies particularly to the IC design sub-sector. The enabling factors also comprise effective telecommunications tools and Internet communication.

First, the disintegration of the semiconductor production stages, as discussed earlier, has enabled the separation of these stages organizationally and spatially. Hence, it becomes possible to "slice up the supply chain" in order to pursue the international division of labor in the sector. For instance, the detachability of IC design from IC fabrication created the potential for the emergence of the fabless-foundry business model in the late 1980s. Such a change in the industry, in contrast to the IDM model where every production stage is done in-house often in a single location, makes it possible to geographically disperse IC design and fabrication operations across national borders, enabling the globalization of chip production activities (Ernst 2005: 60–1).[26]

Moreover, the physical features of semiconductor components, namely their high value-to-weight ratio, have permitted the transportation of these miniaturized components over any geographical distance at an affordable cost, thereby fostering the globalization of chip production operations. Arguably, the relative ease of shipment of chips due to their light weight and small size partly accounted for the earliest relocation of the assembling business to lower-labor-cost regions of the world (Flamm 1985: 48–9). The same physical feature of chips has also enabled the globalization of fabrication activities.

The third enabling factor refers to the increasing automation in IC design activities. Computerization of IC design has enabled the transmission of design data through computer networks to anywhere in the world (Leachman and Leachman 2004: 206). The introduction of automation, in turn, has enabled companies to subdivide the complex design flow across national borders through intra-firm or inter-firm arrangements (Brown and Linden 2005: 301).

The fourth enabling factor concerns the increasing complexity in IC design. The continuous advance of chip technology has resulted in the creation of enormous areas of "silicon real estate" for complicated chips, especially SOCs, which could potentially be designed. However, design automation alone often cannot accomplish the complex design tasks necessary to keep pace with Moore's Law, and firms are compelled to recruit more IC designers than are available in a single country. This development has fostered offshore IC design activities led by US companies and their counterparts elsewhere. Any design team working on leading-edge chips has been compelled to recruit and retain highly skilled design engineers across national borders. As Ernst (2005: 65) put it, "Such design talent is scarce everywhere, and hence SOC design teams need to recruit and retain them wherever they exist."

However, this factor sometimes inhibits firms from globalizing their design activities. This is because the level of complexity of the chips involved is such that to manage multiple-site design flow across national boundaries is more costly than to have the overall design functions retained within the firm's headquarters or at least within national boundaries (Brown and Linden 2005: 303–4; Fabless Semiconductor Association and Industry Directions Inc. 2007: 7, 12).

The fifth and final enabling factor is related to high-bandwidth telecommunications and Internet communications. The introduction of automation in IC design alone will not make it possible to transmit computerized design data. As mentioned earlier, the availability of effective telecommunications and Internet communications has facilitated the globalization of IC design activities. Moreover, the availability of Web-based supply chain management systems has contributed to the success of foundries concentrated in the Asia-Pacific region. This has not only fostered the growth of fabless sector in the industry but also enabled an increasing trend among IDMs to outsource part of their fabrication needs to the foundries – all contributing to the increasing geographical dispersion of semiconductor production activities in the global economy (Leachman and Leachman 2004: 222–3).

Pull factors

Against the backdrop of the enabling factors identified above, a complex combination of push and pull factors has encouraged semiconductor companies to globalize their production activities. The five pull factors pertaining to host countries include cost reduction, location-specific resources, market access, government policies, and political risks.

First, cost reduction is one of the most important explanations for the globalization of semiconductor production activities, according to various scholarly studies, industry surveys, and seasoned industry players I interviewed. According to George Scalise, President of SIA, "Cost is always the key driver" behind offshore semiconductor investments. "If you can't produce a lower cost somewhere, then you have to have some other very compelling reasons," he observed in 2004.[27] A survey by Fabless Semiconductor Association (FSA) in 2007 (Fabless Semiconductor Association and Industry Directions Inc.: 3, 7–8) also regarded cost reduction as a critical outsourcing driver: "Cost reduction continues to be the most significant driver behind outsourcing operations." About 40 per cent of respondents identified cost reduction as the most successful outsourcing objective, compared with 28 per cent in an FSA survey two years ago.[28]

The desire to reduce production costs has been the primary driver behind the globalization of assembly activities. The opening of assembly subsidiaries in Southeast Asia that began in the 1960s resulted in substantial part from firms' decision to exploit the low-cost high quality labor in the region (Flamm 1985). The desire to reduce costs has also propelled firms to engage in offshore investments in fabrication and design. In their respective studies, Brown and Linden (2005: 301–2) and Ernst (2005) concluded that cost reduction primarily results in the increasing design offshore investments and outsourcing activities in Asia. According to Ernst (ibid.: 56), firms are attracted by supply-side forces to shift part of their design functions to Asia, particularly the lower cost of hiring chip design engineers in the region, which is typically between 10 and 20 per cent of the cost in Silicon Valley.

The second pulling factor is location-specific resources, including engineering and technical talent and critical infrastructure. Flamm's study (1985: 48–9) identified the supply of skilled science talent as one of the major determinants of research-related chip location decision. Because the high-end semiconductor technological advances pertaining to product and process technology are knowledge-intensive and skill-intensive, access to a rich pool of supply of skilled and seasoned engineers is critical to retaining the competitiveness of a chip firm. In their respective studies, Brown and Linden (2005) and Ernst (2005) observed that the availability of engineering talent was one of the major explanations for the increasing rise of the Asia-Pacific region as a favorable destination for offshore design investments and third-party outsourcing. "Asian design engineers, especially those from the emerging giant economies of China and India, represent an important source of supplemental engineering talent," argued Brown and Linden (2005: 309). Henderson's study (1989: 118, 127–30), in the context of Europe, concluded that the supply of high-quality engineering and technical labor in Europe explained why US and other companies had invested in wafer fabs in Europe, although market access was the initial impetus behind these investments. Based on a survey with ten samples of chip-makers in the USA, Taiwan, and Japan, Leachman and Leachman's study (2004: 226) identified the supply of engineering and technical talent as the No. 2 factor that chip-makers would consider when deciding where to locate their fabs. In the same study, location-specific resources that may entice firms to globalize their production activities also include the supply of basic infrastructures in host countries, such as the quality of water supply, the reliability of utilities, and local transportation infrastructure.

The third pulling factor is market access. The desire to gain access to an enlarging market has driven chip firms to internationalize their production activities because to be physically close to their customers may help to increase their market share. As discussed earlier, market penetration, instead of cost reduction, was the initial impetus behind investments by US firms in fabs in Western Europe and Japan. Scalise also identified market access as a driving force, second to cost consideration, behind the increasing shift of chip production activities to China. As he observed in 2004:

> In today's globalized world, market access should not be an issue. I don't think it is in a major way with regards to China, but it certainly has some influence. It's easier to access the market in China as it is going to be such a large market.[29]

According to Ernst's study, the need to be close to their customers in the growing Asia-Pacific market has driven design firms to set up subsidiaries in the region, particularly in view of the increasing need for IC designers and process engineers to communicate in close geographical proximity in order to complete their tasks.

> Global firms emphasize the need to relocate design to be close to the rapidly growing and increasingly sophisticated Asian markets for communications, computing and digital consumer equipment, to be able to interact with Asia's lead users of novel or enhanced products or services.
>
> (Ernst 2005: 55)[30]

The fourth pulling factor includes the government policies of host countries pertaining to tax advantages, trade, and foreign investment. As Flamm (1985: 54) points out, "National policies have had important effects on the international patterns of specialization in the production of semiconductors." In the context of the globalization of semiconductor assembly activities that dated back to the 1960s, "Barriers to direct investment together with trade restrictions determine whether a foreign firm chooses to export, or to invest behind a tariff wall" (ibid.: 68). Dicken (2007) also identifies the actions of national governments as one of the two factors that shape the globalization of the industry.[31] Moreover, Henderson's analysis (1989) of the offshore investments in Western Europe by US chip companies behind the high EEC tariff wall attests to the importance of government policies in host countries to sectoral globalization. Although tariff barriers to semiconductors are almost non-existent today in a globalized economy, government policies of host countries continue to affect firms' decisions about offshore investments. For instance, respondents in Leachman and Leachman's study (2004: 226) viewed tax advantages as the top consideration when deciding where to invest their fabs. Government policies of host countries can also accelerate the establishment of critical infrastructure in the countries in question in order to attract foreign firms to set up shop there.

The fifth pulling factor involves political risks in host countries. Flamm's study (1985: 68–9) argues that chip firms do consider political constraints in host countries when assessing whether to extend their production activities to a foreign land. In their analysis of the globalization of Taiwanese high-tech industries, including the semiconductor sector, Berger and Lester (2005) contend that political factors pertaining to host countries, aside from economic factors, also affect firms' decisions about where to globalize their production activities. The decision by UMC, the world's No. 2 foundry headquartered in Taiwan, to invest in fabs outside the politically volatile Taiwan is a case in point. Jackson Hu, CEO and president of UMC, admitted in 2005 that the relative political stability in Singapore and Japan, in contrast to that of Taiwan, had attracted his company to invest there. After all, some of UMC's customers had preferred gaining access to the firm's foundry services outside Taiwan, in a location where political risks are relatively low.[32]

Push factors

Push factors related to countries of origin and changes in the fast-moving and competitive industry have further accounted for the globalization of chip production activities. These considerations include the rising costs of labor and other factors of production, the lack of location-specific resources, government policies, rising political risks as well as limited domestic market in countries of origin.

First, respective studies by Ernst (2005) and Brown and Linden (2005) conclude that the comparatively high cost of hiring IC design engineers in Silicon Valley has pushed the US design firms to shift part of their design activities to Asia, where cheaper qualified engineers are available. Second, they also contend that the lack of sufficient number of qualified design talent at home has further driven US firms to seek talent beyond their national borders.

Third, the US tariff and trade policies affected the way the globalization of assembly activities unfolded in the early years of the industry (Flamm 1985; Hung *et al.* 2006: 367). Under the related US tariff regulations at the time:

> Products could be manufactured in the USA, exported for further processing, and then returned to the USA without paying full import duties. Such products received duty free treatment on the components that were originally manufactured in the USA, and only the value added portion of the products was subject to tariffs.

The fourth point concerns the UMC case mentioned earlier. The Taiwan-based foundry, as discussed earlier, decided to set up offshore operations in Singapore and Japan because of political risks pertaining to the home country. Some of its customers had regarded the prospect of an armed conflict across the Taiwan Strait as detrimental to their business interests, thereby urging the chip-maker to geographically diversify their fabrication services.

Fifth, the limited domestic market can also drive firms to globalize their production activities, as will be discussed in Chapter 5 regarding the Taiwan–China case.

Aside from the aforementioned push factors pertaining to firms' countries of origin, determinants of the semiconductor globalization also include changes in the industry that have elevated the importance of specialization/core competence and time-to-market considerations. Because of the increasing importance of vertical specialization in the sector, the consideration of specialization and core competencies are, at times, outsourcing drivers. In a FSA study (Fabless Semiconductor Association and Industry Directions Inc. 2007: 7–8), 33 respondents said specialization/core competencies are important in their outsourcing decisions, ranked as the No. 2 driving force next to cost reduction. Moreover, the time-to-market factor also affects firms' outsourcing decisions. This factor was ranked as the No. 3 determinant of outsourcing activities by firms surveyed in the same study. About 11 per cent of respondents identified this element as their most successful outsourcing objective. In particular, this figure in the study represents an almost 400 per cent increase from the 3 per cent who indicated it was their most successful outsourcing objective two years ago. This sharp increase in the importance of time-to-market factor attests to the increasing pressure faced by chip firms to beat their rivals in a fiercely competitive sector.

The military globalization of the semiconductor industry

Whereas the previous section analyzes the economic globalization of the semiconductor industry, this section examines the globalization of the production of chips for military uses, which account for a small proportion of worldwide total consumption today. It is argued that although the national security significance of these military semiconductors may justify the homebound production of these components, the pertinent production has also become partly globalized due to commercial firms that are qualified military IC suppliers. Thus, the phenomena that pertain to the globalization of military production and defense industrial base (Buzan and Herring 1988: 42–6; Held *et al.* 1999: 89, 138; Brooks 2005: 6, 80–128; Kirshner 2006b: 21) also apply to the semiconductor industry. However, the scale and depth of the globalization of the production of chips for military uses have not been comparable to that of commercial ones, as will be illustrated below by the case of the USA.

Varied degrees of globalized production of US military chips

Semiconductors for US military systems vary in the degree to which their production is globalized (Table 3.4). To begin with, as commodity products churned out by the mainstream commercial chip sector, COTS items used in US military systems are highly globalized in production to a degree similar to that discussed in the section on the economic globalization of semiconductor production.

The US QPL and QML approaches: emerging multinational influence

Moreover, other types of US military chips have become increasingly globalized in production over the past few decades. This can be shown by analyzing the evolutionary QPL and QML approaches.

The QPL system, which the Pentagon introduced in the 1960s to control the sources of military semiconductors, confined military chips to exclusively onshore production. Conventionally, Washington preferred autarkic arms production, favoring domestic sources for critical systems, materials, and components. These components included chips for US defense systems, given the significance of advanced semiconductor devices in improving the qualitative edge of information-based modern weaponry (Kanz 1991). Since the 1960s, Washington thus established mil-spec requirements, standards, and procurement rules in order to control the sources of military semiconductors. The first initiative was the QPL system.

Since then, the Pentagon has changed its definition of military chip quality, reliability, and production location (onshore vs. offshore) requirement levels in order to eradicate chronic weaknesses in the relatively rigid QPL system. These

Table 3.4 Varied degrees of globalized production of semiconductors for US military end-uses

Types of chips for US military end-uses	Degrees of globalization of production	Notes
COTS	Highest	• Growing proportion in military uses • Hard to trace the actual sources of COTS in military systems
Military Grade Advanced ICs (QML-38535)	High (Feb. 2007 version of QML*: 72%)	• Increasing degree of globalization • Emergence of the Asia-Pacific region: Japan's fall vs. Taiwan's rise
Military Grade Advanced ICs by Accredited Trusted IC suppliers	Non-globalized	Only foreign influence concerns TSMC processes and design rules
Military Grade Advanced ICs by government-related agencies	Non-globalized	Eroding home-bound supply base

Sources: Kanz 1991; Perry 1994; Carlson 2005; Defense Science Board Task Force 2005; Defense Supply Center – Columbus 2007; Trusted Access Program Office n.d.

Note: *As for the level of radiation hardness of these chip products, the list includes both rad-hard and non-rad-hard items. The technology type and product type of these devices also vary. The former encompasses mainstream CMOS technologies, bipolar, and other types of technologies. The latter covers a vast array of items ranging from microprocessors, microcontrollers, FPGAs, DSPs, A/D converters, and memory chips.

problems include higher costs, constrained supply, unrealistic device qualification rules, on/offshore manufacture, and assembly (Defense Science Board 1989: A15–A16).[33]

A major reform was the initiation of the QML system. The QML approach, detailed in *MIL-I-38535, General Specifications for Integrated Circuits (Microcircuits) Manufacturing*, represents a new grade of military chip products. It aims to control product quality and reliability: adding a manufacturer and its product to the QML demonstrates that a product can meet the specified performance, quality, and reliability requirements that the military needs for its weapon platforms (Defense Supply Center – Columbus 2000).

The QML, which eventually replaced the QPL approach, outperforms the QPL in four major ways. First, it focuses on the supplier's quality and process control system, qualifying parts by technology rather than by part type as does QPL. Second, it is more flexible and faster than QPL because the qualification establishes the Pentagon's recognition of a manufacturer's ability to provide a good product without imposing detailed tests and procedures for each product. Third, it is far more convergent with better process control practices and better industry quality than QPL by exercising concepts well accepted by major commercial chip firms because they help to standardize quality management in their mainstream businesses (Sheppard 1990; Kanz 1991; Pecht *et al.* 1997). Despite initially being listed by the Pentagon as an ordinary military standard and specification, the QML subsequently went beyond its mil-spec status by embracing the best commercial practices. This, in the eyes of TI as a seasoned QML firm, "went a long way in providing an IC standard that offers the best of both the military and commercial worlds" for IC manufacturers and defense contractors (Texas Instruments n.d.).

In November 1989, the most important reform in the QML system occurred as the Pentagon merged the QPL into the QML program, and more importantly, revised MIL-I-38535 to scrap the provision for onshore production. In May 1993, the US military permitted offshore assembly and testing of certified QML-38535 ICs. In 1995, it permitted overseas fabrication of these military ICs (Cohen 1997). Since then, the QML reform eliminating the provision that they be produced exclusively onshore has paved the way for these devices to be assembled, tested or fabricated offshore (Pecht *et al.* 1997). This has deepened the globalization of production for this category of military chips. By 2000, the Defense Supply Center – Columbus (DSCC) data (2000) showed that there were eight certified manufacturing lines in Japan run by high-volume commercial manufacturers which performed various operations to support QML-38535 companies. As homebound and autarkic production of military graded QPL/QML chips for the US military applications has become unsustainable, offshore production of these devices has increased over time. The USA, in spite of being the world's leading semiconductor player, has given the green light to the globalization of the production of these military chips for US defense uses.

Deepening multinational influence in MIL-PRF-38535 military supply chain

In 2007, the QML-38535 US military chip supply chain became even more globalized, as evidenced in the following analysis of the February 2007 version of the QML of products qualified under performance specification MIL-PRF-38535 (Defense Supply Center – Columbus 2007).[34]

Before we discuss the deepening multinational influence on the military supply chain in question, it is important to note that analysis of the 2007 document on QML-38535 firms deepens our understanding of the US military chip supply chain for two reasons. First, the list identifies manufacturers qualified to produce advanced ICs that can withstand harsh environments, thereby meeting the US military demands. As the document states, "Only those products that have been manufactured, assembled, and tested on the certified/qualified lines herein can be supplied as qualified microcircuit QML devices" to the US military. Second, the document emphasizes that the information contained in the QML in question reflects the actual manufacturing lines, materials, and technologies of the particular test samples. Hence, the analysis of the document can shed light on the globalization of the US military chip production activities involved.[35]

An examination of the list shows that both US and foreign firms are among the 41 chip firms that have been granted full QML certification/qualification status, with their certified/qualified supply chain partners located either onshore or offshore. Thirty-six of these companies are US-owned, whereas five are foreign-owned.[36] The latter include STMicroelectronics (STM), the BAE Systems, Semelab, PSI Technologies, and Millennium Microtech.[37]

The technology type and product type of these qualified devices also vary. The former encompasses mainstream CMOS technologies, bipolar, and other types of technologies. The latter covers a vast variety, ranging from microprocessors, microcontrollers, FPGAs, DSPs, A/D converters, and memory chips. In terms of the radiation hardness level of these products, the firms in question churn out both rad-hard and non-rad-hard items.

More importantly, a quantitative analysis of the data indicates that 72 per cent of the 39 companies analyzed have internationalized their production activities for the components, whereas 28 per cent have not.[38] The globalizing group of firms has either invested overseas or subcontracted their work to foreign-owned companies. These firms rely on 195 certified manufacturing lines scattered in 16 foreign countries, which Washington considers as allies (Figure 3.4).

The geographical distribution of these foreign-based productions supporting QML-38535 firms shows the growing importance of the Asia-Pacific region in the related US military chip supply chain. Eight of these 16 countries are in the Asia-Pacific region, seven in Europe and one on the American continent. Among the top five countries which house the largest number of certified manufacturing lines in question, which include the Philippines, France, Taiwan, Japan, and Singapore, four are in the Asia-Pacific region. The 73 lines housed in the Philippines concentrate on assembly and test activities. The 24 lines in France are involved in design, fabrication, mask development, test, and assembly operations. The 18 lines that Taiwan houses concentrate on all of the production stages except IC design.

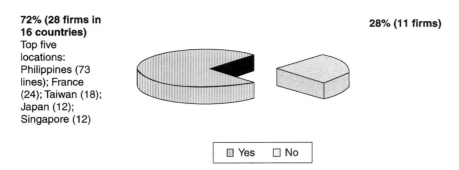

72% (28 firms in 16 countries)
Top five locations: Philippines (73 lines); France (24); Taiwan (18); Japan (12); Singapore (12)

28% (11 firms)

▥ Yes   ☐ No

Figure 3.4 The percentage of QML-38535 firms that have globalized their production activities outside US soil (n = 39).

Source: Based on pertinent analysis of data in Defense Supply Center – Columbus 2007.

Singapore-based 12 operations encompass all of the production stages, whereas Japan-based 12 lines are involved in all but test activities.

The top five countries that support the largest number of QML-38535 firms through their production activities are also located in the Asia-Pacific region. They include the Philippines (14), Taiwan (10), Japan (6), Singapore (6), and Thailand (6).

The importance of the Asia-Pacific region to the supply chain in question can be further evidenced by the fact that the region accounts for 73 per cent of the total 195 certified offshore manufacturing lines. This is followed by Europe and Middle East (26 per cent), and North America excluding the USA (1 per cent). Chip production operations in the Asia-Pacific region is responsible for 93 per cent of the overall offshore assembly operations, 67 per cent of the mask development tasks, 69 per cent of the test operations, and 62 per cent of wafer fabrication activities. However, in the sub-sector of chip design, the region accounts for 22 per cent of the overall certified offshore operations, whereas Europe and the Middle East are responsible for 70 per cent of these production activities (Table 3.5).

Although the Asia-Pacific region, following its emergence as a key player in the economic globalization of the semiconductor industry, has become increasingly significant in the globalization of the US military chip production activities, firms in the region have engaged in the varied degrees of semiconductor production activities for the US defense end-uses.

Several key observations follow. First, although Japan was key to the US military chip supply chain in the late 1980s and the early 1990s, Japanese support for QML-38535 firms is waning. Taiwanese and Singapore firms, however, outnumber the Japanese ones. For example, Taiwan houses nine of the 18 certified fabrication lines in the region, accounting for 50 per cent of the certified fabrication capacity in the region available to the US military. Japan houses seven lines, comprising

Table 3.5 Geographical distribution of the certified manufacturing lines located outside the USA in support of QML-38535 firms

Region	Number of certified mfg. lines	Design	Wafer fabrication	Mask development	Test	Assembly
Asia-Pacific region including Japan	142 (73%)	5 (22%)	18 (62%)	6 (67%)	34 (69%)	79 (93%)
Europe and the Middle East	51 (26%)	16 (70%)	11 (39%)	3 (33%)	15 (31%)	6 (7%)
America excluding the USA	2 (1%)	2 (8%)	0 (0%)	0 (0%)	0 (0%)	0 (0%)
Total	195	23	29	9	49	85

Source: Based on an analysis of data in Defense Supply Center – Columbus 2007.

39 per cent of the total capacity. Singapore houses two lines, accounting for 11 per cent of the total.

Second, Korea and Taiwan have dominated the region's fabrication capacity since the 1990s, but Korea, unlike Taiwan, does not house any certified manufacturing lines to fabricate QML qualified chips for US defense end-uses. Instead, it only runs low-end assembly and test operations. Third, the region, as mentioned earlier, is relatively insignificant in providing QML 38535 firms with certified IC design activities in comparison to Europe and the Middle East. As the data indicate, two certified IC design lines are located in Japan and Singapore respectively, whereas Xilinx in India operates one. Fourth, despite the perceived emergence of China and India in the mainstream commercial chip industry, as discussed earlier, neither countries have played any noticeable role in the transnationalization of the US defense chip industrial base. Whereas India houses one certified line for IC design activities, China has none.

The final observation concerns the role of Taiwan in supporting QML-38535 firms. The 18 certified manufacturing lines on the island are run by five local Taiwanese firms, one US subsidiary and one US–Taiwanese joint venture. They undertake various production tasks excluding IC design: wafer fabrication (nine lines), mask development (three), assembly (four), and testing (two). These offshore production activities support ten QML-38535 companies. They include Actel, Aeroflex-Colorado Springs, Analog Devices, Cypress Semiconductor, National Semiconductor, QuickLogic, TI, Vertical Circuits, Xilinx, and Pyramid Semiconductor. Because of Taiwan's dominance in chip fabrication capacity in the region, the island houses half of the certified fabrication lines in the region. TSMC alone runs six of them; it provides foundry services to Analog Devices, Cypress Semiconductors, National Semiconductor, QuickLogic, Aeroflex-Colorado Springs, and Pyramid Semiconductor. UMC runs two lines for Actel and Xilinx respectively. Winbond Electronics, an IDM, operates one line for Actel (Table 3.6).

It is noteworthy that the analysis of the DSCC document in question has helped us systemically trace US–Taiwanese business links at the firm level within the QML system because it complements other sources of data that, albeit useful, are sketchy and partial. These sources include US government documents, technology journals and company websites that have identified some of the business links (Ciufo 2004; Fabula and Padovani 2004; O'Neil 2004; Defense Science Board Task Force 2005: 24).[39] They also include interviews with Taiwanese chip industry players who have either downplayed or initially denied the business links in question. My interview with a senior executive at TSMC is a case in point. When first asked if TSMC had any military or aerospace customer, he said, "We do not have any military customer." I then cited a statement by the Applied Wave Research (AWR), TSMC's EDA partner, which described TSMC "as a pure-play semiconductor wafer foundry that serves customers targeting wireless, optical networking, power management, storage, aerospace/defense and other high-performance applications." In response, he said, "We do not know to whom our customers sell the chips." When asked if chips fabricated at TSMC foundries had been used for

Table 3.6 Outsourcing activities in Taiwan supporting QML-38535 firms for the production of military ICs

Company	Rad-hard level	Outsourcing productions in Taiwan	Location, line, and flow	Upstream–downstream customer relationship

Actel	N/A	Wafer fabrication	(1) Location: Hsinchu, Taiwan
			    Line: Winbond Electronics
			    Flow: 0.045 μm, 0.6 μm, 0.8 μm CMOS FPGA
			(2) Location: Hsinchu, Taiwan
			    Line: UMC
			    Flow: 0.15 μm, 0.25 μm CMOS and 0.22 μm Flash/CMOS FPGA
				Actel–Winbond Electronics (fabrication)
				Actel–UMC (fabrication)

Aeroflex-Colorado Springs	1 Meg	Wafer fabrication	Location: TSMC* Line: 0.18 μm, 0.25 μm, 0.35 μm Flow: Antifuse Logic	Aeroflex-Colorado Springs–TSMC (fabrication)
Analog Devices	100 K	Wafer fabrication	Location: TSMC; Hsinchu, Taiwan Line: CMOS, Bi-CMOS Flow: ADI-0088	ADI–TSMC (fabrication)
Cypress Semiconductor	N/A	Wafer fabrication	Location: TSMC, Hsinchu, Taiwan Line: FPGA line Flow: QM Plan 90-00033	Cypress Semiconductor–TSMC (fabrication)
National Semiconductor	100 K	Wafer fabrication	Location: TSMC, Hsinchu, Taiwan Line: 8″ (Bipolar CMOS) Flow: Q only	National Semiconductor–TSMC (fabrication)
Pyramid Semiconductor	—	Wafer fabrication and mask development	*Wafer fabrication* Location: Hsinchu, Taiwan Line: TSMC Mask Flow: TSMC ISO 9000 Flow *Mask development* Location: Hsinchu, Taiwan Line: TSMC Fab 2A Flow: TSMC ISO-9000 Flow	Pyramid Semiconductor–TSMC (fabrication; mask development)

QuickLogic	N/A	Wafer fabrication	Location: Hsinchu, Taiwan Line: 0.35 μm CMOS Flow: TSMC military	QuickLogic–TSMC (fabrication)

Texas Instruments	N/A	Assembly and test	*Assembly*
			(1) Location: Taipei Taiwan
			    Line: Plastic
			    Flow: 07-012-1601, 1901
			(2) Location: Taipei Taiwan
			    Line: Special product assembly (hermetic)
			    Flow: 07-0012
			Test
			Location: Taipei, Taiwan
			Line: Special product test
				Texas Instruments–Texas Instruments Taiwan (assembly; test)

Vertical Circuits	N/A	Mask development	(1) Location: Hsinchu, Taiwan
			    Line: Fupo Electronics
			    Flow: wafer remetalization
			(2) Location: Hsinchu, Taiwan
			    Line: Fupo Electronics
			    Flow: wafer masks
				Vertical Circuits–Fupo Electronics (mask development)

Xilinx	—	Wafer fabrication and assembly	*Wafer fabrication*
			Location: UMC, Taiwan
			Line: 8" and 12", 0.35/0.25/0.18/0.15/0.13 μm
			Flow: Per UMC
			Assembly
			(1) Location: Unitive, Taiwan
			    Line: BUMP
			    Flow: Per Unitive
			(2) Location: SPIL, Taiwan
			    Line: BUMP
			    Flow: MFC0082
				Xilinx–UMC (fabrication)
				Xilinx–Unitive (assembly)
				Xilinx–SPIL (assembly)

Source: Based on an analysis of data in Defense Supply Center – Columbus 2007.

Note: *The location is not specified in the data; it is assumed here that the location is in the firm's headquarters in Hsinchu, Taiwan.

military end-uses because a journal article (Ciufo 2004) identified the partnership between TSMC, as a foundry service provider, and Aeroflex, as a rad-hard memory chip supplier to the US military, he replied, "I need to check whether we have fabricated the chips in question." He added that even if TSMC did fabricate chips for military applications, these devices would account for a small proportion of the total.[40] As such, the DSCC document provides us with the most concrete hard evidence available that delineates the comprehensive US–Taiwanese inter-firm business ties in the area of military chip production. As the analysis of the document has shown, pertinent Taiwanese chip firms have become part of an increasingly globalized US defense chip industrial base.[41]

Some of my interviewees in the Taiwanese industry, who were willing to disclose their business ties with QML-38535 companies, have verified part of the customer ties between the US and Taiwanese firms in the QML-38535 military chip supply chain. In one case, a senior executive at UMC admitted that some of the Xilinx FPGAs fabricated at UMC might end up in US military systems. According to him, the standard practice is that Xilinx, upon receiving these chips from UMC, would conduct a series of stringent military grade tests in order to pick items that pass these tests for the US defense/aerospace end-uses.[42] In another instance, a senior director at TSMC revealed that some of the chips fabricated at TSMC were used in US military systems. As he recalled, a US customer, who was a subcontractor for the Pentagon, asked TSMC to write a report for the US defense authorities explaining why the customer chose TSMC, instead of IBM, as its foundry partner.[43]

To sum up the analysis of QML-38535 data, the globalization of the production of these chips for the US military end-uses is deepening. As more than 70 per cent of the qualified QML-38535 firms have transnationalized their manufacturing activities, commercial companies located in an increasing number of foreign countries have contributed to the US military chip supply chain in question.

A case of non-globalization: the Trusted IC supplier program

The least globalized production of military chips involves that of confidential chips for use in US mission-critical defense systems. The production of these devices is, for the most part, regulated through the Trusted IC supplier program, and represents a case of non-globalization because offshore firms are excluded from pertinent production activities.

As detailed in Chapter 2, Washington initiated the Trusted Foundry program in 2004 and the complementary trusted IC supplier accreditation scheme thereafter in order to reach part of the objectives of the 2003 Wolfowitz memo. The two initiatives emerged because of the depleting domestic supply of classified chips for US military end-uses. The erosion of the onshore supply base resulted from two factors. The first is the withdrawal of giant commercial chip firms (i.e. AMD, Motorola, Intel, and Philips) from the QML system in the 1990s (McHale 1996; Pecht *et al.* 1997). The second refers to the non-viable operations of US government-owned, government- or contractor-operated or dedicated facilities (i.e. those by NSA and Sandia) that had supplied these classified chips to meet the US military need (Defense Science Board Task Force 2005).

In 2004, IBM became the Pentagon's first partner under the Trusted Foundry program. By November 2007, seven additional firms had become accredited trusted IC suppliers. They include IBM, the Trusted Solutions Business Unit of National Semiconductor, BAE Systems, Honeywell International, Northrop Grumman, HRL Laboratories, and Intersil. Semiconductor technologies available at these seven onshore suppliers include rad-hard CMOS and compound semiconductor technologies.

The scheme is a case of non-globalization because it excludes the participation of offshore firms from related production operations. The exclusion is due to two considerations. First, the criteria for a trusted domestic supplier would be more achievable and affordable than those for a foreign-based one. Second, Washington intended to mitigate security concerns over the potential untrustworthy offshore supply of these sensitive chips. The scheme thus contrasts with the QML approach which permits certified foreign-based commercial chip companies to take part in the production of given chips for US military end-uses.[44]

Determinants of the military globalization of the semiconductor industry

What, then, accounts for the military globalization of the industry in the US context? The answer is two-fold: performance and cost considerations. To begin with, military grade chips made in a homebound and autarkic environment in the USA, such as at the government-controlled NSA and Sandia facilities, may, at times, fail to meet performance requirements. Hence, commercial firms with better performance and technological capabilities are accepted as qualified manufacturers of these components once they pass related auditing tests. Moreover, the onshore military chip industrial base in the USA was weakened in the 1990s because several giant commercial firms dropped their military-certified product lines. This development coincided with the emergence of competitive and viable semiconductor companies outside US soil, particularly in the Asia-Pacific region. If better quality production services are only available offshore instead of onshore, it is reasonable for the Pentagon to permit their certified commercial suppliers to globalize their production activities so long as security measures, through pertinent auditing procedures, are in place. Such logic partly accounted for the Pentagon's decision in the 1990s to scrap the onshore requirements for certain military chip productions. Besides, the pursuit of cost reduction further drives qualified commercial suppliers, such as QML-38535 firms, to globalize their production activities, because the majority of these companies are commercial entities. Hence, cost consideration drives not only the economic globalization of the semiconductor industry but also the globalization of military chip production. Both of these factors combined explain why the production of QML-38535 military chips, which is conventionally homebound and non-globalized because of their importance in national security terms, has become increasingly internationalized.

The globalization of production of military chips in the US context, as analyzed above, echoes the arguments put forward by Brooks and Kirshner in their respective studies of the globalization of defense-related production activities. In his study, Brooks (2005) hypothesizes that states, including great powers, have to pursue significant internationalization in weapons production instead of autarkic production if they aspire to effectively pursue cutting-edge military technology to enhance their capabilities. Although autarky remains a conventionally desirable approach to defense production, states have to build up strong gains in terms of reduced cost and enhanced quality of the globalization of weaponry production in order to obtain state-of-the art military technology. After all, the rise in costs, complexity and scale of developing new military technologies has made it impossible for any states to "go it alone" in defense-related production unless they dare to run the risk of trailing behind in the military technological race.

To test his hypothesis, Brooks examines the experience of the USA during the last two decades of the Cold War. He concludes that Washington could no longer effectively pursue autarky for its defense-related production during this period. Instead, the USA "strongly pursued globalization in defense-related production" because of the perceived gains from pertinent production internationalization. This put Washington at a relative advantageous position vis-à-vis Moscow in the related areas. Consequently, Moscow's "isolation from the geographic dispersion of MNC production had an independent, negative effect on Soviet ability to remain competitive with the United States in key dual-use industries and defense-related production during the 1980s" (ibid.: 126).

Kirshner, in his study, identifies the encroachment of the market spheres in the realm of security, as illustrated by the commercialization and globalization of the defense industry. Hence, the functions of military security are gradually contracted to private sector players or even subcontracted to foreign suppliers. He concludes that the desire by defense contractors to pursue the same economic incentives (i.e. pressures for efficiency and economies of scale) faced by the producers of consumer goods has driven the globalization of defense-related production.

In brief, the globalization of military semiconductor production, as an integral part of the internationalization of defense-related production activities, has deepened over time because of cost and performance considerations, with the exception of the production of confidential chips for mission-critical systems in the US context.

Conclusion

To conclude, political and economic forces combined have accounted for the varied degrees of economic and military globalization of semiconductor production activities. Some of the manufacturing operations in the supply chain have become deeply globalized. They include fabrication, packaging, and testing segments of the industry for mainstream commercial end-uses. They also, of a lesser degree, concern the production of certain types of military graded chips; for instance, 38 QML-38535 firms in the USA have relied on certified manufacturing lines located in 16 foreign countries in order to supply pertinent semiconductors to the US military. Others have remained homebound and non-globalized. The most telling examples are the accredited trusted IC suppliers of semiconductors for end-uses in mission-critical US military systems. To a lesser extent, the rule of non-globalization also applies to IC design activities in the mainstream commercial chip sector and the tangential defense one. Hence, the web of the globalization of semiconductor production is complex. Dicken's description of the globalization of the industry, cited at the beginning of the chapter, applies to a substantial part of the sector, albeit not all. Because the migration of the Taiwanese chip industry to China, as part of the sectoral globalization processes, unfolds amid the uneasy political and security relations across the Strait, it is essential to examine the development of the industry in China and its links to national security and globalization. It is to this theme that we shall turn in Chapter 4.

Notes

1 These production activities can take place in close proximity to each other, or they can be geographically dispersed. Additionally, they can be integrated under one firm or run by different companies. See Leachman and Leachman 2004: 204–5.
2 Interview, 3 November 2005, Hsinchu, Taiwan. See also interview with Michael G. Hadsell, vice-president and managing director-Asia, Toppan Photomasks, 16 September 2005, Shanghai, China.
3 There is, however, disagreement over what actually constitutes IC design.
4 These include salaries and benefits and hardware and software allocation.
5 The underlying project for the study is assumed to be a digital logic design, the industry's typical product. To echo the trend of increases in design integration density, the representative chip is assumed to contain more transistors as the feature dimension of chips shrinks over time. For a survey of some 300 IC design firms in North America, Europe and Asia that indicates the average time spent on each stage of hardware IC design activities, see Electronic Engineering Times 2006: 11.
6 The study normalizes the engineering hours required to design a logic chip with one million transistors at each generation. This is based on the assumption that the project is increasingly complex, which reflects the trend in the industry: two million transistors at 0.35μm, five million at 0.25μm, 20 million at 0.18μm, 40 million at 0.13μm, and 80 million at 90nm.
7 Interview, 20 July 2005, Taipei, Taiwan.
8 Typically, a piece of processing equipment in the fab is useful for three or four generations of process technology. Each generation of technology involves a 50 per cent reduction in the minimum feature size from the previous one. Hence, each succeeding generation requires replacement of between 25 and 35 per cent of the equipment used in the previous generation. See Leachman and Leachman 2004: 205. This observation is supported by the case of TSMC subsidiary in Shanghai. See interview with the president of TSMC (Shanghai), 26 September 2005, Shanghai, China. At the time of the interview, about 60 to 70 per cent of the process equipment in the fab, originally set up for 0.25 micron process technology, was useful for the next generation of technology (i.e. 0.18 micron).
9 No comparable data are available for the 12-inch fab.
10 Interview, 8 December 2004, San Jose, CA, USA.

11 See also Henderson 1989: 45; Brown and Linden 2005: 282–5; Hung *et al.* 2006: 367. In the successive waves of investment, many of these US companies involved employed a greater number of workers in Asia than they did at home. See Scott and Angel 1988.
12 Aside from the market consideration, the cheapness and high quality of engineering and technical labor also propelled foreign companies to invest in wafer fabrication plants in Europe. See Henderson 1989: 118, 127–30.
13 Presentation by Marco Mora, chief operating officer of SMIC, at SEMICON China, 15 March 2005, Shanghai, China.
14 Interview, 12 January 2005, Dallas, TX, USA.
15 Interview, 29 June 2005, Hsinchu, Taiwan.
16 US-based Chips & Technologies (C&T), the world's first fabless design house, used the excess semiconductor capacity created by some of the Japanese IDMs to pioneer the fabless business model in the early and mid-1980s. Cirrus Logic, Xilinx, and Altera soon followed. See Hurtarte *et al.* 2007: 3.
17 See also interview with Klaus C. Wiemer, former president of TSMC, 12 January 2005, Dallas, TX, USA.
18 The same factor explains why the Taiwanese fabless design houses, following the establishment of pure-play foundries at home, shifted their fabrication business from Japanese IDMs to these local contract manufacturers. See interview with Klaus C. Wiemer, former president of TSMC, 12 January 2005, Dallas, TX, USA.
19 Interview, 12 January 2005, Dallas, TX, USA.
20 See also interview with Yoshio Nishi, the director of Center for Integrated Systems at Stanford University and former senior fellow at TI, 6 December 2004, Palo Alto, CA, USA.
21 Interview, 23 September 2005, Shanghai, China.
22 Interview with the president of a fabless design firm, 7 September 2005, Beijing, China.
23 Interview with the chief of the design firm's subsidiary in Beijing, 9 September 2005, Beijing, China.
24 The respondents in the survey included 367 chip designers and a few respondents who did not indicate nationality.
25 The analysis below is also based on various interviews with industry players.
26 Although Ernst categorizes the vertical specialization as a "push" factor in his analysis of the determinants of the shifting chip design activities to Asia, I think it is more appropriate to categorize it as an "enabling" factor.
27 Interview, 8 December 2004, San Jose, CA, USA.
28 The survey is based on the triangulation of quantitative online questionnaires with 106 valid industry respondents across the supply chain and qualitative interviews with industry leaders.
29 Interview, 8 December 2004, San Jose, CA, USA.
30 The market share of the overall Asia-Pacific region, chiefly including Taiwan, China, and South Korea, is impressive: it increased from 2 per cent in 1985 to 26 per cent in 2006 at the expense of Japan.
31 The second factor is the evolving corporate strategies of major chip firms.
32 Interview, 11 July 2005, Hsinchu, Taiwan.
33 The same study also urged the Pentagon to make commercial acquisition the core of procurement reform in order to exploit resources available in the commercial sector to benefit the defense sector.
34 All devices are in the military temperature range −55 to +125 Celsius, unless otherwise specified on the list.
35 However, the analysis of the data on QML-38535 firms has its limitations. First, the information on firm operations in four cases is incomplete, and these cases are thus removed from the quantitative analysis of the geographical distribution of offshore production operations. These cases involve QP Semiconductor's unspecified IC design center in Europe, Austin Semiconductor's assembly operations, Rochester's assembly

operations and testing operations in Malaysia that support IDT. Second, several pieces of information in the document are obsolete and are thus removed from the analysis. For instance, although the document still included Hato, Japan, as TI's certified/ qualified fabrication line, the company website stated that its fab in Hato was closed in late 2000. Third, and finally, the proportion of QML-38535 ICs in relation to the overall semiconductor consumption by the US military is unspecified. See Defense Supply Center – Columbus 2007: 50, 79, 84, 90.

36 The 2007 list does not include several giant commercial chip firms, such as Intel, the first company to be granted QML certification, Motorola, AMD, and Philips, because they dropped their military-certified QML product lines in the 1990s. See McHale 1996; Cohen 1997; Pecht *et al.* 1997.

37 The first three companies are European-owned. STM is headquartered in Switzerland, whereas BAE Systems and Semelab are headquartered in the UK. It is the US subsidiary of the BAE Systems in Virginia that enjoys full certification/qualification QML status. PSI Technologies and Millennium Microtech are specialized in assembly and test operations. The former is headquartered in the Philippines, with its qualified manufacturing lines located in the Philippines. The latter is headquartered in the Cayman Islands, with its qualified manufacturing lines located in Thailand.

38 Out of the 41 firms on the list, two (i.e. Rochester and BAE Systems) are discarded in the analysis. It is unclear in the document whether Rochester's assembly operations, which are run by "subcontractors," are onshore or offshore. In the case of BAE Systems, the document describes the firm's mask development center as at an "outside location," without specifying whether it is onshore or offshore.

39 Ciufo's report identifies the TSMC-Aeroflex partnership. O'Neil, Director of Military and Aerospace Product Marketing at Actel, revealed that RTAX-S, rad-hard Actel FPGAs, had access to advanced process technologies at UMC. Fabula and Padovani of Xilinx revealed their company's access to advanced process technologies through its partnership with UMC and IBM.

40 Interview, 29 May 2005, Hsinchu, Taiwan.

41 A handful of Taiwan-based firms have also become certified partners for the qualified suppliers of discrete and hybrid microcircuits for end-use in the US military systems. For instance, General Semiconductor Taiwan is a certified offshore partner for the US-based Microsemi Lawrence, a QML-19500 firm supplying discrete semiconductors to the US military. In the area of hybrid microcircuits, International Rectifier Aerospace & Defense, Interpoint Taiwan Corporation (ITC), and VPT Incorporated are the three certified QML-38534 firms that have relied on three qualified manufacturing lines in Taiwan. Universal Scientific Industrial Company Ltd. (USI), Taiwanese subcontractor for International Rectifier Aerospace & Defense, in Nantou, Taiwan, operates the first line. The second is operated by ITC in Taoyuan, Taiwan. Established in 1973 by the then General Instrument Microelectronics Taiwan, ITC is now a subsidiary of US-based Crane Aerospace Group. Most of ITC products are military grade parts, and its chief customers include Boeing, Lockheed Martin, Raytheon, BAE, and Northrop Grumman. Delta Electronics, a Taiwanese electronics company, runs the third line for VPT in Kaohshiung, Taiwan. See Defense Supply Center – Columbus 2007; 2008.

42 Interview, 11 July 2005, Hsinchu, Taiwan.

43 Interview with a TSMC director who was requested by the US customer to write the report, 25 August 2005, Beijing, China.

44 However, the only foreign influence on the scheme concerns the case of the Trusted Solutions Business Unit of National Semiconductor located in Maine because the CMOS process technology at the 0.25 μm and 0.18 μm in the unit is based on and compatible with TSMC processes and design rules.

4 The PRC semiconductor industry, national security, and globalization

> Not merely to science, the semiconductor industry is also very important to China's economic development and defense security.
>
> (Zhongyu Yu, president of CSIA[1])

> The semiconductor industry in mainland China is bitter history . . . During their retirement farewell parties, pertinent forerunners often shed tears lamenting their failure to see the take-off of the sector.
>
> (A veteran with 37 years experience in the PRC defense chip industry[2])

Introduction

The semiconductor industry is crucial to national power and security, as discussed in Chapter 2, and this chapter explores such a theme in the context of the PRC. It examines the extent to which the evolution of China's chip sector is relevant to its national security amid the backdrop of increasing sectoral globalization by analyzing the dynamic interplay of the Chinese industry and China's domestic and international environment. After a brief introduction, the chapter is composed of four sections. The second section identifies the link between the Chinese industry and China's economic, technological, and defense security through the lens of pertinent policy-makers and senior industry executives. The third and fourth section examine, respectively, the historical evolution of the Chinese defense and commercial chip industrial base by scrutinizing related policy initiatives and key players and the links between the industrial developments and China's national security. The fifth and final section briefly summarizes the major arguments of the chapter.

The link between the PRC semiconductor industry and national security

Beijing's acknowledgement of the importance of semiconductors to the country's national power and security has been long-standing.[3] It is argued that economic, technological and defense security considerations have propelled China to pursue a strong domestic chip industry. In recognition of the importance of semiconductors to its power and security, Beijing has endeavored to increase its indigenous chip capability to foster its economic competitiveness and independence, to pursue its technological innovation, to deepen its military modernization, and to mitigate vulnerabilities because of foreign dependency.

Senior Chinese political leaders and seasoned industry executives have long desired to build a strong domestic electronics production base to advance Chinese economic and defense interests based on their understanding of the semiconductor–security interconnection (Hu 2001; Jiang 2001: 83, 125,180).[4] As Zhang Xuedong, former vice-chairman of the Commission of Science, Technology, and Industry for National Defense (COSTIND), stated in 1999 (Editing Committee on China Electronics Industry in Fifty Years), "The development of electronics and information industries gives service to the building of the national economy and national defense."[5] Yu Zhongyu, president of CSIA, highlighted the centrality of the chip industry to China's economic and military modernization as follows: "Not merely to science, the semiconductor industry is also very important to China's economic development and defense security."[6] According to another seasoned Chinese engineer, who spent more than 37 years in the sector, "Microelectronics is strategic to China because it is related to our national security, including defense security and economic security, although today's IC market is dominated by civilian end-use."[7] In a 2002 *neibu* study (Task Force on China's IC Industry), researchers at a think tank under the PRC Ministry of Science and Technology also stressed that the IC industry is important to the Chinese economic security, information security as well as defense security.

The link with economic security

Economic security drivers behind the Chinese aspiration to build a viable domestic chip industry have revolved around three arguments put forward by Chinese policy-makers and chip industry players. First, a strong indigenous sector increases China's economic security because it mitigates any potential vulnerability due to its overt dependence on a foreign supply of semiconductors.

The following data have substantiated the predominant foreign penetration of the Chinese chip market in recent years. China's IC consumption–production gap, which refers to the difference between IC consumption and IC industry revenues, has continued to increase in spite of Beijing's efforts to contain it. This annual gap increased from US$5.9 billion in 1999 to US$68 billion in 2008 (PricewaterhouseCoopers 2010: 59). Moreover, joint ventures outperformed state-owned enterprises (SOEs) in sales revenue and profits between 1998 and 2003. Their sales revenue accounted for an average of 84 per cent of the total on an annual basis over the period analyzed (Table 4.1) (National Bureau of Statistics *et al.* 2004). Besides, China's trade deficit reached US$92.5 billion in ICs in 2006 (World Trade Organization 2007: 39). In 2007, China, for the first time ever, consumed more than one-third of the semiconductors in the global market (PricewaterhouseCoopers 2008). In the same year, it remained the fastest growing market for Asian economies' exports of ICs and other intermediate goods (World Trade Organization 2008: 39). In

Table 4.1 Comparison of state-owned enterprises and joint ventures in the Chinese IC manufacturing sector in terms of sales revenue and profit, 1998–2003 (100 million yuan)

		1998	*1999*	*2000*	*2001*	*2002*	*2003*
Number	SOEs	17	22	25	22	21	29
	JVs	84	102	106	94	107	140
Sales revenue	SOEs	6.66	32.35	54.07	40.66	37.46	49.82
	JVs	89.33	108.17	216.82	177.79	260.09	422.50
		(84.5%)	(78%)	(83%)	(80%)	(84%)	(91%)
	Total	105.45	138.33	260.27	223.29	309.19	464.62
Profit	SOEs	−1.68	−2.82	4.16	−15.16	−9.19	−7.27
	JVs	0.53	1.89	16.08	−5.81	−5.48	7.27
	Total	−0.63	0.54	17.69	−4.02	−2.04	10.33

Source: National Bureau of Statistics *et al.* 2004.

Note: SOE stands for state-owned enterprise, whereas JV stands for joint venture. The number in brackets under the joint ventures' sales revenue each year represents the percentage share in the total revenue in the same year.

2009, the global semiconductor giants continued to dominate the Chinese market; the top 10 suppliers in terms of sales revenue were all MNCs accounting for 53 per cent of the market share (PricewaterhouseCoopers 2010: 14).

Foreign dominance in China's chip market had thus propelled Jiang Zemin (2001: 125), then Chinese President and Central Military Commission (CMC) Chairman, to lament, in 1999, foreign firms' "monopoly" of the domestic market, which he viewed as a threat to their indigenous counterparts. Wang Qinsheng, Chair of HED and Chair of CIDC, considered a strong Chinese domestic IC industry "important to Chinese economic security" because it would enable the country to supply homegrown chips for its critical national economic and defense systems. As she put it, "If we allow others to make our systems, they will know our systems inside out. The same applies to our usage of foreign-made chips. All of these are threats."[8] An official echoed Wang's observation. She envisaged that if the Chinese economic infrastructure were exposed to security threats at the chip component level, it would compromise China's economic security: "We need to be in control of our own fate not merely for the defense sector, but also for the critical part of our economic and government sectors."[9]

Briefly, the consideration of China's IC consumption–production gap, its dependence on imports of core semiconductor technologies for critical national economic infrastructure, and the influence of foreign rather than local chip players in generating a majority of the industry sales revenue have propelled Beijing to accelerate the formation of a strong indigenous chip sector in its pursuit of economic security.

Second, China's strong chip industry increases its economic security because semiconductor technology is critical to its national technological innovation. As semiconductor technology has enabled technological innovations in other aspects of the economy, as detailed in Chapter 2, Chinese policy-makers have held a similar view. According to Zhou Ruojun, a deputy director-general at the Chinese Ministry of Commerce, China considered semiconductors to be a core technology in the development of its technological innovation: "If software is the brain, IC is the heart. It is the core technology." She envisioned the development of software and IC hardware technologies in China as helping to foster the growth of domestic electronics and information industries.[10] China's 10th Five-Year Plan also regarded the domestic IC sector as strategic in economic terms because it would facilitate the country's move towards informationalization, which, in turn, would drive industrialization (Task Force on China's IC Industry 2002).

Third, Chinese policy-makers have viewed a viable domestic chip sector as a contributor to Chinese economic growth and competitiveness. The report of the 16th National Congress (China Semiconductor Industry Association 2002) regarded the IC sector as a pillar of Chinese economic growth. As an official put it, the domestic chip sector was emerging as a "force multiplier" for the GDP growth of the Chinese economy.[11] According to an experienced industry player, a viable domestic chip industry would improve the Chinese economy in two ways: (1) by spurring new sectors (i.e. bio-tech and MEMS) in the economy through combining a diverse array of fields; and (2) by revamping traditional industries through supplying them with semiconductors which would improve their productivity.[12] The above three arguments surrounding the linkage between chips and economic security have thus reinforced China's resolve to upgrade its semiconductor capability.

The link with technological and defense security

Aside from recognizing the economic security importance of a strong indigenous chip industry on the PRC soil, Chinese senior officials, military planners, and industry players have also identified the significance of such an industrial base to the country's technological and defense security based on four sets of arguments.

The first involves the perceived importance of a strong local chip industry as the fundamental building block of Chinese defense modernization, which would empower the PLA conventional war-fighting capability. The second argument focuses on the offensive side of IW that the PLA can potentially launch against its enemies by exploiting homegrown advanced semiconductor goods and technologies. It highlights the significance for the PLA of exploiting a viable domestic chip industrial base to launch its asymmetrical warfare against its technologically superior enemies, thereby empowering its unconventional war-fighting capability. The third argument emphasizes the defensive side of IW from the PLA perspective. It stresses the long-standing risks and vulnerabilities that the PLA has faced due to its dependence on unreliable foreign supplies of critical chip goods and technologies, and the extent to which a strong indigenous chip industry can help mitigate these threats and vulnerabilities engendered by foreign dependency. The fourth argument highlights the PRC's recognition of the dual-use nature of semiconductor goods and technologies, the general trend of spin-on and its policy towards the gradual use of COTS in its military systems with the aim of integrating its rising commercial chip industrial base into its military industrial base, if appropriate. Each of the above arguments will be elaborated as follows.

Basis of PLA modernization and conventional military capability

First, a strong local chip industry would underpin Chinese defense modernization, contributing to the PLA's conventional war-fighting capability. This argument was reinforced by the Chinese acknowledgement that the backward domestic chip industry had undermined the modernization of the PLA. It was also accentuated after the First Gulf War when the Chinese realized that US semiconductor superiority had underpinned its military predominance.

Since the 1950s, China's attempts to develop a domestic electronics industry have had a clear military agenda. In 1952, the Second Ministry of Machine Building was founded to be in charge of China's conventional weapons development. Soon after this institutional initiative, Red-Army-General-turned minister Zhao Erlu identified aeronautics and electronics as the two "weakest links" to be fortified in order to wage the Korean War and meet China's long-term defense requirements. His ministry thus gave precedence to creating basic research institutes in electronics, optics, aeronautics, and special materials in an attempt to build China's defense industry (Zheng and Li 1989: 117; Lewis and Xue 1994: 75). In 1956, a 12-year national plan of scientific development pinpointed the defense importance of semiconductors to China and identified semiconductor technology as one of the four fields critical to Chinese national security, which Beijing vowed to take emergency measures to develop (Simon and Goldman 1989; Zhang and Zhang 2007: 22). Thereafter, China's long journey to develop its electronics industry has had a clear military agenda. In 1960, an enlarged CMC meeting vowed to give priority to missiles and nuclear bombs, as well as the development of electronics and radio technology (Bao 1989: 261). In September 1963, the Fourth Ministry of Machine Building was established to be in charge of electronics and telecommunication industries.

After the Cultural Revolution, many Chinese leaders realized that the low capabilities of the country's electronics industry had inhibited its military technology. Some contended that the elevation of electronics would advance China's defense modernization by contributing to both strategic and conventional weapons R&D (Feigenbaum 2003: 84–6). Similar lamentations for the low level of the indigenous microelectronics sector and its negative impact on Chinese military capability have remained unabated (Yang and Liao 2003).

Technology journals published by the PLA and conglomerates in the Chinese defense industry have recorded instances of problematic electronic components, churned out by the domestic defense production base, which damaged various Chinese military systems, thus sabotaging Chinese defense security. Three examples follow.

First, a failure rate of more than 40 per cent in various rockets during their assembly and range tests from September 1971 to January 1978 resulted from problems with home-made electronic parts and components. Thus, various defense system manufacturers called for reforms of China's defense electronic production base in the late 1970s (Cao and Gao 1989: 445). Second, according to a specialist at China's military standardization center, a faulty electronic part manufactured by a production unit within the Chinese defense industry caused the explosion of Long March-2E, a China-made launch vehicle, during its launch in January 1995. In the wake of the rocket's explosion, the Chinese aerospace industry implemented a military standard concerning destructive physical analysis (DPA), which involves a series of stringent tests designed to ensure the high reliability of electronic components for military/aerospace end-uses. However, the result of a subsequent DPA of imported and homegrown electronic devices selected between October 1996 and January 1997 was another blow to Beijing: the quality of indigenously made devices trailed far behind that of imported ones (Zhao 2000: 9).

The third example concerned a bottleneck at the IC level that had disrupted the development of *Dongfeng-31*, China's long-range ground-to-ground missile. The missile had its successful test launch in August 1999. The launch occurred three weeks after Taipei said it should be treated as a separate state in any negotiations with Beijing. The reported range of the missile is in the 5000-mile category and it is capable of carrying a 1500-lb weapon. Crucially, the launch marked a milestone in China's military modernization program: it demonstrated the PLA's ability to deploy a more robust deterrent in the form of long-range mobile and solid fueled missile (Wu and Li 2000).[13] Nevertheless, according to the missile's chief engineer, the insertion of a large number of ill-designed ICs in the system, along with other technical hurdles, had caused heat dissipation problems jeopardizing the reliability of the missile (ibid.). Although the problems were eventually solved, the case demonstrated that badly designed ICs by the domestic defense sector had inhibited the PLA's missile modernization ambition (Yang and Liao 2003). To remove the aforementioned problems, Chinese policy-makers resolved to overhaul China's chip capability.

The first argument also gained prominence after the First Gulf War because of its two-fold lesson for China: US semiconductor superiority accounted for its military supremacy, and China should improve its semiconductor capability in order to modernize its military (Luo *et al.* 1992: 36; Li 1993; Mulvenon 1999: 178–9; Xu 2004; Zhang and Zhang 2005: 93–100). As Jiang Zemin (cited in Zhang 1995) proclaimed after the war, "Military electronics has a bearing on national security . . . [and] must be given first place." Since the latter half of the 1990s, one of the PLA's priority objectives during its third phase of modernization has been to digitalize its forces (Gao and Guan 2004; Ling and Wang 2005; Wang 2005), with semiconductors viewed as indispensable to the digitalization process.[14]

Lu Xinkui (2001), the vice-minister of the Chinese Ministry of Information Industry, stressed that IT and IT industries would increasingly become instrumental to the Chinese defense industrial base and that they would continue to be pillar industries in supporting Chinese military reform, augmenting the PLA's capacity to win regional wars under high-tech conditions.[15] Such wars include any

Taiwan Strait contingency. In his 2003 report to the 16th Party Congress, Jiang Zemin further urged the Chinese military to accomplish the dual tasks of mechanization and IT application in order to accelerate its transformation and to leapfrog ahead in its modernization goals. He instructed the PLA to adopt transformational IT-related technologies – underpinned by semiconductors – in place of conventional mechanized systems to pursue "military changes with Chinese characteristics" (Jiefangjun Bao (PLA Daily) 2003; Ling and Wang 2005; Department of Defense 2009: 11). Similarly, Chinese President Hu Jintao (Xinhuanet 2007) proclaimed in his report to the 17th Party Congress:

> To attain the strategic objective of building computerized armed forces and winning IT-based warfare, we will accelerate composite development of mechanization and computerization, carry out military training under IT-based conditions, modernize every aspect of logistics, intensify our efforts to train a new type of high-caliber military personnel in large numbers and change the mode of generating combat capabilities.

Neibu publications have also continued to recognize the importance of semiconductors to the modernization of the PLA, emphasizing China's military motivation behind the build-up of a strong domestic IC industry. (Such an accentuation has often been lacking in public official statements, which tend to stress the economic motivation only.) In a *neibu* study (Task Force on China's IC Industry 2002), researchers at a government think tank stressed that China's effort to build a strong chip industry would advance Chinese defense security. As the proportion of IC in modern military systems would continue to increase, the technological level of these systems would rise accordingly. As the First Gulf War and the Kosovo War in the 1990s emerged as high-tech wars, the two sides in each of these conflicts competed militarily through the comparison of their information power, weaponry clout, and semiconductor capability. Hence, China should increase its military power by improving its chip capability, the study concluded. Xu Shi-liu (2004: 6), the head of the 24th Research Institute, a key military microelectronics center in China now a part of China Electronics Technology Corporation (CETC), argued that China should concentrate on its chip technology to modernize its military because it was IT that had spurred the RMA in the West in recent years.[16]

Asymmetrical warfare and unconventional war-fighting capability

The second argument has centered on the belief that a viable Chinese chip production base would strengthen the PLA's unconventional war-fighting capability in the area of offensive IW attacks. Such a prospect constitutes the implementation of asymmetrical warfare that some PLA planners have advocated in recent years, the notion that the inferior can defeat the superior by selecting measures of offensive operation other than strictly defined military means (Qiao and Wang 2002). From the Chinese perspective, the PLA, by utilizing a greatly improved domestic chip industrial base, would upgrade its capability to launch asymmetrical warfare against its technologically superior enemies, thereby advancing its unconventional war-fighting power.

Secondary materials have documented Chinese recognition of the centrality of semiconductors to the conduct of IW, which the PLA or its enemies can launch (Cheng *et al.* 1997). For instance, chipping, a semiconductor-targeted tactic detailed in Chapter 2, is not foreign in PLA writings (Qiao 1994; Wang and Xu 2003; Fu and Li 2004): some have advocated that China should use chipping to destroy US and Taiwanese information systems.

Mitigation of vulnerability due to foreign dependency

The third argument has focused on the belief that a strong local chip sector would help upgrade the PLA's defensive IW capability and mitigate vulnerabilities caused by foreign dependency.

Information and network security concerns, which are identified by the State Council Informatization Office (2005: 55–60) as the top problems Beijing has faced in establishing an information-based country, also apply to defense-related fields. PLA strategists (Mulvenon 1999: 178; Xu and Fang 2001; Ma 2003: 238; Fu and Li 2004; Zhang 2004: 303) have highlighted the reported US move to insert viruses in their enemies' information systems during the First Gulf War and the Kosovo War, although some of the claims were unsubstantiated. Others (Li 1993; Zhang and Guo 2004), in view of similar attacks launched against the PLA, have called for improvements in the Chinese capability to design and fabricate chips for critical military systems in order to increase its defensive IW capability and to minimize any risks these attacks may incur.

Still others (Guo and Gu 1996; Wang 1998; Zhang and Guo 2004) have amplified long-standing risks the PLA has faced because of its dependency on foreign supplies of critical chips. These risks include backdoor devices, supply cuts and counterfeit components. Engineers from the Central Research Institute of Huajing Microelectronics Group (Luo *et al.* 1992), one of China's major military semiconductor operations, argue that China's dependency on foreign-sourced military ASICs has threatened its security. They thus urge Beijing to ensure that onshore production of military ASICs would meet the demands of the PLA: "We should independently establish our military ASIC industry so as to fabricate these military components at home."

In a *neibu* study, the Chinese Ministry of Science and Technology (Task Force on China's IC Industry 2002) also highlights the necessity for China to advance its domestic chip production base to churn out quality ICs for critical defense and economic systems in order to mitigate foreign dependency risks. "The design and production activities of critical ASICs for important national defense systems and systems relating to national economic security should be fundamentally based at home," it concludes.

A senior member of the Chinese defense industry expressed a similar concern in 2005: "If you fully depend on others for semiconductors, you will fear that others may cut your supply at critical times even if the supply is unproblematic during peacetime." Hence, China should run its domestic semiconductor operations for the design and fabrication of critical military chips.[17] Moreover, researchers at a government think tank (Xu and Fang 2001) regard the lack of homegrown CPU technology as an information security threat to China. They view the 2001 debut of Arca-1, China's first homegrown 0.25-process, 32-bit CPU, as an opportunity to cement China's information security: the country could equip its critical military and infrastructure systems with indigenously made or designed CPUs.

Spin-on, dual-use, and the move towards COTS

The fourth and final argument has been based on the recognition by PLA strategists and defense industry players of the dual-use nature of semiconductors and the trend of spin-on in developed countries.[18] Such recognition has propelled them to support the build-up of a strong domestic commercial chip sub-sector in order to benefit the Chinese military.

They have urged the PLA to recognize the dual-use nature of many technologies (including semiconductors) and the position of civilian firms as technology drivers in the Chinese economy as a basis to facilitate spin-on in order to advance the military's digitalization endeavors (Xu 2004; Zhang 2004: 312; Editing Committee on Studies of Jiang Zemin's Thoughts on Defense Technology Industry Construction 2005: 94–5; Ling and Wang 2005: 19). In their study published by the PLA publishing house, Ling and Wang (2005: 19) argue:

> The civilian sector contains abundant resources including various dual-use IT technologies and COTS goods available for military end-uses. So long as these IT goods and technologies are governed under a unified set of regulations and standards, civilian enterprises should manufacture IT products for military end-uses.

The head of the 24th Research Institute (Xu 2004) also urges the Chinese defense establishment to institutionalize the trend of spin-on to benefit the military microelectronics community. He regards such a policy initiative feasible due to China's recent progress in developing a vibrant civilian chip industry, with some of the local firms already acquiring technologies close to or on a par with those of their counterparts in leading semiconductor countries. China's success in spin-on, he envisages, could enable the military to use existing goods and technologies in the domestic civilian market. To complement spin-on, he also urges Beijing to revamp its relatively backward defense chip industrial base to develop chip technologies for future military systems.[19] According to a senior member of the Chinese defense chip sub-field in 2005, the sub-sector had lagged far behind its commercial counterpart in terms of equipment and technology for the production of semiconductors. He thus urged Beijing to ensure that the domestic commercial chip sub-sector be permitted to supply its qualified products to help drive the PLA modernization.[20]

To sum up this section, the aforementioned national security considerations have driven Beijing to introduce various policies over recent decades to build a strong domestic chip industry. These drivers have thus shaped the pertinent policy environment in China for the development of the sector, which we will explore in the next two sections.

Chinese defense chip industrial base: from autarky to gradual opening-up

This section examines the development of China's defense chip industrial base and its relevance to China's national security. Although the defense and commercial chip industrial arenas overlap at times,[21] it is possible to differentiate the two segments (Figure 4.1). The former is relatively autarkic in its production system compared to the latter, and is technologically weak compared to the latter especially since the latter's take-off since 2000. This section focuses on the defense chip industrial base by exploring three inter-related themes. The first analyzes the evolution of the traditionally autarkic defense chip industrial base, its key players and its relevance to China's national security. It is argued that in spite of various hurdles, pertinent state-owned defense operations had manufactured semiconductors to satisfy part of the PLA defense needs, thereby serving, at least partially, as a security asset to Beijing. The second segment examines the formulation of a Chinese military specification, standard, and certification system pertaining to semiconductors and its relevance to China's national security. It is contended that the implementation of the rules-based system has been a security asset and liability to China. The third and final part examines the recent reform of civil–military

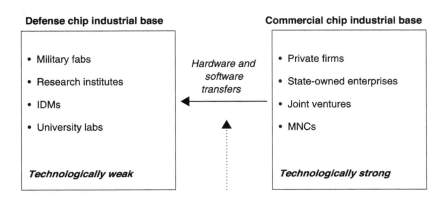

Figure 4.1 The PRC chip industrial base and spin-on reform.

Sources: Interviews and secondary data.

integration that has expanded the Chinese defense chip industrial base to include non-state-owned enterprises and its impact on China's national security.

Origin, key players, and national security linkages

China started to develop its electronics industry because of the aforementioned national security considerations with initial assistance from the Soviet Union. In the early 1950s, semiconductors as a branch of science and technology were introduced to the PRC. In 1956, the Chinese Academy of Sciences (CAS) pioneered a course on theories and practices of semiconductor. The origin of semiconductor technology research in the PRC dates back to 1956 with the establishment of the 13th Research Institute.[22] The institute is the earliest established, large-scale, and specially equipped comprehensive semiconductor research institution in China, and has strong technical potential today (Pillsbury 2005: 47). Soon after its inauguration, the institute produced China's first germanium transistor. In July 1959, China applied its first homegrown semiconductor in its military systems (Wang and Zhang 1989: 1–16). In 1960, the Institute of Semiconductors of CAS was founded to meet the needs of the aforementioned 12-year national plan of 1956, which identified semiconductors as an area critical to China's national security that Beijing should take urgent measures to develop. In 1965, the 13th Research Institute produced China's first homegrown IC, only three years after the debut of the world's first IC.[23] Just as Moscow contributed to the establishment of China's defense industry in the 1950s (Zheng and Li 1989: 118; Editing Committee on China Electronics Industry in Fifty Years 1999; Shambaugh 2004: 226), it also helped Beijing develop its defense chip industrial base. In 1958, Soviet experts began to aid No. 784 Factory, one of the major Chinese electronics industry players. Moreover, Moscow helped Beijing fabricate semiconductors for 109 Computer Model-III, a large all-purpose digital computer, at a captive plant attached to the Institute of Physics of CAS. As Chinese semiconductor scientist Wang Shouwu, who was involved, recalled (Chinese Academy of Sciences 2000: 36–7; Zhang and Zhang 2007), the computer with CAS-fabricated transistors had been crucial to China's development of atomic bombs from 1967 until the early 1980s. In 1960, however, the Russian experts withdrew from China because of the Sino-Soviet split.

Since the 1950s, China has gradually established a largely autarkic indigenous production system for semiconductors tailored to meet Chinese military and aerospace objectives; the system comprises state-owned research institutes, academies, factories and university labs.[24] Many of these were located in remote regions of China contributing to the construction of a self-sufficient defense industrial base from 1964 to 1971 under the Third Front initiative put forward by Mao Zedong. Their location aimed at reducing vulnerability to potential attacks by the USA or the Soviet Union (Naughton 1988; Lewis and Xue 1994: 94; Shambaugh 2004: 227). In spite of sporadic restructuring and downsizing efforts over past decades, the structure of the Chinese defense chip industrial base has largely continued to the present day. In 2005, the chairman of a leading Chinese

IC design house, who also led the Very Large Scale Integration (VLSI) design project under the National 863 Program, admitted that the internal source of chip supplies to the PLA would rely on many fabs in China, fabs with their own designs and facilities not controlled by outsiders.[25] Another seasoned industry player also stressed in 2005 that "these old-fashioned factories are still in existence" today.[26]

Such an internal source of military semiconductor supplies is complemented by the foreign source of supplies amid stringent, albeit flawed, export controls by the West, according to a military journal published by the Electronics and Information Base Department (*Dianzi Xinxi Jichu Bu*) of the PLA's General Armaments Department (GAD) (Wang 1998). China has thus developed a dual-track strategy to acquire ICs for its military end-uses from both domestic and international sources echoing, in some ways, the partial globalization of defense chip production activities in the case of the USA as analyzed in Chapter 3. Although it is difficult to estimate the ratio between imported and indigenously made military ICs for PLA end-uses due to a lack of available data, the same study revealed that China had imported a majority of space-qualified electronic parts and components for end-uses in its military aerospace and aviation systems. Because of the end of the Cold War and the ensuing loosening of pertinent export controls by the West, China has found it easier than before to purchase high-end military electronic parts and components from the West in general and the USA in particular (Guo and Gu 1996).[27]

Key players and contributions to the PLA modernization

The following section identifies key players in the traditionally enclosed Chinese defense chip industrial base and analyzes the extent to which they have contributed to PLA modernization, based on analyses of interview data and secondary materials including *neibu* documents. It is argued that pertinent Chinese defense producers have contributed to China's security by supplying semiconductors to the military in spite of various technological, economic, and political constraints common to China's ossified state industrial system.[28] Between 1986 and 1990, for instance, 22 major Chinese military systems benefited from the insertion of 530 million indigenously made electronic parts and components, including semiconductors (Ye 1992: 7).[29]

First, seven state-owned research institutes, factories, and universities in the Chinese electronics industry won the annual progress award in military science and technology in 1990 because of their work on semiconductor discrete components and ICs (China Machinery and Electronics Industry Yearbook Editing Committee 1991: IV-12-IV-13). These institutes included the Nanjing Electronic Devices Institute (five awards), the 13th Research Institute (four awards), the 24th Research Institute (four awards), the University of Electronic Science and Technology of China (one award), the Xi'an University of Electronic Technology (one award), the Jinan Semiconductor Device Institute (one award), and the Beijing Semiconductor Device No. 3 Factory (one award). They won the prize because of their work in 16 types of semiconductor goods and technology including GaAs ICs, semiconductor switches, diodes, field effect transistors, etc. In 1991, eight institutes won the same award (China Machinery and Electronics Industry Yearbook Editing Committee 1992: IV-14-IV-15). They included the 24th Research Institute (seven awards), the 13th Research Institute (five awards), the 55th Research Institute (four awards), Huajing (three awards), the Xi'an University of Electronic Technology (two awards), the 47th Research Institute (one award), State 970th Factory (one award) and Jinan Semiconductor Device Institute (one award). They won the prize because of their work in 23 types of semiconductor goods and technology including rad-hard ICs, GaAs ICs, semiconductor switches, memory chips, transistors, diodes, etc.

As the aforementioned institutes were part of the Chinese defense chip industrial base in the early 1990s, at least eight of them continued to operate in the 2000s, appearing on the June 2002 version of QPL, which will be analyzed below. They included the 13th Research Institute, the 24th Research Institute, the 55th Research Institute, the 47th Research Institute, Huajing, the State 970th Factory, and the Jinan Semiconductor Device Institute and Beijing Semiconductor Device No. 3 Factory.

Second, a *neibu* publication (Hua 1996) identified military semiconductor devices developed by China's state-run research institutes and academies (Table 4.2) during the 8th Five-Year Plan period from 1991 to 1995 and analyzed the features of these defense operations and their link to China's national security as follows.

The technological level of these Chinese defense institutions during this period was behind that of its leading Western counterparts by a few generations in areas such as CPUs, DSPs and radiation-hardened ICs for military end-uses. Specifically, the gap was about 15 years in the area of hybrid ICs. Despite this, China managed to insert some of the home-made semiconductors in its information-based military systems during the period, ranging from satellites, missiles, avionics, to communication gadgets. Hence, China's defense security was arguably enhanced in two ways. On one hand, the homegrown miniaturized and better functioning chips, compared to their immediate predecessors, helped reduce the weight and size of the PLA systems in question and improve the reliability and performance of these military gadgets. On the other hand, these components were substitutes for pertinent imports, which were, at times, unavailable because of export controls.

A telling example concerns the DSPs. China had attempted to indigenously design and fabricate these devices for the PLA end-uses because of the fear that insertion of foreign-designed and -manufactured components into its military systems might undermine its national security. In 1995, China successfully made its first homegrown DSP (TMS320C25), paving the way for insertion of these chips in its military systems from 1996 to 2000. Crucially, my interview with an industry insider in 2005 substantiated the fabrication of these homegrown DSPs. In the late 1990s, the interviewee led an American EDA company's branch in China, which had sold software tools to the state-run Chinese manufacturer of these DSPs. According to the indigenous chip-maker running a 6-inch wafer fab,

Table 4.2 Key players in the PRC defense chip industrial base and their contributions to Chinese defense security, 1991–95

Products	Period	Key players	Contributions to PRC defense security
Hybrid ICs	8th Five-Year Plan period (1991–95)	43rd Research Institute, 13th Research Institute, 24th Research Institute, No. 771 Research Institute, No. 214 Research Institute, Beijing Semiconductor Device No.1 Factory	• Technology level of some of the products comparable to that in the West in the late 1980s and the early 1990s • Insertion of these hybrid ICs in PLA systems helped reduce the weight and size of the systems and increase their reliability and performance
GaAs VHSICs	First national program during 6th Five-Year Plan period (1981–85); R&D during 7th (1986–90), and 8th Five-Year Plan period	13th Research Institute, 55th Research Institute, Shanghai Institute of Metallurgy of CAS	Exact end-use not specified
DSPs	R&D during 8th Five-Year Plan period	Not specified	• 1995: debut of first homegrown DSP (TMS320C25) • Debut of first homegrown 1.5μm-process, CMOS DLM DSP • Insertion of these DSPs in PLA systems during 9th Five-Year Plan period (1996–2000)
CPUs	R&D and production of military-use, 16-bit CPU began at the end of 7th Five-Year Plan period	Not specified	R&D and production of military-use, 16-bit 80C86 series of CPU began during 7th Five-Year Plan period, and completed during 8th Five-Year Plan period
A/D, D/A Converters	8th Five-Year Plan Period	24th Research Institute, No. 777 Factory, No. 4433 Factory, No. 771 Research Institute, No. 214 Research Institute	• About 20 types inserted in PLA systems • Insertion of bipolar 10-bit A/D converters (SAD571) in military radars, missiles, avionics, and communications systems
Rad-hard technology and ICs	R&D began in the 1980s	13th Research Institute, 24th Research Institute, No. 771 Research Institute, Beijing Semiconductor Device No. 3 Factory	• 1990: 49 types of rad-hard electronic devices passed military qualification tests • 44 of them subsequently inserted in PLA systems • Rad-hard CMOS ICs inserted in military satellites achieving import substitution and overcoming security hurdles due to Western blockade • 1995: another 35 types passed military qualification tests

Source: Based on data in Hua 1996.

the Chinese military decided not to purchase DSPs from TI because of concerns that these components, if supplied by a US firm, might be tampered with by backdoor devices that would cause the Chinese military systems to be ineffective. Instead, the domestic chip-maker fabricated comparable semiconductor devices for the PLA.[30] Based on this telling account, one can argue that the Chinese defense chip industrial base had successfully accomplished import substitution in the case of DSPs, thereby advancing China's defense security.

Third, a *neibu* publication in 2002 further identified qualified Chinese military semiconductor manufacturers who had passed a new set of certification and screening tests in order to supply the PLA. This set of data, combined with analysis of related publicly available information, deepens our understanding of these Chinese defense players and their contributions to China's defense security.[31]

According to the June 2002 version of the QPL, 182 indigenous manufacturing lines churned out 1,195 types of military electronic parts and components (including semiconductors) from April 1992 to 31 May 2002. During this period, 24 qualified state-owned local chip-makers supplied 226 types of QPL semiconductors – comprising 122 discrete components and 104 ICs – to the Chinese military. Of the 24 suppliers, 15 were state-owned factories, eight were research institutes, and one was state-owned-factory-turned enterprise (Table 4.3). Many of these defense operations have been in existence for decades. In particular, the eight research institutes, with their history dating back to the 1950s, are now under a diverse array of defense industry conglomerates.[32] Six of them are part of CETC under the leadership of the Ministry of Information Industry. They are the 13th, 18th, 24th, 47th, 55th, and 58th Research Institutes. One is the No. 771 Research Institute[33] of the China Aerospace Corporation. The remaining one is No. 214 Research Institute of China North Industries Group Corporation, the largest weaponry-manufacturing group in China. Table 4.4 further identifies QPL producers in question in terms of the military standards that apply to their products, with the names and locations of their qualified manufacturing lines. Additionally, the third version of QML identified six certified manufacturers running a total number of eight hybrid IC production lines for PLA end-uses (Table 4.5) (Yu 2002).[34]

Although the exact military end-uses of these qualified semiconductors are often unclear to researchers because vague descriptions of the pertinent end-uses are common in the official data,[35] there are notable exceptions such as the case of No. 214 Research Institute. In December 2007, the institute (No. 214 Research Institute of China North Industries Group Corporation) detailed its role as a QML and QPL producer in advancing China's defense security in a press release. Five types of the ICs produced by the institute had met pertinent military standards, passed a series of stringent tests and then become the building blocks of a newly designed long-range precision strike weapon deployed by the Chinese Ground Forces. Consequently, improvements at the IC component level reduced the weight and size of the military system in question while increasing its reliability.

Table 4.3 Military semiconductor suppliers on the June 2002 version of QPL by types of operation

Types of operation	Player(s)
State-owned factories	State 746th Semiconductor Plant
	No. 1 Factory of State 774th Plant
	State 777th Plant
	State 798th Plant
	State 871st Plant
	State 873rd Plant
	State 877th Plant
	State 970th Plant
	Beijing Semiconductor Device No. 3 Factory
	Beijing Semiconductor Device No. 5 Factory
	Hebei Shijiazhuang Radio No. 2 Factory
	Jinan Semiconductor Device Institute (Jinan Semiconductor Corp.)
	Liaoning Fuxin Transistor Factory
	Shenyang Feida Semiconductor Device Factory
	Harbin Transistor Factory
Research institutes	13th Research Institute, CETC
	18th Research Institute, CETC
	24th Research Institute, CETC
	47th Research Institute, CETC
	55th Research Institute, CETC
	58th Research Institute, CETC
	No. 771 Research Institute, China Aerospace Corporation
	No. 214 Research Institute, China North Industries Group Corporation
State-owned-factory-turned enterprise	China Hua Jing Electronics Group Co. (former State 742nd Plant)

Source: Based on data in Yu 2002.

Four case studies and the diagnosis of the defense chip industrial base

Crucially, the following four case studies shed further light on the strengths and weaknesses of the industrial base. These cases include the 13th Research Institute, the 45th Research Institute, the 24th Research Institute and the No. 771 Research Institute. The first case concerns the 13th Research Institute, the first comprehensive semiconductor research institution in China. So far, the institute has pioneered various semiconductor projects in China. It was home to China's first germanium transistor in 1956, its first IC in 1965 and its first GaAs microwave power field effect transistor in 1980. It also pioneered the study of GaAs VHSIC in China between 1980 and 1990 (Yuan 1992). In 1990 and 1991, it won the annual progress award in military science and technology because of its work on GaAs ICs, GaAs microwave field effect transistors, and rad-hard ICs; between 1991 and 1995, it

Table 4.4 Military semiconductor suppliers on the June 2002 version of QPL

Qualified manufacturer	Military standard	Qualified manufacturing line
State 746th Plant	GJB 33–85	Silicon low power junction field effect transistor
No. 1 Factory of State 774th Plant	GJB 33A-97	Silicon high frequency and microwave power transistor
State 777th Plant	GJB 597A-96; GJB 597–88	Analog IC
State 798th Plant	GJB 597–88	Thick-film hybrid IC
State 871st Plant	GJB 597A-96; GJB 597–88	100 mm bipolar digital IC
State 873rd Plant	GJB 33–85; GJB 33A-97	Silicon voltage regulator diode; silicon switch regulator diode; silicon voltage reference (including low-noise types) diode; transient voltage suppressor diode
	GJB 33A-97;GJB 33–85	Silicon switch rectifier diode
	GJB 33–85; GJB 33A-97	NPN epitaxial planar power transistor
State 877th Plant	GJB 33A-97	Glass passivated diode
	GJB 33–85	Silicon low frequency high power transistor
State 970th Plant	GJB 33–85	Microwave PIN diode
Harbin Transistor Factory	GBJ 33–85	Current switch diode
Liaoning Fuxin Transistor Factory	GJB 33–85	Large power high frequency transformer transistor
Beijing Semiconductor Device No. 3 Factory	GJB 597A-96; GJB 597–88	CMOS IC
Beijing Semiconductor Device No. 5 Factory	GJB 597A-96	Integrated regulator
	GJB 33A-97; GJB 33–85	PN silicon unijunction transistor
Shenyang Feida Semiconductor Device Factory	GJB 33A-97; GJB 33–85	PNP silicon high frequency low power transistor, and switch tube
Hebei Shijiazhuang Radio No. 2 Factory	GJB 33–85	Bipolar NPN transistor for low-to-medium power
Jinan Semiconductor Device Institute (Jinan Semiconductor Corp.)	GJB 33A-97; GJB 33–85	Schottky diode
China Hua Jing Electronics Group Co.	GJB 33A-97	PNP silicon transistor
	GJB 33A-97; GJB 33–85	Bipolar NPN transistor for low-to-medium power
	GJB 33A-97	Bipolar NPN power transistor for low-to-medium power

Institute	Standard	Product
13th Research Institute	GJB 33–85; GJB 33A–97	Silicon microwave transistor; GaAs microwave power field effect transistor
18th Research Institute	GJB 2443–95	Thermoelectric cooling module
24th Research Institute	GJB 597A–96; GJB 597–88	Bipolar high frequency low noise wide-band amplifiers; bipolar analog IC
47th Research Institute	GJB 597A–96; GJB 597–88	CMOS IC
55th Research Institute	GJB 33A–97; GJB 33–85	GaAs microwave power field effect transistor
	GJB 33A–97	Silicon microwave diode
	GJB 597A–96	1.0–1.5 µm CMOS IC
No. 214 Research Institute	GJB 597A–96	3 µm CMOS IC
No. 771 Research Institute	GJB 597A–96	Four-Inch CMOS
	GJB 597A–96; GJB 597–88	Analog IC

Source: Based on data in Yu 2002.

Table 4.5 Military semiconductor suppliers on the third version of QML

Qualified manufacturer	Military standard	Class	Manufacturing line	Location	Note
Feiyu Microelectronics	GJB 2438–95	H1	Thin-film hybrid IC	Beijing	1993: received Certificate of Conformity of Quality System Certification of Military Product Supplier from GAD
43rd Research Institute	GJB 2438–95	H1	Thin-film hybrid IC	Hefei, Anhui	Earliest research institute engaged in hybrid microelectronics technology production, research, and development
	GJB 2438–95	H1	Thick-film hybrid IC		
No. 214 Research Institute	GJB 2438–95	H1	Thick-film hybrid IC	Bangbu Anhui; new R&D center in Suzhou	2007: five types of ICs for end-uses in a long-range precision strike weapon by Ground Forces
No. 771 Research Institute	GJB 2438–95	H1	Thin-film hybrid IC	Xian, Shanxi	2000: received Certificate of Conformity of Quality System Certification of Military Product Supplier from GAD
	GJB 2438–95	H1	Thick-film hybrid IC		
Qingdao Semiconductor Research Institute	GJB 2438–95	H1	Thick-film hybrid IC	Qingdao, Shandong	ICs for end-uses in China's first man-made satellite, weather satellites, and manned spacecrafts
Xijing Electronic Corp.	GJB 2438–95	H1	Thick-film hybrid IC	Xian, Shanxi	• 1970: No. 895 Factory set up as part of the Third Front initiative • 1985: founding of Xijing Electronic Corp. • 2001: Xijing restructured as Chuang Lian Electronic Component (Group) Co. Ltd.

Sources: Yu 2002; Pillsbury 2005: 50–1; No. 214 Research Institute of China North Industries Group Corporation 2007; No. 771 Research Institute n.d.; various company websites.

was involved in the R&D of hybrid ICs, GaAs VHSICs, and rad-hard technology (Hua 1996). Moreover, it was home to two qualified manufacturing lines that provided the PLA with silicon microwave transistors and GaAs microwave power field effect transistors (Yu 2002).

On the sidelines of an industry show in 2005, seasoned engineers from the institute proudly showed me a colorful booklet detailing in-house GaAs products and technologies. According to an engineer from the institute's 3-inch GaAs process technology fabrication center, "Our GaAs technology has reached the level that the US accomplished in the mid-1990s. We are one of the world's leading players."[36]

Despite this, export controls by the West and China's fragile industrial infrastructure have constrained the institute's development, constraints that have also troubled other members of the domestic defense chip industrial base. Although some of the Chinese defense chip operations had obtained controlled semiconductor materials and equipment through "third countries" or "individual channels," another seasoned engineer from the institute argued:

> The US tried hard to sabotage China's efforts to develop its defense industry, and its export controls of semiconductor materials and equipment is a good example . . . When we visited Europe, Europeans revealed that the US had pressured them not to sell products to us.[37]

Even worse, China's weak industrial structure has further constrained these Chinese defense chip production hubs, including the institute, to pursue self-sufficiency, a long-term strategic aspiration for the PLA. According to another member of the institute, "It will not work if we depend on ourselves. Our basic industries are bad. We need to depend on other countries for semiconductor equipment and materials."[38]

The second case concerns the 45th Research Institute, the only indigenous state-owned institution which has continued to work in the area of lithography equipment since the 9th Five-Year Plan period, with customers including domestic military and commercial chip-makers (Zhang 2002). As of 2005, it was under the helm of a Chinese expert with a PhD degree from the Georgia Institute of Technology.[39]

The institute, however, has been fighting an uphill battle because of various constraints. Lack of resources has been a major challenge. In March 2005, an engineer from the institute revealed that his employer had spent three years developing its latest stepper, which would have its test-run in small-scale defense chip operations in southern China. In spite of this accomplishment, he admitted that resource constraints had put him and his colleagues under tremendous pressure. "Whereas leading semiconductor equipment makers in the West can afford to spend a long time developing a new stepper, we are requested to produce ours within two to three years," he admitted.[40] Besides, the relatively weak domestic industrial infrastructure and the gap in technology and investment between the institute and the global giants have further hindered its development (Zhang 2002).

The third case concerns the Sichuan-based 24th Research Institute, a key provider of analog ICs for military and commercial end-uses in China. Between 1986 and 1990, it pioneered China's military foundry projects, running small-volume military ASIC experimental lines (Wu 1991; Chen 1992; Guo and Liu 1992). In 1989, it fabricated military-grade analog semiconductors for end-uses in Long March-2E (LM-2E), an indigenously made launch vehicle which had its first successful test flight on 16 July 1990 reaching the desired low Earth parking orbit with the 50-kg Pakistani Badr piggy-back satellite.[41] In 1990 and 1991, it won the annual progress award in the area of military science and technology because of its work on CMOS analog ICs, rad-hard ICs, and the development of computer-aided design (CAD) for rad-hard ICs (China Machinery and Electronics Industry Yearbook Editing Committee 1991: IV-12-IV-13; China Machinery and Electronics Industry Yearbook Editing Committee 1992: IV-14-IV-15). Between 1991 and 1995, it was involved in the R&D and/or production of hybrid ICs, A/D, and digital-to-analog (D/A) converters, and rad-hard ICs (Hua 1996). Moreover, it housed two qualified manufacturing lines that had provided the PLA with bipolar, high frequency, low noise, wide-band amplifiers, and analog ICs (Yu 2002). In 2003, China's National Laboratory of Analog Integrated Circuits was founded under the umbrella of the institute (Xu 2004). The institute also supplied semiconductors for gadgets in China's spaceships – from Shenzhou I to V – as part of the *Hangtian* project (Zheng 2007).[42] When China's first manned spaceship, Shenzhou V, had its debut in October 2003, the spaceship was equipped with a variety of semiconductors supplied by the institute (Lei 2003).

In spite of its contribution to the Chinese military modernization, senior researchers at the institute (Wu 1990; Xu 1991) have identified a major weakness in the Chinese defense chip industrial base that has sabotaged the institute's development. This pitfall refers to the gap between China and its counterparts in leading semiconductor countries in the areas of reliability and fabrication technology pertaining to military analog ICs.

However, the recent restructuring of the institute has enabled the defense unit to narrow the technological gap in question. Southwest Integrated Circuit Design Co. Ltd. (SWID), an IC design house which sprang from the institute, used the 0.35μm 3P3M SiGe BiCMOS process technology at the Taiwan-based TSMC, the world's leading pure-play foundry, to fabricate a type of radio frequency ICs for end-use in GPS. Test results indicate that the device has satisfied the need for a GPS/GLONASS compatible receiver (Yang *et al.* 2005).

The fourth and final case is the No. 771 Research Institute. Founded in 1965, it is China's flagship state-owned unit specializing in research, development and manufacturing of airborne computers and semiconductors for end-uses in missiles and satellites. It has contributed to the Chinese military modernization by supplying space-qualified and military-grade ICs and IC-based computers for end-uses in China's missiles and spacecraft. In 1965, it made China's first bipolar microcomputer using small-scale ICs (No. 771 Research Institute 2005). In 1971, it fabricated China's first P-Channel Metal-Oxide-Semiconductor (PMOS) micro-computer using medium-scale ICs.[43] These computers became the building blocks of China's solid propellant strategic missiles (No. 771 Research Institute n.d. b). In 1977, it debuted China's first 16-bit microcomputer using large-scale ICs.[44] In 1980, it manufactured China's first CMOS, application-specific rad-hard micro-computer using large-scale ICs. Between 1991 and 1995, it was engaged in the development of hybrid ICs, A/D, D/A converters, and rad-hard ICs (Hua 1996). Furthermore, it has been qualified to make QPL products, including analog and digital ICs, by using its in-house 4-inch CMOS process technology (Yu 2002). As a QML military product supplier, it has run two certified manufacturing lines to fabricate thin-film and thick-film hybrid ICs. Its military-certified CMOS IC manufacturing line could produce an average of 5,000–10,000 wafers per month (No. 771 Research Institute 2005). When China launched its second manned spacecraft, Shenzhou VI, on 12 October 2005, the spacecraft was equipped with 22 IC-based space-qualified computers from the institute (Du 2005). Briefly, the institute is expected to nurture China's ongoing manned space flight program and pertinent military modernization projects whereby the size, weight and reliability requirement of the navigation and guidance systems for the rockets is achievable only with continuous miniaturization.

In the mid-1990s, however, the institute attempted to initiate a technological transfer from Taiwan in order to support China's *Hangtian* project but to no avail. The head of the institute contacted a senior engineer, who then worked for a Taiwanese IDM, during his trip to China in order to lure the Taiwanese firm to provide the institute with rad-hard chip design service. Although the Taiwanese in question declined the deal, the anecdote arguably highlighted the perceived technological weakness of the institute at that time.[45]

In sum, the analysis of the four cases has demonstrated that since its inception in the 1950s the traditionally enclosed Chinese defense chip industrial base has contributed to Chinese military modernization through the supply of semiconductors for end-uses in various indigenous military and aerospace systems. The system has thus contributed, at least partially, to China's defense security. According to an official-turned-industry player, the objective of the Chinese indigenously made military semiconductors is applicability, not to catch up with the state-of-the-art technology. In her words:

> Semiconductors for our military end-uses do not need to be extremely high-tech. At the Ministry of Electronics Industry, I realized that the US military was still using high-tech goods, which had lagged behind the state-of-the-art by more than 10 years. We are no different from the US when it comes to defense applications . . . Even when we trailed far behind the West, we made homegrown semiconductors for military end-uses, including *Dongfanghong* 1, China's first man-made satellite launched in 1970.[46]

However, the Chinese defense chip production system has faced daunting challenges such as its technological weakness compared to its counterparts in leading semiconductor countries, export controls, and the lack of sufficient investment in the national industrial infrastructure.

The gradual build-up of a regulatory and military standards-base regime

To govern its defense industrial base and to ensure the quality and reliability of homegrown defense components (including semiconductors) and systems, China has endeavored to establish a military standards-based regime since 1950. The following section reviews the three pertinent phases (Zhang 1986), namely, 1950–76, 1977–82 and 1982–present,[47] and identifies major policy initiatives. These include: (1) the "Seven Specializations" (*qizhuan*) system introduced in the late 1970s; (2) the establishment of China's national military standards for military products in the first half of 1980s; and (3) the decision to move towards a comprehensive national military standards-based regime in 1989. The section will further analyze the strength and weakness of these military modernization endeavors, and their impact on China's defense security.

1950–76: Dispersed management, the Russian influence and the Cultural Revolution

In the beginning, China replicated the Russian model in order to build its military standardization system from scratch. However, this effort was plagued by the lack of a centralized management system, inconsistency in rules and standards, and the difficulty in replicating a foreign system incompatible with the level of techno-logical development in China. Even worse, China's increasing technological capability soon made the crude military standardization system obsolete (Xie 1985; Zhang 1986).[48] Following the withdrawal of the Russian experts in 1960, China was compelled to build a military standardization system without any external input. A national-level bureau in charge of national standardization was established in 1964, but it became defunct because of a governmental reorganiza-tion in the following year. In 1966, the Cultural Revolution started, further disrupting China's plan to establish a fully-fledged military standardization system. Hence, the aforementioned efforts during the first phase had contributed little to China's defense security.

1977–82: The "Seven Specializations" scheme

During the second phase, which began in 1977 and ended in 1982, strong official support for the establishment of a centralized management system resulted in the birth of the "Seven Specializations" (*qizhuan*) approach in 1978 (Zhang 1986; Cao and Gao 1989; Wang 2005: 20). This scheme was designed to improve the quality and reliability of homegrown military electronic parts and components.

Major developments during this phase are summarized as follows. In April 1978, the Seventh Ministry of Machine Building, in charge of the development of China's ballistic missiles, urged the CMC to improve the quality and reliability of homegrown military electronic parts and components by identifying more than 400 pertinent issues in its report to the CMC. For instance, problems in the indigenously made electronic devices accounted for more than 40 per cent of failures in various rockets during their assembly and range tests from September 1971 to January 1978. In response to mounting pressure from the Chinese military-industrial complex, the CMC instructed the Fourth Ministry of Machine Building in charge of electronics and telecommunication industries to introduce related reforms.[49] Consequently, a joint meeting held by the ministry and the CMC in December 1978 finalized the introduction of the "Seven Specializations" (*qizhuan*) scheme. In 1979, the State Council and the CMC further identified the need to strengthen the Chinese military standardization system as one of PLA's modernization objectives.

The *qizhuan* system comprised four parts and was implemented from 1981 to 1990, overlapping with some of the initiatives introduced during the third phase. The first was to select and control the types and production sites of these military gadgets. The second was to identify measures to ensure the reliability of products; these included "special authorization, special technology, special personnel, special equipment, special materials, special screening and special checks" (*zhuanpi, zhuanji, zhuanren, zhuanji, zhuanliao, zhuanjian, zhuanka*). The third involved the information feedback system established between users and producers. The fourth was to analyze failure patterns common among malfunctioning electronic parts and components in order to systematically overcome pertinent technological hurdles (Cao and Gao 1989: 445–6).

Key players in the *qizhuan* system included the CMC, the Ministry of Electronics Industry, the Ministry of Space Industry, 81 factories, and research institutes as well as a few universities.[50] The agency in charge was the Bureau of Defense Industrial Base (*Jungong Jichuju*) of the Ministry of Electronics Industry with support from the CMC. Between 1981 and 1985, Beijing invested over 100 million renminbi (RMB) to revamp 26 pertinent production lines. In total, there were 158 certified manufacturing lines across China (ibid.; Bureau of Defense Industrial Base 1996: 16), churning out military products such as diodes, transistors, connectors, and capacitors (China Machinery and Electronics Industry Yearbook Editing Committee 1989: V-24).

Nevertheless, the impact of the scheme on China's defense security is mixed. The system is a security asset in three major ways (Liu *et al.* 1986: 79; Cao and Gao 1989: 446–7; Bureau of Defense Industrial Base 1996: 16; Wang 2005: 20). First, it helped China partly realize its pursuit of self-reliance in the arena of defense electronic components. Between 1981 and 1990, 158 certified *qizhuan* manufacturing lines produced about 30 million electronic parts and devices for end-uses in more than 30 major military systems. According to Cao and Gao (1989), two officials from the Ministry of Electronics Industry, these indigenous producers provided the designer of a domestic telecommunication satellite, which completed its successful test flight on 8 April 1984, with more than 97 per cent of its key electronic parts and devices (including ICs and discrete components). Moreover, they manufactured more than 90 per cent of components for China's space carrier rockets.

Second, producers under the scheme helped improve the reliability of the military systems in question by supplying devices of improved quality as the building blocks of these defense gadgets. The improvement in the guidance system of the *Dongfeng-5* missile, China's first ICBM, was a case in point.[51] Screening tests eliminated 73.1 per cent of components manufactured by the non-*qizhuan* producers for end-uses in the missile's guidance system, whereas similar tests removed merely 5.5 per cent of those produced by *qizhuan* manufacturers (ibid.). Another example is related to ship inertial navigation systems. Whereas the mean time before failure increased to 1,532 hours for systems equipped with *qizhuan* components, it remained a dozen hours for those with non-*qizhuan* devices. Third, the improvement in the quality and reliability of devices manufactured under the system helped shorten the production time needed for pertinent military systems.

However, the scheme was flawed in three ways, thereby reducing China's security. First, the program was economically unsustainable. In the words of a seasoned Chinese engineer who had had first-hand experience of the project, "There are many stringent regulations governing the manufacture of defense products, such as *qizhuan*. If you are qualified to make a cup by using a certified *qizhuan* line, the line would be discarded as soon as you finish producing the cup."[52] Second, the fast development rate of electronics technology sometimes made *qizhuan* manufacturing lines obsolete, resulting in their inability to produce quality parts which would meet the continuously rising demand for high-quality electronics from the designers of PLA weapons systems. Third, the production yield of pertinent manufacturing lines at times was very low, and the quality and reliability of finished products consequently had to rely on post-production screening procedures (Chen 1992: 40). Hence, the aforementioned weaknesses made the system a security liability for China, resulting in Beijing's subsequent decision to forgo the scheme.

1982–present: Towards a comprehensive national military standard system

The third phase began in 1982 when Beijing started to imitate the US MIL system (Ye 1992: 8), which was discussed in Chapter 2, promulgated new laws and regulations, and reorganized institutions, in order to establish a comprehensive national military standardization system because of national security considerations (Zhang 1991; Bureau of Defense Industrial Base 1996). As Xie Guang (1985), the deputy director of COSTIND, pointed out in the 1980s, the lack of a comprehensive military standardization system weakened the PLA's modernization and damaged China's defense security: "We are a country with millions of soldiers and yet we have not fully established our military standardization system."

Major developments during this phase are summarized as follows. In 1982, COSTIND was established as the major agency responsible for military standardization (Zhang 1986: 4). In 1983, COSTIND set up a study unit specializing in military standardization. In 1989, the unit was upgraded to become the center of military standardization, which was attached to the No. 301 Research Institute of

China Aerospace Industry (Yang 1989). In 1984, the State Council and the CMC issued No. 8 Document detailing measures to manage the military standardization system (Kong 2003: 249). In the same year, COSTIND began to set up professional technology committees for military standardization, and these committees comprised representatives from defense R&D units, defense suppliers and customers, and professional communities. Committee members would review pertinent technological issues in their capacity as government consultants. In 1985, two committees were set up: one for military electronic parts, and the other for military electronic components including discrete devices and ICs (ibid.: 223). In the same year, Beijing announced its first national military standard for military electronic parts and components, namely, the general regulations (i.e. GJB 33-85) governing the quality requirement of semiconductor discrete devices for military end-uses (Wang 2005: 17). In 1988, the Ministry of Electronics Industry began to implement the related regulations through the launch of pilot projects at selected sites (Bureau of Defense Industrial Base 1996: 18; Wang 2005: 18).

Another milestone came in 1989 when a national-level meeting about the "Seven Specializations" scheme concluded that China should discard the model and move instead towards the establishment of a national military standardization system in order to manage the production of military electronic parts and components. The meeting proposed the following policy changes: to certify manufacturing lines; to ensure that electronic parts and components for military end-use would be produced by certified and qualified manufacturing lines; to appraise various components in order to establish QPLs; to conduct periodic appraisal of QPL goods; and to ensure the stable supply of quality QPL products (ibid.).

Hence, COSTIND initiated the "Mars Project" (*jinxing jihua*) between 1991 and 1995. The project aimed to impose national military standards to regulate the production of military electronic parts and components in order to ensure the quality and reliability of these devices (Du and Xie 1997: 9).[53] Concurrently, the Bureau of Defense Industrial Base of the Ministry of Electronics Industry, with help from COSTIND, continued to launch pilot projects at selected production hubs in order to implement the related regulations. According to an official account, by 1996, the system had covered an increasing number of defense electronic parts and components (Bureau of Defense Industrial Base 1996: 18).

With the reorganization of GAD and COSTIND in early 1998 (Jencks 1999; Mulvenon and Yang 2002: 276–8, 304–8; Shambaugh 2004: 143–6), a number of weapons R&D, and scientific research organizations and tasks came under the aegis of GAD.[54] Since its inception, GAD has taken over the role of the old COSTIND in supervising the establishment of a national military standardization system at the central government level (Liu 2004: 2).[55] Moreover, its Military Product Quality Assurance System Certification Committee (*jungong chanpin zhiliang baozheng tixi renzheng weiyuanhui*) issues certificates to qualified military product producers, with the Zhongguo Xinshidai Certification Center authorized to carry out certification work on the ground.

In October 2002, measures to manage professional technology committees for military standardization were promulgated by GAD, the General Staff Department (GSD) and the General Logistics Department (GLD) under the CMC. To implement these measures, GAD restructured the 24 military standard committees originally under the aegis of COSTIND. In 2004, GAD inaugurated the Military Electronic Parts and Components Standardization Technology Committee; it integrated the two committees that COSTIND had set up in 1985, for military electronic parts and components respectively (ibid.). By 2005, China had largely completed its national military standards for military semiconductors (Table 4.6) (Wang 2005).

As part of its efforts to modernize the PLA, Beijing has continued to upgrade its national military standardization system in recent years. According to Liu Chenghai (2004: 2), the deputy minister of GAD's Electronics and Information Base Department, the PLA has regarded the standardization of military electronic parts and components as crucial to its modernization projects. In his words:

> Electronic parts and components are the core parts of electronics-based information-dependent systems. Whether or not we can develop and manufacture high-class, high-quality, and highly reliable electronic devices will directly affect the development and production of our electronics-based information-dependent military systems.

As Cheung (2007a) pointed out in front of the United States-China Security Review Commission, the establishment of a common and comprehensive technical standards and military specifications regime has been one of the major tasks for the Chinese defense industry following the 1998 reorganization. After all, China's desire to pursue technological leapfrogging would depend on the upgrading of its weapons systems, which would have to rely on reliable and high-quality standardized parts and components as their building blocks.

However, the impact of China's efforts to establish a standards-based regime since the 1980s to govern its defense production system on its national security has been mixed. On the one hand, the system introduced so far has enhanced

Table 4.6 Chinese military standards and quality classification of semiconductors for military end-uses

Types of parts and components	Military standards	Quality classification (from high to low)
Semiconductor ICs	GJB 597A-96	S, B, B1
Hybrid ICs	GJB2438A-2002	K, H, G, D
Semiconductor discrete components	GJB 33A-97	JY, JCT, JT, JP

Source: Wang 2005.

China's security in two major ways. First, it has outperformed its predecessor, namely the "Seven Specializations" approach, in helping enhance the technological level of the Chinese defense industrial base and to ensure the quality of military goods in question (Bureau of Defense Industrial Base 1996: 18). Second, following the gradual formulation of the system, Chinese defense system producers have preferred to use homegrown qualified electronic parts and components on the Chinese QPLs rather than using their foreign counterparts. For instance, the China Aviation Industry Corporation (AVIC) decided in the 1990s to refrain from using imported military electronic parts and components if comparable alternatives could be found on the indigenous QPLs (ibid.: 17). This policy, if implemented, would reduce security threats to China because of its dependence on foreign-made military devices.

On the other hand, the implementation of the new policy has sabotaged China's defense security because of three major challenges. First, the Chinese authorities have been slow in completing related work since the 1980s. Zhang Bin, a senior member of the State 877th Plant, which is a longstanding player in the Chinese defense chip industrial base, has criticized Beijing for being tedious in setting up regulations that specify the quality requirements of military discrete devices. A telling example concerns GJB 33-85, the first Chinese national standard for defense electronic parts and components (Wang 2005: 17). Although Beijing announced the general regulations governing the quality requirement of pertinent semiconductor components in 1985, it only managed to finalize 39 related rules from 1987 to 1990. In contrast, Washington concurrently announced and implemented some 600 regulations governing the use of semiconductors in US military systems (Zhang 1991). By the end of 2001, China had announced more than 500 military standards for electronic parts and components (Yu 2002: 200). In 2004, a GAD official (Liu 2004: 3) estimated that more than 600 were still pending. Worse, the number is expected to rise as the PLA continues to upgrade its information-based military systems.[56]

Equally slow has been the speed of certifying pertinent manufacturing lines, to complete the QPL and the QML, and to issue certificates to qualified military producers. By 1996, the Ministry of Electronics Industry had certified 21 manufacturing lines within the indigenous defense industrial base and consequently included 30 products in the QPL. These operations accounted for merely 20 per cent of a total number of 106 lines which the ministry had planned to certify (Bureau of Defense Industrial Base 1996: 16). Worse yet, the June 2002 version of the QPL included only 1,195 types of military electronic devices produced by 182 certified lines from April 1992 to 31 May 2002, and the third version of the QML included merely six certified manufacturers running eight hybrid IC production lines (Yu 2002). Besides, in spite of being longstanding players in the Chinese defense chip industrial base, Feiyu Microelectronics, and No. 771 Research Institute, in their capacity as QML producers of hybrid ICs, had to wait until 1993 and 2000 respectively to receive pertinent certification from GAD.

The second problem involves the difficulty in sustaining the operation of qualified manufacturing lines for the production of QPL goods. According to two members of the 4th Research Institute of the Ministry of Electronics Industry (Du and Xie 1997), some qualified QPL suppliers could not survive because of their small-volume production models and limited demand from the PLA: "Although QPL was established following a conventional appraisal method, it was unsustainable, thereby disrupting the prospect of QPL goods as the priority choices for indigenous decision-makers for military procurement matters. Consequently, the appraisal and qualification system became meaningless."

The third challenge stems from the chronic lack until the 1990s of a consistent yardstick to measure the reliability of indigenously made military electronic parts and components. While the USA began to introduce a reliability indicator for some of its military electronic parts and components as early as 1955, China lagged behind the USA for at least 35 years. Between 1986 and 1990, the Chinese defense industrial base introduced over a thousand new electronic parts and components for the PLA end-uses every year without providing any reliability indicator for these devices. It was not until 1991 that Beijing started to introduce pertinent reliability indicators (Ye 1992: 10). Other problems that have sabotaged China's efforts to establish a comprehensive national military standardization system include latent defects in qualified parts and components, inconsistency in regulations and technical standards, the incomplete quality classification of qualified parts and components, as well as competing yardsticks for product appraisal and acceptance inspections (Wang 2005: 20). Hence, the aforementioned three major pitfalls have made the implementation of the system a liability to China's defense security.

Civil–military integration and the opening-up of defense industrial base to civilian firms

The following section further examines another milestone in the restructuring of the Chinese defense chip industrial base, namely the policy to open the traditionally enclosed production system to non-state-owned enterprises, and its impact on China's defense security. As part of China's strategy to integrate its defense industry into a broader civilian economy, this initiative comprises the following elements: the strategy on civil–military integration introduced during the 10th Five-Year Plan; the backing of the strategy by the 16th Party Congress, the 2004 defense White Paper, and the 17th Party Congress; as well as a policy to permit certified civilian firms to take part in weapons production (including that of military semiconductors) announced in 2005. Detailed analysis of these changes and their impact on China's defense security is to follow.

Policy initiatives

China has recently pursued a two-pronged approach to modernize its defense industry. One pillar of the approach aims at internally re-engineering the defense industry by breaking down bureaucratic barriers and nurturing an entrepreneurial institutional culture in the sector; the other encourages the integration of the defense industry into a broader civilian economy with the aim of establishing a dual-use technological and industrial base that serves both civilian and military needs. The implementation of the latter has benefited from the overhaul of the defense industry that began in the late 1990s. The structural reform of the sector aimed at establishing a small inner circle of dedicated defense prime contractors, which is complemented by a large supporting base of secondary sub-contractors (Cheung 2007b).[57] Since then, the reform has gradually created an opportunity for the involvement of civilian firms with no prior participation in the defense sector.

In 2001, China released its 10th Five-Year Plan stressing the aforementioned second pillar of the defense reform.[58] It did so by introducing a new set of 16-character guiding principles for China's economic and military modernization. The principles were as follows: "*Junmin Jiehe* (Combining Civil and Military Needs), *Yujun Yumin* (Locating Military Potential in Civilian Capabilities), *Dali Xietong* (Vigorously Promoting Coordination and Cooperation), *Zizhu Chuangxin* (Conducting Independent Innovation)" (Xinhuanet 2001). These principles have replaced the original ones that Deng Xiaoping set out in 1982 for the reorganization of the defense industry. Deng's famous dictum was as follows: *Junmin Jiehe* (Combine the Military and Civilian), *Pingzhan Jiehe* (Combine Peace and War), *Junpin Jiehe* (Give Priority to Military Products), *Yimin Yangjun* (Let the Civilian Support the Military). In the 1980s and the 1990s, Beijing reformed the defense industrial base according to Deng's emphasis on spin-off and defense conversion (*jun zhuan min*). Consequently, various defense sectors had been retooled for civilian production thereby nurturing the civilian economy. Hence, there was scant attention to spin-on initiatives with the major exception being the establishment of the 863 Program in 1986, which Cheung (2009) regards as "the first significant effort by the authorities to genuinely pursue coordinated civil-military R&D." However, the notion of *Yujun Yumin* embodied in the new dictum has encouraged spin-on[59] thereby deepening the integration of the defense industry into a broader civilian economy with a rich array of technological and industrial capabilities that the PLA can absorb in order to modernize its forces (Cheung 2009).

Subsequently, the 16th Party Congress in 2003, the 2004 China Defense White Paper and the 17th Party Congress in 2007 endorsed the notion of *Yujun Yumin*. The Party Congress in 2003 passed a policy document in support of the construction of a new dual-use technological and industrial base. The document, entitled the "Decision of the Chinese Communist Party Committee on Several Issues in Perfecting the Socialist Market Economy," called for the building of an innovative "*Junmin Jiehe, Yujun Yumin*"-based system and the "mutual promotion and coordinated development of the defense and civilian technologies" (Xinhuanet 2003; Xue and Chen 2004: 24). This elevated the *Yujun Yumin* guiding principle to the strategic outline for the future construction of a dual-use economy (Cheung 2007a; 2007b). The 2004 Chinese defense White Paper (The State Council Information Office 2004) identified the same 16-character principles as part of the strategic guide on defense industry reform. It also included provisions on PLA armaments procurement, which had been introduced in October 2002, with respect to spin-on. These regulations "supported state-owned enterprises outside the military industry and private high-tech enterprises to enter the market of military products."[60] In 2007, the Chinese President Hu Jintao proclaimed to the 17th Party Congress that China will "combine military efforts with civilian support, build the armed forces through diligence and thrift, and blaze a path of development with Chinese characteristics featuring military and civilian integration" (Xinhuanet 2007).

To further implement the *Yujun Yumin* scheme, COSTIND organized exhibitions and conferences in order to lure non-state-owned firms to take part in defense production activities. One of the first events took place in Beijing in the Spring of 2004, attracting the participation of nearly 150 firms, many of whom were civilian entities (Zhongguo Junzhuanmin (Defense Industry Conversion in China) 2004). Some pushed the government to offer incentives, such as tax preferential treatments, to civilian firms in order to attract them to engage in defense production activities (Xue and Chen 2004). Others urged the authorities to reform the current military standardization system by adopting mature commercial standards in order to remove obstacles that prohibited the participation of civilian enterprises in military production activities (Ye 2003: 36; Yu 2004: 10). The government was also urged to finalize the related regulations governing the infusion of civilian resources into defense production operations (Liu 2004: 36; Wu 2004; Qian 2005: 38). After all, the lack of these regulations had resulted in, for instance, the failed attempt by Jiangsu Changdian Advanced Packaging Technology Co. Ltd., China's leading indigenous semiconductor packaging and testing firm, to engage in military production activities. In 2004, the firm had been contacting members of the defense establishment over the previous two years urging them to allow the company to take part in defense-related manufacturing activities, but to no avail (Wu 2004).

In 2005, the State COSTIND decided to permit certified civilian firms to engage in weapons production activities, thereby opening up the Chinese defense production system to non-state-owned enterprises. On 27 May 2005, an official from the State COSTIND announced (*Keji Ribao* (*Science and Technology Daily*) 2005) that Beijing would soon issue new licenses for weapon development and production, some of which would be given to civilian firms, including semiconductor companies. Furthermore, he identified the arena of ICs as a key segment of the defense industry to which technologically capable civilian firms would be expected to contribute. He tipped a Beijing-based IC design house, led by five Chinese PhD returnees from the USA, as the government's targeted partner. Because the company had designed cutting-edge ICs to be readily deployed in the Chinese military systems, including fighter aircrafts, warships, and missiles, it was highly likely that the firm would become one of the first private enterprises to receive the pertinent licenses from the government.

On 30 July 2007, the State COSTIND (Commission of Science Technology and Industry for National Defense 2007) specified the procedures that private firms should follow in order to become certified weapons producers. These include the need to apply for a quality assurance certificate, a security clearance, and a permit to research on and to produce specific weaponry listed in a special COSTIND catalogue.[61]

What, then, accounts for the formulation of the *Yujun Yumin* policy intent on institutionalizing the trend of spin-on? There are three explanations for the policy change; they include: (1) the chronic deficiency of the Chinese defense industry; (2) the attractive resources embedded in a fast-expanding Chinese civilian economy; and (3) Beijing's recognition of the dual-use nature of many technologies and the dominant trend of spin-on.

First, the long-standing deficiency of the Chinese defense industry had forced Chinese policy-makers to revamp the domestic defense industrial base. The chronic weakness of the sector is three-fold. These weaknesses include the technological backwardness of many of the indigenously made weapons systems, the historically long R&D and production timelines for most locally built military gadgets, and China's increasing reliance on foreign countries for the supply of major weapons systems. Thus, the industry has persistently failed to meet the needs of the military (Medeiros *et al.* 2005: 8–11).

Whereas the defense industry has had a troubled existence, the more market-oriented sectors of the Chinese economy have grown at an impressive speed, accumulating technological assets readily exploitable by the Chinese military in its march towards modernization. In particular, much of the IT-related technology and knowledge lies in the civilian economy and is spearheaded by high-tech civilian firms, and the technological capability of these companies is often better than that of their counterparts in the conventional defense industrial base. Hence, Chinese policy-makers find it logical to push for an increase in the effective use of the civilian industrial base in order to nurture Chinese military needs.

Finally, Beijing's desire to increase the use of COTS items in domestic defense systems further accounts for the policy change. Such a desire results from the Chinese government's recognition of the increasing trend of spin-on in developed countries and the dual-use nature of most technologies including semiconductors (Sha *et al.* 2001; Ye 2003; Chen 2004; Liu 2004; Yu 2004: 9; Qian 2005: 38; Zhang *et al.* 2005), as discussed earlier in the chapter.[62]

The policy change, driven by the aforementioned three factors, is expected to open the channel through which resources can be transferred from the civilian sector in China's fast-growing economy to the domestic defense industry. This, in turn, is expected to maximize the national technological base upon which China can further modernize its military (Drewry and Edgar 2005).

Defense security repercussions

To what extent is the new policy package conducive to China's defense security? A preliminary assessment, particularly in the area of semiconductors, leads us to conclude that the initial impact of the policy on the PRC defense security is mixed.

So far, a few civilian firms in China's commercial semiconductor sector have already become part of the Chinese defense chip industrial base. Shanghai Fudan Microelectronics, which has direct connections with Fudan University, is a case in point. With funding from the Chinese military, the company began a project on the development of a 32-bit CPU, which was similar to the Intel 80386 microprocessor. A junior engineer at the firm even consulted an experienced Taiwanese IC designer about the project.[63] By late 2002, the company had successfully debuted Shenwei I, a 32-bit embedded CPU, which represented the highest level of China's homegrown microprocessor technology at the time (Li 2002). The CPU subsequently won a first-class national defense technology award from the government (Shanghai Fudan Microelectronics n.d.). In October 2006, Fudan University's publication revealed that the university had recently passed a second-class security clearance check by local COSTIND officials and was in the process of getting two other compulsory certificates in preparation for its participation in defense-related scientific research and production activities (Department of Science and Technology 2006).

From the case of Fudan Microelectronics, it is logical to infer that so long as supply meets demand, other members of the Chinese commercial chip industrial base will engage in defense-related design and production activities in order to satisfy the needs of the Chinese military. Domestic IC design houses with ambivalent links to the PLA are particularly likely to do so. A prime suspect is Hisilicon Technologies, which is a spinout from Huawei Technologies Ltd., China's leading high-tech giant, founded by a former PLA officer. Some of them may transfer technologies to the defense sector in the same fashion that Intel sold its microprocessor technology to Sandia in 1999 to help build less expensive and better functioning radiation-hardened chips for nuclear weapons.

However, implementation of the policy is not without its challenges. First, the decrease in radiation hardness in COTS items readily available on the commercial chip market as a result of semiconductor technology advances may challenge the designers of defense systems and sub-systems in their decision to insert COTS directly into their military systems, as discussed in Chapter 2. Second, it remains unclear whether the designers of China's defense systems and sub-systems, after having for a long time been confined to a technologically weak and enclosed production system, would have enough in-depth understanding of technological issues pertaining to COTS insertion, which would be crucial to their decisions to use COTS in Chinese military systems. Third, the problem of confidentiality, and the compatibility of commercial and military standards may further affect pertinent policy implementation. Fourth, indigenous Chinese civilian firms largely lack innovative capability and depend overtly on foreign technologies. Once these companies are engaged in defense production activities to serve the PLA, the latter may thus deepen its dependence on foreign technologies (Zhong 2005; Song and Niu 2006; Zhang and Li 2006).

The Chinese commercial chip industrial base: gradual take-off amid sectoral globalization

So far, we have analyzed the Chinese defense chip industrial base and its link to China's national security. The following section examines the Chinese commercial chip industrial base; it explores the three developmental phases of the industry since the 1960s, reviews the current state of the sector amid increasing trends of sectoral globalization, and analyzes the national security implications of the evolving industrial base.

Historical overview: policy initiatives and major players

Below is an analysis of major initiatives and key players during the three phases in the development of the Chinese commercial chip industrial base: the 1960s to 1978, 1978 to 2000, and 2000 to the present day. In the late 1960s, Beijing initiated the industrialization of semiconductor technology, thereby paving the way for the establishment of an indigenous commercial chip industrial base. Until 2000, related official efforts had been unsuccessful. As an experienced industry player recalled in 2005, "The semiconductor industry in mainland China is bitter history... During their retirement farewell parties, pertinent forerunners often shed tears lamenting their failure to see the take-off of the sector."[64] Since 2000, however, the sector has started to take off.

1960s–78: Self-reliance

Self-reliance was the primary feature of the first phase, which began in the 1960s and ended in 1978. During this period, Beijing attempted to establish a domestic semiconductor industrial base, chiefly for military purposes, in an enclosed economy with few international connections. In the early 1970s, China imported seven second-hand semiconductor manufacturing lines from Japan in spite of the Coordinating Committee for Multilateral Export Controls (COCOM) regime during the Cold War, which was created to restrict sensitive exports (including semiconductor goods and tools) to the then Eastern bloc (Hu 2001: 382). Because the industry focused largely on producing semiconductors for military end-uses, few manufacturing lines churned out semiconductors for commercial end-uses, with the exceptions being radios. The major chip players during this phase were all state-owned factories, research institutes, and university laboratories. Industrial policies intent on fostering the development of the commercial chip industrial base were absent.

The industry during this period was severely handicapped because of various internal and external constraints. These included the lack of capital, lack of technological know-how and human resources, problems in following the path of a planned economy to develop a high-tech sector, the Cultural Revolution, and export-control regulations under COCOM. For instance, China failed to produce semiconductors by operating second-hand manufacturing lines from Japan in the 1970s because the Chinese related engineers lacked sufficient know-how (ibid.).

In addition, the Cultural Revolution disrupted the development of the industry in three major ways in spite of Zhou En Lai's 1972 order, which helped retain part of the continuous development of Chinese electronics sector (Editing Committee on China Electronics Industry in Fifty Years 1999: 10). First, the turmoil deepened the technological gap between the PRC and leading countries in the sector. Second, it took a substantial part of engineering manpower away from the sector. By the time some of the skilled engineers returned to work, they had already passed their prime. An anecdote about a 1979 encounter in Silicon Valley between a Taiwanese semiconductor engineer and his PRC counterparts, who had resumed their careers at the end of the Cultural Revolution, illustrated the point. As the Taiwanese engineer recalled:

> In a hotel where I stayed, I spotted a group of Chinese running around in their sweatshirts one evening . . . Whereas we were in our 30s, they were in their 50s. The team comprised three engineers from a semiconductor unit in Shanghai and a Chinese Communist Party (CCP) secretary. They came to National Semiconductor to learn some process technology.[65]

Third and finally, the foolhardy political order from the top sabotaged the pace of the industrial development. Chen Boda, then a member of the Politburo, contended that China should manufacture semiconductors in unit-established jails, commonly known as cowsheds (*niupeng*), where dissidents were imprisoned. He intended to mobilize the public to develop semiconductor technology through mass movement, with the sheer denial of stringent clean room requirements for modern-day fabs (Chinese Academy of Sciences 2000: 36). Consequently, industrial development was disrupted. Recalling the statement vividly, an experienced semiconductor engineer summarized the plight the Cultural Revolution had inflicted upon him and his colleagues: "We scientists who knew the nature of semiconductors found life really hard."[66]

1978–2000: Opening up and catching up

The second phase, from 1978 to 2000, was characterized by China's access to a wider range of international resources and the initiation of pertinent industrial policies (such as the Sino-Foreign cooperation model) to foster the take-off of the sector, albeit with mixed repercussions.

Following Deng's open-door policy in 1978, the domestic sector gained wider access to international resources. By the end of the 1970s, China had imported 24 fabrication lines with second-hand equipment (Hu 2001: 382). In 1979, a delegation from the Institute of Electrical and Electronic Engineers (IEEE) in the USA visited China in order to understand the development of the domestic sector. According to Bill Spencer, chairman emeritus of International SEMATECH, who was part of the delegation, a Chinese semiconductor facility visited by the American delegation was equipped with foreign manufacturing tools. To his astonishment, an ion implanter at the site was similar to the ones used at Sandia, the leading American microelectronics lab for nuclear weapons, where Spencer worked at the time. As he recalled:

> It looked exactly like the ion implanter we had at Sandia . . . it had been copied by a Japanese company and sold to China. This supports the discussion about

Japan and Europe selling equipment to China even though I am sure these things were embargoed in 1979. They were buying exactly what we would use from Japan.[67]

In spite of the partial availability of imported manufacturing tools, China failed to upgrade its commercial chip industrial base in the initial years of the second phase because of the lack of technological know-how, the absence of appropriate infrastructure, as well as other factors (ibid.: 382). One episode during the IEEE's visit to China in 1979 was illustrative of the challenges the industry faced at the time. One day when the delegation was visiting a domestic semiconductor facility, an unexpected power cut hit the production hub, and the Chinese engineers at the site immediately opened the window of the factory. "Semiconductors are made in clean rooms, so the air conditioning is always on, the air flow is always on, and the temperature is controlled very carefully," Spencer recalled. "So they didn't have a very good concept of the climate required."

The 6th Five-Year Plan (1981–85), announced in the middle of the second phase, contained two pertinent policy initiatives. The first emphasized the need for the Chinese electronics industry to move towards "*Junmin Jiehe* (Combining Civil and Military Needs)." A follow-up of Deng's emphasis on defense conversion,[68] this initiative encouraged defense industry players to produce non-military goods for the civilian market. Consequently, members of the Chinese defense chip industrial base began to shift part of their production to commercial end-uses. A certain degree of spin-off ensued, with the defense chip industrial base shifting its technological expertise to its then primitive commercial counterpart. This move towards defense conversion on the supply side was targeted at the rising domestic demand for semiconductors because of the Chinese economic reforms. By the end of the 1980s, 97 per cent of the electronics industry revenue was for civilian end-use, whereas only 3 per cent was for defense. Arguably, the electronics sector thus demonstrated an impressive ability to transform themselves into producing mainly non-military goods compared to the rest of the Chinese defense industry (Lou 2003: 142–4; Medeiros *et al.* 2005: 6). The same observation applied to the semiconductor segment of the electronics industry.

The second pertinent initiative in the 6th Five-Year Plan promoted the Sino-Foreign cooperation model to develop the domestic commercial chip industrial base in the wake of the failure of the self-reliance model in the preceding decades. Project 907 marked the implementation of such a model involving two schemes: Toshiba's transfer of 3-inch wafer process technology to the State 742nd Factory, located in Wuxi,[69] and the establishment of Hua Yue Microelectronics Co., Ltd. in Shaoxing (China Semiconductor Industry Association 2002: 153; Howell *et al.* 2003). In particular, the first scheme deserves our attention. The state-run Chinese factory not only imported semiconductor equipment from Toshiba but also sent hundreds of engineers to Toshiba's headquarters in Japan for a nine-month training program from 1980 to 1981. According to a member of the Chinese team in Japan, the project was important to China in two ways. First, it implemented China's open-up policy in the area of semiconductors, thereby departing from the closed-door and self-reliance approaches in the preceding decades. Second, it enabled the Chinese factory to create a certain economy of scale in its production of semiconductors for end-use in TVs, thereby outperforming its competitors in the local commercial chip sector.[70]

However, duplicative and poorly coordinated promotional policies introduced during the 6th and 7th Five-Year Plan periods still failed to substantially upgrade China's indigenous semiconductor capability.[71] By the end of the 1980s, the Chinese commercial chip industrial base still lagged far behind that of the global state-of-the-art operations (Howell *et al.* 2003: 23).

During the 8th and 9th Five-Year Plan periods from 1991 to 2000, Beijing initiated the 908 Project and the 909 Project to further implement the Sino-Foreign cooperation model in an attempt to establish large firms in the sector (Fang and Cao 2002). Under the 908 Project, a newly formed state-owned China Huajing Electronics Group attempted to establish a 6-inch wafer fab with equipment purchased from Lucent, and 18 IC design houses were set up in 1994 as Huajing's potential customers.[72] However, the project turned out to be unsuccessful. The IC design houses, including Beijing IC Design Center (BIDC), failed to provide Huajing with any orders.[73] According to Yu Xie Kang, former vice-president of the group, "they could only do reverse engineering."[74] Nor was Huajing a success. The firm began as a 100 per cent state investment scheme in 1990, but it was not until 1998 that it began its operation. The delay resulted from the interference by the government, bureaucratic delays, and the chronic inefficiency common to China's SOEs. The 908 Project was to build China's first 6-inch wafer manufacturing line operating on 0.8–1.2 micron process technology, producing 12,000 pieces of wafer per month. By the time the line was in operation, it was already lagging behind the state-of-the-art operators,[75] thereby putting the firm in a disadvantageous position vis-à-vis its competitors. Consequently, the volume of production was low, trailing far behind its original target (ibid.: 24).[76]

So what accounted for the failure of Huajing? According to Yu, the project failed because of the central planning approach, which involved countless bureaucratic delays, unwise business decisions, and uneconomic measures in running the firm.[77] Yu's view was partly echoed by David N.K. Wang, the then executive vice-president of Applied Materials in 2005. Based on his visit to Huajing in the late 1990s, Wang argued that although Huajing's development was not constrained by export controls because the firm had managed to import semiconductor equipment from Toshiba and Siemens, it was the nature of Huajing as a state-owned enterprise that rendered the operation a failure. As he recalled:

> Huajing employed 6,000 people, and it ran its own cinema, restaurant, school, and hospital. Its semiconductor output would have required the work of 500 employees in a facility established according to the international standard at the time, but it employed 6,000 people there. So it was uncompetitive.[78]

To save its moribund production line, in 1999 Huajing leased the line to Hong Kong-registered CSMC for two years. At the time, CSMC was newly founded by

Peter Cheng-yu Chen, an experienced Taiwanese industry player. Describing his move to take over the line as "bold," Chen was proud to have successfully transformed the state-run operation into China's first pure-play foundry, a leading firm in the 6-inch league. Before the line was leased to the Taiwanese team, it produced 6,000 pieces of wafer per month; by the end of 2004, its volume of production had increased ten-fold.[79]

In spite of the failure of the 908 Project, Beijing continued to beef up its efforts to develop its indigenous commercial chip industrial base by introducing the 909 Project. With a planned budget of US$1.2 billion, the scheme was China's largest strategic project in the field of semiconductors to that date, aiming to boost domestic semiconductor production in view of increasing local demand for semiconductors (Ministry of Science and Technology 2000). The project was composed of three elements: (1) the establishment of Hua Hong Microelectronics for process and product development and production; (2) the establishment of the General Research Institute for Non-Ferrous Metals for wafer technology implementation; and (3) the support of selected IC design houses and institutes for design and development activities (Dunn 1997).

On 29 March 1996, the State Council approved a plan to construct a semiconductor manufacturing facility in Shanghai. The following month, Shanghai Hua Hong Microelectronics Co. Ltd. was chartered to undertake the 909 Project as an IDM. In July 1997, the firm formally launched a joint venture with NEC, namely Shanghai Huahong NEC Electronics. An investment project of US$1 billion, Huahong NEC had 28.6 per cent of its capital contributed by NEC and its affiliate NEC (China) Co. Ltd., and the remaining 71.4 per cent by Shanghai Hua Hong Microelectronics (NEC Corporation 1997; Fang and Cao 2002). The joint venture was to design, manufacture, and market memory and logic chips using 0.5–0.35 micron process technologies transferred by NEC, in what would be China's first 8-inch wafer fab. It also aimed to have a maximum capacity of 20,000 8-inch wafers per month.

In February 1999, the construction of its chip fabrication lines was completed, and produced chips of 0.35–0.24 micron, an upgrade from the originally planned 0.5 micron. "The factory was a copy of NEC Hiroshima fab . . . Over 300 Chinese engineers went to Japan to learn NEC process technologies and recipes," recalled Zou Shichang, the then vice-chair of the joint venture.[80] In 2000, the joint venture began to sell its memory chips, ranging from 64Mb to 128Mb, on the international market (Ministry of Science and Technology 2000). By the end of 2000, it was churning out 20,000 pieces of 8-inch wafer per month. However, by the end of 2001, the market performance of its 128Mb memory chips failed because of the IC market downturn, resulting in internal strife, with high-ranking executives debating the firm's future.[81] Consequently, the firm decided to switch to foundry operations.

Even worse, the 909 Project was faced with various daunting tasks. First, NEC had irritated its Chinese partner by dominating the operation of the joint venture due to its technological and managerial advantages.[82] As a former senior executive recalled, "The Japanese couldn't possibly give us the real technologies. They failed to wholly abide by the technology transfer contract. This is because they represented Japan and we represented China. And this was not limited to Huahong NEC."[83] Second, the eight design houses failed to perform as Huahong NEC's potential customers.[84] Initially established as part of the 908 Project, these firms were designated under the 909 Project to advance their design activities to provide the joint venture with orders. However, they failed to deliver because of their technological constraints. "They lacked experienced team members," admitted a former top executive of the joint venture.[85]

To sum up the second phase, Beijing's initiatives to upgrade the Chinese commercial chip sector from the late 1980s to the late 1990s gave rise to several national champions such as Huajing and Huahong NEC, but the performance of these companies was mixed. By the late 1990s, Chinese semiconductor firms still lagged behind global state-of-the-art participants, relied heavily on the import of semiconductor equipment, supplied less than 20 per cent of the domestic market, and their market share outside China was insignificant (Howell *et al.* 2003: 27). Worse yet, the trade deficit in the area of ICs continued to soar from 1996 to 2001 (Figure 4.2) (Lou 2003: 183).

Despite this, the overall industry concurrently made progress in terms of technology, output, sales revenue, and the formulation of the supply chain. Whereas IC manufacturing technology in China was at least 12 years behind the global state-of-the art in the late 1980s, it dropped to 6–8 years by the late 1990s. Moreover, while China produced 97 million ICs in 1993, it churned out 560 million of the 5.7 billion ICs it consumed in 1995 (Cliff 2001: 12–13). In terms of IC revenue, the amount reached 18.6 billion RMB in 2000, which satisfied 19.1 per cent of the domestic market and accounted for 1.2 per cent of

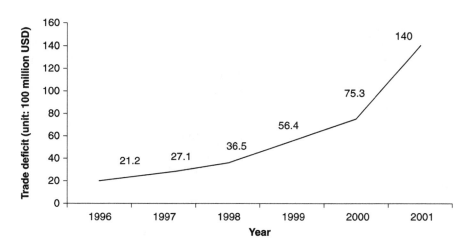

Figure 4.2 The PRC's increasing trade deficit in the area of ICs, 1996–2001.

Source: China's Electronics Industry Yearbook data from 1996 to 2001, cited in Lou 2003: 183.

the global total (Task Force on China's IC Industry 2002). From 1996 to 2001, China's semiconductor production capacity rapidly expanded, in terms of production volume and sales revenue (Lou 2003: 181). Most important of all, a preliminary semiconductor supply chain was in place in China by the late 1990s. Indigenous players were involved in IC design, fabrication, packaging and testing, the supply of semiconductor material and equipment, and the development of computer-aided IC design tools (Wang *et al.* 2002: 133–4). Despite this, the majority of these local players still trailed far behind the global giants.

2000–present: Accelerating catch-up efforts through liberalization policies

The third phase, beginning in 2000, has been characterized by marked changes in the official Chinese strategy to enhance domestic semiconductor capability. Beijing discarded heavy state intervention in market operations, favoring, instead, incentive policies and measures commonly found in capitalist countries. This policy shift was first embodied in the 10th Five-Year Plan for the period from 2001 to 2005 and in sectoral policies implemented pursuant to the Plan. Subsequently, it was included in the 11th and 12th Five-Year Plans for the period from 2006 to 2015. The new approach shapes the current policy environment for the sector and will be summarized as follows.

The 10th Five-Year Plan put forward a national strategy to push for informationalization in order to facilitate industrialization. Information-related sectors thus became the focal points in China's strategic industries, among which the IC industry enjoyed the top priority (Task Force on China's IC Industry 2002). The plan aimed to ensure that China would meet 60 per cent of domestic demand for ICs by 2005 and that Chinese companies would "gradually design and develop their own IC products (including CPU)." Aside from promoting IC design and fabrication sectors, it also stated that the government would "support the development and production of materials used in the manufacturing of ICs and electronic components" (Ministry of Information Industry 2000; Xinhuanet 2001).

The plan represented a substantial departure from its predecessors with regard to the promotion of high-tech industries and China's accession to WTO in 2001 partly accounted for such a policy change. To implement the plan, Beijing started to discard traditional policy tools, including reliance on state-owned enterprises to lead major developmental efforts and the prohibition on 100 per cent foreign ownership of enterprises in strategic sectors. Instead, the state provided broad industrial policy tools and incentives, such as subsidies, tax incentives, and the promotion of new businesses, while leaving specific investment, commercial, and R&D decisions to individual firms (Howell *et al.* 2003: 37–9). Before its WTO accession, Beijing had pursued a market-in-exchange-for-technology strategy to lure foreign investment to flow into China. Nevertheless, WTO entry constrained such a strategy, and Beijing had to turn to the industrial policy tools that were common in capitalist economies to promote the development of its high-tech sectors, including semiconductors.

In June 2000, the State Council published Circular 18, a seminal document defining Beijing's policy toward the semiconductor industry. The objective of the Circular was to make China a leading design and manufacturing base for ICs by 2010, and to ensure that it would "meet the majority of domestic market demand" for ICs and to export "a certain amount" within 10 years. To achieve this, the document specified a set of promotional measures to be made available to qualifying firms (both foreign and domestically owned) in the IC industry. These measures involved value-added tax (VAT), taxation holidays/reductions, foreign currency retention, and the authorities' shift from relying on government bank loans to obtaining capital investment directly from overseas or local financial markets. The document was a powerful tool Beijing used to lure the influx of FDI into advanced IC design and fabrication segments of the local industry. Since October 2004, however, China has either scrapped or revised the provisions in the Circular concerning the controversial issue of VAT refunds because of a pertinent memorandum of understanding (MOU) between China and the USA. Washington had argued that the provisions were WTO-inconsistent and subsequently brought the case in front of WTO's dispute settlement mechanism. In July 2004, the signing of the MOU settled the bilateral dispute (China Semiconductor Industry Association and China Center of Information Industry Development 2005: 20).

The 11th Five-Year Plan, which covered 2006 through 2010, continued to identify the development of a domestic IC industry as Beijing's priority (Fang 2005). The plan aimed to increase China's IC sales revenue to about 8 per cent of the global total by 2010. It also called for the development of five IC design companies, each worth US$375 million to US$624 million in revenue, and 10 additional ones each worth US$125 million to US$375 million in revenue (PricewaterhouseCoopers 2007: 40). In the 12th Five-Year Plan, which covered 2011 to 2015, however, Beijing's semiconductor industry policy shifted away from the pursuit of capacity and output value growth toward a focus on improving R&D capabilities and firms' global competitiveness. Moreover, a policy that emphasized market mechanisms replaced that of direct government investment.

Although enabling, push and pull factors, as detailed in Chapter 3, have often accounted for the deepening globalization of semiconductor production activities, Beijing's move towards liberalized policies since 2000 has certainly attracted an explosive amount of FDI targeting at semiconductors, dubbed the "silicon gold rush" (Geppert 2005), to penetrate almost all aspects of the Chinese chip supply chain. Consequently, Beijing has raised its semiconductor capability at an unprecedented pace, further integrating its industry into the global supply chain. SMIC, China's flagship foundry today, is a case in point. Founded in 2000 with tremendous assistance from Taiwan, the firm began to operate the first 12-inch fab in Shanghai in 2005, and became the world's third largest pure-play foundry in terms of sales by the first half of 2007 (PricewaterhouseCoopers 2010: 9).

Overview of the current industry

The following aspects of the current Chinese commercial chip sector will be examined: market size and production capacity, three major sub-sectors in the supply chain (i.e. IC design, fabrication, packaging and testing), and the role of foreign investment amid the trend of deepening sectoral globalization.

Market size and production capacity

The size of China's semiconductor market has grown at a notably greater rate than the worldwide market over the past few years. In 2004, China's semiconductor consumption grew by 41 per cent to reach US$41.8 billion, whereas the global market grew by 28 per cent. In 2009, its consumption declined only 2.5 per cent from US$104 billion in 2008, whereas the worldwide market decreased by 9 per cent. Hence, its explosive semiconductor market growth since 2000 has substantially increased its share of the global total. While it accounted for 7 per cent of the worldwide total in 2000, its share reached over 20 per cent in 2004, surpassing Europe and the Americas for the first time. In 2005, its chip market reached US$40.8 billion in terms of overall consumption, making it the world's largest IC market for the first time. In 2009, its share of the US$226.3 billion worldwide semiconductor market reached 41 per cent. Figure 4.3 summarizes global semiconductor consumption by region from 2003 to 2009, indicating China's increasing share of the overall market over the period covered.

China's growing role in the Asian semiconductor market is equally impressive. In June 2007, *Gartner Dataquest* predicted that China's dominance of the Asian market would increase over the next four years. Whereas China accounted for more than half of the regional market outside Japan by early 2007, it would represent 63 per cent of the US$203 billion regional total by 2011 (Hopfner 2007).

Three factors have contributed to China's insatiable demand for semiconductors in recent years. The first is the continuous transfer of worldwide electronic system production to China, thereby increasing its share of the global total from 17 per cent in 2004 to 33 per cent in 2009 (PricewaterhouseCoopers 2010: 10). In 2009, less than 66 per cent of semiconductors China consumed were used in the manufacture of electronic products for export. The second factor is the fast-growing semiconductor end-use markets in the world's most populous country, with its economy growing at an explosive rate in recent years. In 2005, for instance, computing, communications, and consumer end-use markets accounted for 95 per cent of the Chinese market, whereas automotive and military/industrial ones represented the remaining 5 per cent of the total (Figure 4.4). In 2009, more than one-third of semiconductors China consumed were used in the manufacture of electronic products for domestic consumption. According to PricewaterhouseCoopers, the portion of semiconductors used for these products for domestic consumption is expected to increase further because of increasing middle-class consumption, China's economic stimulus package and other government initiatives. For instance, two of the programs that constituted Beijing's economic stimulus package,

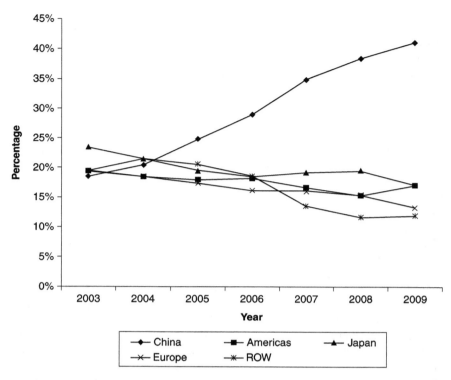

Figure 4.3 Worldwide semiconductor market by region, 2003–09.

Source: PricewaterhouseCoopers 2010: 9.

announced in 2008, could account for about US$50 billion in additional domestic chip consumption; they were the "Electronics Go to Farmers Subsidy Program" and the "Home Appliance Replacement Subsidy Program" (ibid.: 53, 76). The third driving force is the above-average semiconductor content of the electronic systems which have been produced in China, which averaged 25 per cent compared to the global average of about 19 per cent (ibid.: 11).

In terms of production capacity, the local industry has grown steadily since the mid-1990s (Figure 4.5 and Figure 4.6) but it has not expanded fast enough to narrow the gap between the domestic demand and supply of semiconductors (Figure 4.7). In recent years, China's semiconductor sales revenue has met less than 25 per cent of the domestic demand (China Semiconductor Industry Association and China Center of Information Industry Development 2005: 3). Worse, its IC consumption–production gap has continued to widen at an accelerating rate: it increased from US$5.9 billion in 1999 to US$68 billion in 2008. According to CSIA, China's IC market is expected to grow to US$119 billion by

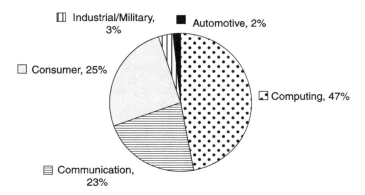

Figure 4.4 China's semiconductor market by application, 2005.

Source: CCID, *Gartner Dataquest*, CSIA, 2006 cited in PricewaterhouseCoopers 2007.

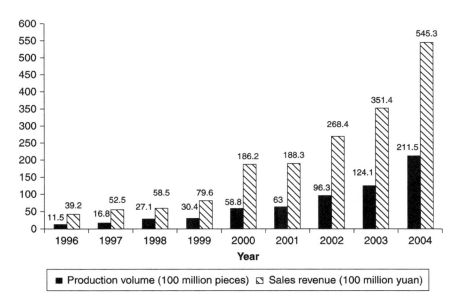

Figure 4.5 Volume production and sales revenue of China's semiconductor industry, 1996–2004.

Sources: 1996–2001 data based on China's Electronics Industry Yearbooks from 1996 to 2001, cited in Lou 2003: 181; 2002–2004 data from CSIA and CCID, cited in China Semiconductor Industry Association and China Center of Information Industry Development 2005.

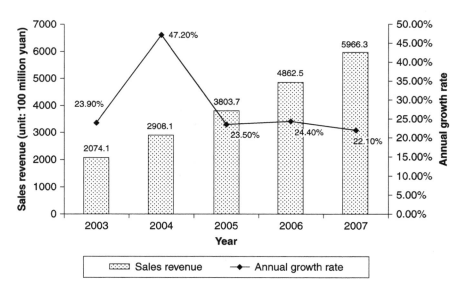

Figure 4.6 Sales revenue and growth rate of China's semiconductor industry, 2003–07.
Source: CCID Consulting 2007.

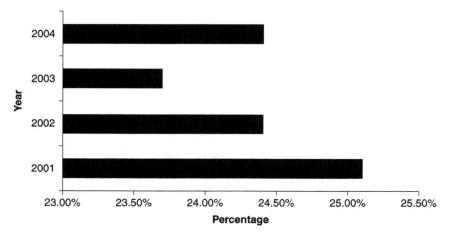

Figure 4.7 The ratio of China's semiconductor sales revenue to domestic semiconductor demand, 2001–04.

Source: China Semiconductor Industry Association and China Center of Information Industry Development 2005.

2012 and its chip industry revenue is forecast to reach US$25 billion. This forecast implies a further widening of China's chip consumption–production gap to US$94 billion by 2012 (PricewaterhouseCoopers 2010: 59).

Measured respectively by sales revenue, production revenue, and the volume of 8-inch wafers produced, China's semiconductor production capacity has not been significant on a global scale, although the ratio has been on the rise. Its semiconductor sales revenue accounted for 5.64 per cent of the worldwide total in 2004, up from 4.98 per cent in 2003, 4.81 per cent in 2002 and 3.64 per cent in 2001 (China Semiconductor Industry Association and China Center of Information Industry Development 2005: 2–3). In terms of production revenue, China's sector accounted for 11 per cent of the global total in 2009, up from 10.7 per cent in 2008, 6 per cent in 2005, 5 per cent in 2004, and just 2 per cent in 2000 (PricewaterhouseCoopers 2007: 19; PricewaterhouseCoopers 2010: 27). Measured by the volume of 8-inch wafers produced, China's semiconductor production capacity is minimal on a global scale. In 2001, the number of 8-inch wafers produced in China accounted for 1 per cent of the worldwide total, which remained the same in 2002 and grew to 2 per cent in 2003 and to 3 per cent in 2004.

Supply chain development amid sectoral globalization: design, fabrication, packaging, and testing

This section examines IC design, fabrication, packaging, and testing sub-sectors in the Chinese chip supply chain amid increasing forces of sectoral globalization identifying major progresses and challenges involved.[86] Before the analysis, it is noteworthy that various sub-sectors have grown at different paces, thereby making varying degrees of contribution to the industry production revenues. According to CSIA and China Center for Information Industry Development (CCID), the ratio of revenue generated by IC design, IC manufacturing and IC packaging and testing was 13:17:70 in 2003, and it became 15:33:52 in 2004 (China Semiconductor Industry Association and China Center of Information Industry Development 2005: 24). If the discrete device sub-sector were included,[87] it would be the largest part of the industry from 2003 to 2005, accounting for 47 per cent of the total production revenue. The second largest sector was the IC packaging and testing, which generated 26 per cent of the total revenue. The third largest sector was IC manufacturing by IC foundries and IDMs; it contributed to 18 per cent of the total. IC design sector was the smallest and yet the fastest growing segment accounting for 9 per cent of the total (PricewaterhouseCoopers 2007: 21). The ranking of the four segments in the Chinese industry remained static from 2003 to 2005, though each sector grew at a different rate. For instance, the revenue share of IC packaging and testing production declined from 36 per cent in 2003 to 29 per cent in 2004 and dropped further to 26 per cent in 2005. By contrast, that of IC manufacturing production increased from 9 per cent in 2003 to 18 per cent in 2004 and 2005, and that of IC design sector increased from only 6 per cent in 2003, to 8 per cent in 2004, and to 9 per cent in 2005 (ibid.). From 2005 to 2009, the distribution of China's semiconductor industry continued to change. In 2009, for instance, IC design sub-sector experienced an exceptional growth in an otherwise negative environment because of the global economic recession. Below we shall begin with an analysis of the IC design sub-sector.

Despite its slow pace of development before the 1990s, China's IC design sub-sector has picked up momentum since 2001, emerging as the fastest growing segment of China's semiconductor industry amid daunting tasks ahead. In 1986, China established its first commercial oriented design house, BIDC.[88] As discussed earlier, the 908 Project and the 909 Project funded the establishment of dozens of IC design companies, but they failed to provide the two respective factories with orders because of their technological constraints. However, these firms have contributed to the gradual formulation of the sub-sector since the late 1990s.

The introduction of Circular 18 in mid-2000 attracted an influx of foreign and domestic capital to the emerging sub-sector, resulting in the increase in the number of IC design companies from 200 in 2001 to 472 in 2009 (Figure 4.8). Since 2001, IC design has remained the fastest growing segment of the Chinese industry. IC design revenues grew from US$178 million in 2001 to US$3.95 billion in 2009, experiencing a CAGR of 47 per cent (PricewaterhouseCoopers 2010: 27). From 2001 to 2005, its CAGR reached 71 per cent.[89] By way of comparison, the world's top 10 fabless design houses grew with a CAGR of 22 per cent from 1994 to 2006.

In terms of ownership, by the end of 2005, about 380 of the 479 IC design companies were domestic firms,[90] and 100 were design units or activities of foreign-invested or subsidiary MNCs. The latter group included the Chinese

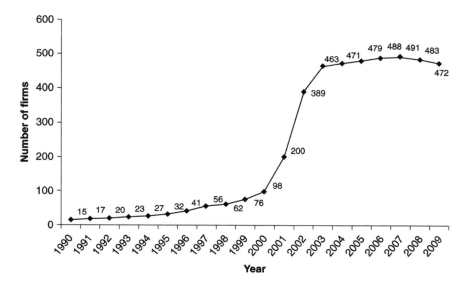

Figure 4.8 Number of IC design firms in China, 1990–2009.

Source: Original data from PricewaterhouseCoopers 2010: 46.

design activities of 18 of the top 25 semiconductor MNCs. In terms of operation types, some were spin-offs from system manufacturers,[91] others were functional departments within IDMs, and still others were pure fabless design houses.[92] Domestic enterprises could be further divided into two broad categories (PricewaterhouseCoopers 2007: 34–5): those domestically focused and those globally focused. The former group consisted of government-connected companies operated by locally educated Chinese with lower levels of technological expertise and experience.[93] These firms primarily relied on exclusionary domestic standards, such as Time Division-Synchronous Code Division Multiple Access (TD-SCDMA)[94] and smart ID cards. The latter group, on the contrary, was composed of globally capable enterprises operated mainly by foreign-educated Chinese with higher levels of technological know-how and experience, firms that would be key to the establishment of a globally competitive sub-sector in China.[95]

Since 2001, the IC design sub-sector has achieved notable qualitative improvements in three aspects. First, performance by leading companies in China has been impressive. In 2005, Actions Semiconductor Corporation (ASC, or Actions) reported US$155 million in revenue, thereby becoming China's first IC design enterprise to break the US$100 million revenue mark.[96] In the same year, Vimicro, then China's second largest fabless design house, reported US$95 million in revenue. Both firms became the first Chinese fabless design houses to complete successful stock market launch at NASDAQ.[97] More importantly, both achieved impressive pen-etration in focused market niches. Actions gained a 30 per cent share of the global market for the non-iPod Moving Picture Experts Group Layer-3 Audio (MP3) player SOCs, and Vimicro gained profits from its share of the PC video camera chip market (PricewaterhouseCoopers 2007: 14, 35). By the end of 2006, China's top six IC design houses respectively had broken the US$100 million revenue mark (Table 4.7) (Electronic Engineering Times Taiwan 2007).

Table 4.7 China's top ten fabless design companies in 2006

Rank	Company	Revenue (US$M)*	Main products
1	Actions	170.4	Consumer IC
2	Vimicro	128.2	Multimedia IC
3	Datang Microelectronics Technology	116.3	IC card
4	HiSilicon Technologies	114.4	Communication IC
5	Wuxi China Resources Semico	106.8	Consumer IC
6	Hanzhou Silan Microelectronics	103.8	Consumer IC
7	Shanghai Huahong Integrated Circuit	83.2	Consumer IC
8	CEC Huada Electronic Design	74.7	IC Card
9	Hsinghua Tonfang Microelectronics	64.1	IC Card
10	Spectrum Communications	40.8	Communication IC

Source: Original data from CCID and IEK, cited in Electronic Engineering Times Taiwan 2007.

Note: *The original data for the revenue in 2006 were in RMB, and an exchange rate of 1US$ = RMB 7.899 was used for these calculations.

Second, breakthroughs in developing China's homegrown CPUs in recent years have further indicated the country's increased IC design capability, though these CPUs are not comparable to their state-of-the-art counterparts made by global giants, including Intel and AMD. Examples of indigenous CPUs abound. In 2001, Arca introduced Arca-1 to the market, China's first homegrown 0.25-process, 32-bit CPU. The product integrated a 32-bit Arca RISC[98] processor core with various other functions on one chip, and it was regarded as a highly integrated CPU suitable for use in network computers, broadband clients, set-top boxes, digital TVs, and industrial PCs (Xu and Fang 2001; Wang 2001b). In 2002, BLX IC Design Co. Ltd., a CAS spin-off, introduced its 266MHz Godson-1A chip to the market (Clendenin 2003; Lemon 2004). In December 2003, the market saw the debut of the United-863 chip, China's first homegrown SOC designed for use with the Linux operating system. The CPU was the brainchild of the Micro Processor Research and Development Center of Peking University with support from the 863 Program. It differed from its predecessors because it was developed from scratch by using Chinese IP. By 2004, 350MHz Arca-2 CPU made it to the global market as Wyse Technology, the USA's biggest network computer maker, decided to use Arca-2 CPUs for a version of its Winterm 1000 line of thin clients that would be sold globally.

Third, China's IC design houses have migrated to the upper end of the techno-logical ladder over the past few years. According to CCID, these firms have improved their design capabilities to achieve finer design line widths from 2005 to 2009. In 2005, the process technology in use by China's 479 IC design houses was as follows: 36 per cent of designs used 0.25 micron or less, 33 per cent used 0.25–0.5 micron, 23 per cent used 0.5–1.5 micron, and 8 per cent used 1.5 micron or more (PricewaterhouseCoopers 2007: 38). During 2009, almost 40 per cent of designs used 0.25 micron or less. In particular, 41 of these firms had design capa-bilities equal to or less than 90 nanometers. At the same time, the percentage of IC design enterprises with low technologies has decreased (PricewaterhouseCoopers 2010: 48). Surveys conducted by *Electronic Engineering Times Asia* (2005; 2006; 2007a) also concluded that IC design firms in China had made technological progress over the past few years. For instance, 64 per cent of respondent compa-nies utilized 0.25 micron or less process technology for digital design in 2005, and the proportion reached 72 per cent in 2006, and 74 per cent in 2007. In 2007, 10 per cent of the firms surveyed used 0.11 micron or less process for digital design, whereas none did in 2006. In 2007, 9 per cent of designs utilized 0.09 micron and less, 1 per cent used 0.11 micron, and 15 per cent relied on 0.13 micron. In addition, 42 per cent of designs used 0.18 micron, 9 per cent utilized 0.25 micron, 14 per cent relied on 0.35 micron, 9 per cent used 0.5 to 1.5 micron, and 1 per cent utilized process technology greater than 1.5 micron.

Despite its aforementioned growth and qualitative achievements, China's IC design sub-sector has faced two major challenges. First, it started from a small base, and currently involves hundreds of different entities, most of which are small, technologically backward and financially fragile amid a continuous trend of sector consolidation. In 2008 and 2009, some firms went bankrupt, and the number of firms consequently decreased from 491 at the end of 2007 to 472 by the end of 2009. According to a senior executive of a government-funded venture capital targeting IC design start-ups, the majority of indigenous firms are weak and small and will be further absorbed by the strong and the big amid fierce market competition.[99]

Second, the market performance of the majority of these companies has been poor. In 2005, only 11 of them achieved revenue of more than US$25 million, and only 30 had revenue of more than US$5 million (PricewaterhouseCoopers 2007: 35). A comparison of China's top 10 companies with their global counterparts in terms of revenue further indicates the weakness of the Chinese sub-sector. In 2003, the revenue generated by the top 10 Chinese fabless firms reached US$327.3 million. This amount was less than the US$346 million concurrently generated by Novatek, the world's 16th largest design house.[100] By the end of 2006, the world's top 10 fabless companies had achieved billion-dollar annual revenue (Hurtarte *et al.* 2007: 5) and the total revenue generated by the world's fabless IC company sales reached US$42.3 billion (McGrath 2007). By contrast, the revenue generated by China's top 10 fabless firms (Table 4.7) reached US$938.6 million, which was less than the revenue generated by Altera, the world's tenth largest IC design company.

In spite of these challenges, some market forecasts remain upbeat about the future prospect of the sub-sector. *Gartner Dataquest*, for instance, predicted in 2007 that China's IC design houses were expected to gradually become competitors when compared with global giants in the next few years. While most IC designers in China would continue to focus on low-end technologies for end-use devices, such as personal digital assistants and digital set-top boxes, and would have a minimal impact on global vendors in the short term, some would expand their capabilities to more sophisticated areas. In particular, an increasing portion of electronic manufacturers in China would follow the examples of Huawei and ZTE to set up their own IC design houses in order to improve margins. By 2011, about 20 per cent would be expected to run their own chip design facilities (Hopfner 2007). In addition, PricewaterhouseCoopers (2010: 49) predicted that a number of China's IC design firms would soon present initial public offerings (IPOs). If so, these firms may attract a large amount of development funds, thereby luring local and overseas high-tech talent to join the sub-sector, thereby further increasing China's IC design capability.

To conclude, China's IC design sub-sector has been the fastest growing segment of the local semiconductor industry in the past few years amid a series of challenges. Nevertheless, its growth in years ahead will depend on a complex array of internal and external factors such as the size of China's IC market, government policies, domestic wafer fabrication capabilities, R&D spending, the availability of competent engineers, and the presence of foreign design enterprises (PricewaterhouseCoopers 2007: 41).

China's semiconductor fabrication activities for commercial end-use began with the endeavors of defense conversion in the late 1970s. In the 1980s and 1990s, Beijing sequentially introduced the 907 Project, the 908 Project, and the

909 Project in order to establish IDMs. In 1999, CSMC debuted China's first pure-play foundry. Since the introduction of liberalized policies in 2000, which favored the establishment of pure-play foundries, the influx of investment into China's IC fabrication sector has resulted in the construction of a few new fabs. By the end of 2005, China was home to 75 fabs in production, which contributed to 18 per cent of the total industry revenue, a 9 per cent increase from 2004 (ibid.: 21–2). By the end of 2009, China had 115 wafer fabs, and its IC manufacturing sub-sector, including pure-play foundries, represented 17 per cent of its total industry revenue.

China's recent progress in upgrading its IC fabrication capabilities can be measured as follows. First, the dollar revenue of its IC manufacturing sector has grown at an average CAGR of more than 38 per cent from less than US$400 million in 2001 to US$5 billion in 2009 (PricewaterhouseCoopers 2010: 41). Second, China-based fabs have climbed up the technological ladder in terms of wafer size. From 2005 to 2009, the proportion of 12-inch wafer fabs in China increased from 10 per cent of its total fabrication capacity to 27 per cent, whereas that of 6-inch or less wafer fabs decreased from 45 per cent to 38 per cent.

Third, pure-play foundries in China had grown rapidly in terms of sales revenue and production capacities before 2009, the year the global sector suffered severely from the economic recession. From 2002 to 2006, sales revenue of these foundries expanded 6.8 times, with a CAGR of 67.2 per cent. Chinese foundries accounted for 4 per cent of the worldwide market in 2002, 12.4 per cent in 2005 and 12.6 per cent in 2006. Between 2002 and 2006, the production capacity of China's foundries expanded 1.5 times, reaching 7.17 million pieces a year, accounting for 27.2 per cent of the global total (Electronic Engineering Times Asia 2007c). Fourth, performance by leading Chinese foundries has been impressive. By the end of 2006, China-based SMIC, Huahong NEC, and HeJian were among the world's top 10 pure-play foundry companies (IC Insights 2006). In particular, SMIC became the world's third largest dedicated foundry four years after its inception, with its production capacity and revenues growing explosively from 2002 to 2004. In 2006, its fab installed capacity accounted for 2.2 per cent of the global total (Solid State Technology 2007a). In 2007, its monthly capacity reached 185,000 wafers, and its sales reached US$1.55 billion by the end of the year. Moreover, SMIC has also been a leading player on PRC soil in terms of process technology. For instance, it introduced 0.35 micron and less to China.[101] As of early 2008, SMIC was running all of China's state-of-the-art 12-inch wafer fabs, with some beginning to produce chips with 90nm process (Lapedus 2008).

In spite of the aforementioned progress, China-based chip producers still lag behind the global giants in terms of leading-edge wafer size and process technologies. Although China's capacity in terms of wafer size and process technology was comparable with worldwide capabilities in 2005, it lacked substantial 12-inch capacity (Table 4.8). Only two of the world's 58 12-inch wafer fabs in 2005 were in China, accounting for 10 per cent of the country's fab capacity. By contrast, 6-inch and 8-inch wafer fabs, respectively, accounted for 31 per cent and 45 per cent of the total. From a process technology standpoint, China's capacity in

Table 4.8 Comparison of China and the world in wafer fab capacity, 2005

	China capacity	(%)	World capacity	(%)
Total	911.8		12,390.8	
By geometry				
≥ 0.7 micron	173.1	19	2,119.9	17
< 0.7 to ≥ 0.4 micron	103.0	11	1,016.3	8
< 0.4 to ≥ 0.3 micron	52.4	6	935.3	8
< 0.3 to ≥ 0.2 micron	89.6	10	1,014.8	8
< 0.2 to ≥ 0.16 micron	35.8	4	549.3	4
< 0.16 to ≥ 0.12 micron	253.0	28	1,934.5	39
< 0.12 micron	181.1	20	4,770.7	39
n/a	23.8	2	50.0	0
By wafer size				
≥ 4 inch	74.1	8	589.5	5
5 inch	57.4	6	785.4	6
6 inch	277.1	31	2,686.1	22
8 inch	413.0	45	5,329.5	43
12 inch	90.0	10	3,000.3	24

Source: SEMI World Fab Watch, 2006 cited in Table 5 PricewaterhouseCoopers 2007.

Notes: (1) Capacity = 8-inch equivalent wafer starts per month in thousands; (2) World Fab Watch probability > 1.0.

2005 was comparable with worldwide capabilities, the exception being a dearth of leading-edge 0.12 micron or less processes. Whereas the worldwide capacity at the node less than 0.12 micron was 39 per cent, China's capacity at the same node was 20 per cent (PricewaterhouseCoopers 2007: 26–7).

CCID reached a similar conclusion concerning China's wafer fab capacity in 2006. Although China-based foundries were almost comparable with worldwide capabilities, there was a dearth of state-of-the-art 12-inch capacity and 90nm or less process technologies in the PRC. In 2006, 6-inch and 8-inch wafer fabs, as well as 0.35 micron to 0.11 micron processes dominated China's foundry operations. Eight-inch wafer fabs and 0.35 micron to 0.11 micron processes, respectively, accounted for 53 per cent and 55.7 per cent of China's overall foundry production capacity. By contrast, 12-inch and 90nm processes represented 4 per cent and 2.5 per cent of the total (Electronic Engineering Times Asia 2007c). In 2009, China's capacity continued to concentrate on the smaller wafer size ranges. Thirty-eight per cent of its capacity was in six-inch or smaller wafers, 35 per cent in 8-inch ones, and 27 per cent in 12-inch ones. From a process technology standpoint, its wafer fabrication capabilities continued to trail behind the global state-of-the-art as of 2009. When fully equipped and ramped, China would merely have 25 per cent of its capacity at the leading-edge node less than 80nm, compared to a worldwide sector distribution of 46 per cent (PricewaterhouseCoopers 2010: 39).

Looking into the near future, the development of China's IC manufacturing sub-sector remains uncertain. According to one forecast (IC Insights 2006), any growth in China's foundry market share would be at a moderate pace because the Chinese foundries have moved from the explosive growth of the start-up period to the more moderate growth rates encountered by established firms. Semiconductor equipment companies selling fab tools to the Chinese market echoed this prediction (Clendenin 2006). Indeed, some industry observers have been pessimistic about China's foundries. For instance, a Shanghai-based SEMI analyst predicted in 2006 that half of China's fab hopefuls would fail because of the lack of partnership and manufacturing expertise in spite of soaring domestic demand for ICs (Lapedus 2006b). Others, however, have been optimistic about China's future foundry operations because of SMIC's expansion of new fabs and China's insatiable demand for ICs. In 2007, CCID forecast that the sales revenues of China's foundries would reach US$6.53 billion by 2011, accounting for 15.9 per cent of the global total. According to PricewaterhouseCoopers (2010: 42), by 2012, China would increase its share of worldwide foundry production by about 21 per cent if it completely equipped and ramped to full capacity at mature yields all of its existing wafer fabrication modules.

In sum, China has successfully put itself on the global map of IC manufacturing following decades of trial and error. Crucially, the viability of China's foundry sector in the future will determine the extent to which local foundry services can help fabless start-ups advance their growth and upgrade their technology, a key to China's economic competitiveness and national security.

The packaging and testing sub-sector has traditionally been the largest contributor to the production revenue of China's IC industry, in spite of its recent decline in revenue share. In 2009, its share of the overall industry revenue dropped to 25 per cent, down from 28 per cent in 2008 and 30 per cent in 2007. The dollar revenue of the sub-sector has grown at a 20 per cent CAGR from less than US$2 billion in 2001 to slightly more than US$ 7 billion in 2009 (PricewaterhouseCoopers 2010: 31).

The value of China's IC packaging assembly and test production increased to represent about 11 per cent of the worldwide total by the end of 2005, up from 9 per cent in 2004. China had 90 pertinent facilities by the end of 2005, accounting for 22 per cent of the worldwide total, and its share of the global total was about 20 per cent by the end of 2009 (PricewaterhouseCoopers 2007: 45; 2010: 53).

The major challenge the sub-sector has faced is two-fold: foreign dominance and local firms' technological backwardness compared to the global giants. First, foreign investment and MNCs have traditionally dominated the sub-sector. At the end of 2005, only about 30 per cent of China's 90 facilities were locally owned. By the end of 2009, however, half of China's 74 facilities were local, whereas the other half was foreign in terms of ownership. Because of production globalization, the majority of the world's leading IDMs have set up packaging and testing facilities in China, either in the form of purely foreign-owned subsidiaries or joint ventures. By the end of 2009, each of the five largest and nine of the 10 largest MNCs in the sub-sector ran one or more facilities in China (ibid.).[102] Second, foreign firms operate with higher-end technologies in their facilities in China, whereas their local counterparts focus on the lower-end market because of their backward technologies (PricewaterhouseCoopers 2007: 45–6). According to a Chinese engineer at a foreign-invested packaging and testing firm operating on PRC soil, "Our technologies are extremely advanced leaving local firms trailing far behind." His observation is supported by other interviewees from the sub-sector.[103]

The future of the sub-sector thus depends on whether Chinese local firms will climb up the technological ladder in order to compete with their foreign counterparts in the domestic and international market. It remains to be seen whether local firms can change this trend in the near future.

The final key aspect of China's current semiconductor industry involves the role of foreign firms in the sector because of sectoral globalization. The involvement of foreign companies in China's industry was developing from limited participation to growing influence before 2000. As early as the 1960s, semiconductor factories in China managed to import second-hand semiconductor production lines and tools from overseas, especially Japan, in spite of export controls under the COCOM regime. In 1984, Applied Materials became the first foreign semiconductor company to set up a joint venture in China.[104] In the 1980s, Toshiba was the key foreign partner for the 907 Project, transferring its 3-inch wafer process technology to China and helping train over 100 Chinese engineers. Such a Sino-foreign cooperation model was implemented further throughout the 1990s. For instance, semiconductor equipment from Lucent, Toshiba and Siemens enabled the 908 Project to go ahead, whereas NEC's technological transfer helped establish China's first 8-inch wafer fab under the 909 Project. Since Beijing introduced a series of liberalized policies in 2000, MNC-led semiconductor production activities have moved to China at an accelerating rate. By the end of 2005, foreign investment accounted for 80 per cent of total investment in the local industry.[105]

Today, foreign semiconductor firms continue to shift some of their operations to China as part of the continuous sectoral globalization. George Scalise, President of SIA, dubbed the trend "a new outsourcing model," with production activities in IC design, fabrication, packaging and testing concurrently internationalized. Excluding Japan, the Asia-Pacific countries, particularly China, have become hot spots in the investment profiles of global companies (Semiconductor Industry Association 2009: 14, 17).

Foreign firms have a presence in China in various forms, such as FDI-related outsourcing activities, joint ventures, research laboratories, and technological licensing. In the area of packaging and testing, major IDMs have dominated the pertinent production activities on Chinese soil. In 2005, they accounted for over 30 per cent of the production revenue generated by China's top 50 chip players. In the area of IC manufacturing, there has been the entrance of pure-play foundries from overseas, chiefly in the form of joint ventures or foreign-owned subsidiaries. The most notable examples include SMIC, TSMC, and HeJian. In addition, four foreign IDMs had some form of invested IC wafer fabrication capacity in China by the end of 2009; they included Hynix, Intel, NEC, and ProMOS (PricewaterhouseCoopers 2010: 42). By the end of 2009, Intel and Hynix became the largest and second largest semiconductor manufacturers on Chinese soil in terms of sales revenue (ibid.: 32).

In the area of IC design, some of the world's leading fabless firms and IP service providers, which are at the high-end segment of IC design sector, have also increased their production activities in China. An illustrative case involves ARM,[106] a provider of leading IP that enables local companies to reduce their time to market and create competitive high-value products and solutions based on the ARM architecture. The firm set up its first office in China in Shanghai in 2002 and opened its office in Beijing in 2004 to leverage strong growth opportunities in this fast-developing market. Since then, it has expanded its strategic ties with several leading China-based semiconductor firms – including Huawei, Haier, Spectrum and SMIC – through various licensing agreements (ARM 2004). Another case involves Xilinx's move to set up its first R&D center in China in 2007 in order to be close to its customers (Manners 2006). However, IC design has been the least globalized segment of the supply chain as compared to IC manufacturing, packaging and testing, as concluded in Chapter 3. Generally, fabless design firms have been reluctant to shift their core operations offshore, including to China.

National security implications

So far, we have argued that the Chinese commercial chip industrial base has recently taken off, though it has still lagged behind its counterparts in leading semiconductor states, in terms of technological capability and market performance. Such an industrial base, as is contended below, continues to be a liability to China's national security in economic and defense terms.

In economic security terms, the industry potentially results in China's economic insecurity because it still cannot provide the majority of ICs needed in the domestic market, resulting in China's overt dependence on a foreign supply of ICs to meet its soaring domestic demand. Worse yet, the enlarging IC consumption and production gap exacerbates China's economic security concerns. This reality thus continues to reinforce Beijing's resolve to take further measures to increase indigenous IC production.

In defense security terms, the industry is a liability to China in two major ways. On the one hand, China's dependence on foreign supplies of ICs for its fast-developing end-use market exacerbates Beijing's fear of supply cuts in times of contingency, as well as its concern over information insecurity due to IC-targeted IW operations masterminded by foreign IC makers and/or their governments.

On the other hand, China's technologically feeble commercial chip industrial base will not effectively help improve Chinese defense capability, despite even following the move by Beijing to lure civilian semiconductor firms to take part in defense production. Unless the industrial base in question ensures its technological advances, it will not become a fully-fledged asset to China's defense security through the successful transfer of technology from the civilian to the defense side following the dominant trend of spin-on.

Despite this, research data indicate that the Chinese military has already tried to leverage the growing domestic IC capability to advance its modernization. The PLA's funding of a CPU project at Fudan Microelectronics is a case in point. In addition, 10 per cent of respondent companies in a survey (Electronic Engineering Times Asia 2006) of China's IC design houses said they were engaged in military electronics design activities while operating in China. In a follow-up survey (Electronic Engineering Times Asia 2007a), 19 per cent of the firms said they still did, a 90 per cent increase from the previous year. Although neither survey specified which military these China-based IC design firms had served, it would be logical to argue that at least some of the military customers were Chinese.

Conclusion

To conclude, the development of China's semiconductor industry has been linked to Beijing's drive to pursue its economic, technological, and defense security in the past few decades. The viability of the chip sector is linked to China's "compounded national strength" and national security.[107] For decades, however, the industry has developed in a relatively enclosed environment with limited success.

On the one hand, the traditionally autarkic defense chip industrial base has partially contributed to China's defense security by supplying small-volume semiconductors to the PLA, in spite of various internal and external constraints. China, over time, has nevertheless developed a dual-track strategy to acquire semiconductors for its defense end-uses from domestic and international sources, echoing, in some ways, the partial globalization of defense chip production activities in the case of the USA. Moreover, efforts to establish a comprehensive national military standard system in order to govern the pertinent defense industrial base efficiently have been an asset as well as a liability for China's national security. Recent reforms of civil–military integration have potentially opened the door for civilian firms to take part in defense chip production activities, thereby transferring resources in the vibrant civilian economy to the defense side, in spite of apparent challenges ahead.

On the other hand, the commercial chip industrial base has failed to take off for decades in spite of state-led endeavors. This has changed since 2000 when Beijing introduced liberalized policies to lure foreign investment into China's strategic sector amid the increasing trend of sectoral globalization. Consequently, there have been notable qualitative improvements in all the major sub-sectors of the commercial chip sub-sector, in spite of onerous challenges ahead. Crucially, some of the commercial IC design local firms have engaged in production activities to satisfy the PLA need indicating the gradual trend of spin-on in the PRC context. As the globalization of semiconductor production activities led by MNCs further accelerates the development of the Chinese commercial chip sector, the Chinese defense establishment is expected to use the increasingly strong domestic commercial chip industrial base to its advantage.

Taiwan has shifted a segment of its IC production activities to China as part of the sectoral globalization, thereby contributing to the empowerment of the strategic Chinese industry. We shall turn to this under-studied theme in Chapter 5.

Notes

1 Interview, 2 September 2005, Beijing, China.
2 Interview, 17 September 2005, Shanghai, China.
3 This section relies on the analysis of the following sources: official public statements, internal (*neibu*) circulars, statements by the PLA, and my interviews with senior players in the Chinese defense and commercial chip industrial base.
4 Hu was the former vice-minister of the Ministry of Machine-Building and Electronics Industry.
5 COSTIND is China's top national defense administrative organization.
6 Interview, 2 September 2005, Beijing, China. Yu joined the 13th Research Institute in the late 1950s, and subsequently became the chief engineer in the Chinese Ministry of Engineering Industry.
7 Interview, 17 September 2005, Shanghai, China.
8 Interview, 30 August 2005, Beijing, China. One successful example she cited involves the design and fabrication of chips for China's national ID card scheme in 2005, with four indigenous design houses and two local foundries chosen to be suppliers. Wang formerly worked at the Ministry of Machinery and Electronics Industry involved in R&D of state projects of computer networks and satellite control systems.
9 Interview, 23 September 2005, Shanghai, China. See also Economic Security Forum 2002: 113–20; Chen 2005: 218–22.
10 Interview, 16 March 2005, Shanghai, China.
11 Interview, 23 September 2005, Shanghai, China.
12 Interview, 17 September 2005, Shanghai, China.
13 The publication was by China Aerospace Science and Technology Corporation (CASC), a member of the Chinese defense industry.
14 For an excellent summary of views on the link between microelectronics and weaponry systems by Chinese military electronics experts, see Hua *et al.* 2000.
15 The journal was published by COSTIND.
16 The institute is also known as Sichuan Institute of Solid-State Circuits.
17 Interview, 17 September 2005, Shanghai, China.
18 As concluded in Chapter 2, technological changes in the overall semiconductor industry have complicated the link between the chip industry, defense power, and security. While the tangential defense chip sub-sector in a country still matters in national security terms if it provides classified chips for critical military systems through relatively enclosed and autarkic operations at home, the mainstream commercial chip sub-sector in the same country may become militarily important if it is technologically superior to the defense chip sub-sector and is able to transfer resources to satisfy national defense needs. Largely, this conclusion holds true in the PRC context.
19 These technologies include CMOS, GaAs, nanotechnology, MEMS, and radiation hardening technology.
20 Interview, 17 September 2005, Shanghai, China.
21 Both spheres overlap because certain local chip players produce semiconductors for military and civilian end-uses. This is especially so under certain conditions, such as the defense sector's efforts to make civilian goods aside from its military products, and Beijing's recent efforts to integrate the military and civilian segments of the national economy.
22 The institute is also known as Hebei Semiconductor Research Institute.

23 For insiders' accounts, see No. 13 Research Institute n.d.; interview with Yu Zhongyu, president of CSIA, who joined the institute after graduating from university, 2 September 2005, Beijing, China.

24 According to the official account of the fifty-year Chinese electronics industry, players in the industry, which is an integral part of the Chinese defense industrial base, are traditionally state-run because of national security considerations. See Editing Committee on China Electronics Industry in Fifty Years 1999: 28.

25 Interview, 2 June 2005, Taipei, Taiwan.

26 Interview, 14 September 2005, Ningbo, China.

27 The authors were members of the Chinese defense industrial base.

28 The lack of economy of scale in the production of military semiconductors in these operations has been the norm rather than the exception. For instance, the State 877th Plant fabricated 10 250W power transistors in a single military order in 1993. See Zhang 1991: 30.

29 The data came from the 5th Research Institute of the Ministry of Mechanical and Electronics Industry.

30 Interview with a former chief of an American EDA company in charge of China business, 29 June 2005, Taipei, Taiwan.

31 The analysis of the data, however, is faced with two constraints. First, not enough systemic information is available on the actual end-use of these indigenously made QPL and QML components in the PLA systems. Second, the PRC has introduced its own QPL and QML systems fairly recently and as such, both lists do not have a comprehensive coverage of all the electronic components that are in actual use in the PLA systems.

32 These conglomerates form the main trunk of China's current defense-industrial complex that operates within a chain of command that proceeds from the enterprises, through ministerial leadership, to the State Council headed by the premier. See Mulvenon 2005: 211.

33 It is also known as Lishan Microelectronics Corporation.

34 Comparing the QPL and the QML data, No. 214 Research Institute and No. 771 Research Institute appear on both lists.

35 For instance, QML firm Qingdao Semiconductor Research Institute has revealed on its website that its semiconductor products have been inserted into China's first artificial satellite, Dongfanghong 1, weather satellites and manned spacecrafts without specifying the exact systems in question.

36 Interview, 17 March 2005, Shanghai, China.

37 Interview, 17 March 2005, Shanghai, China.

38 Interview, 17 March 2005, Shanghai, China. The area of semiconductor equipment is a case in point. For instance, the lack of a technological base and investment had resulted in the failure of pertinent indigenous efforts to develop lithography equipment, a *neibu* survey revealed in 2002. The only exception is the 45th Research Institute, the second case to be explored here. According to an official-turned-industry player, "We used to have loads of equipment, such as steppers, manufactured by our indigenous research institutes, but they don't work now." See interview, 30 August 2005, Beijing, China.

39 Interview with an engineer from the institute, 17 March 2005, Shanghai, China.

40 Ibid.

41 For a review of the Chinese rocket program, see He 2003. The author worked with the space division of the China Great Wall Industry Corp.

42 The *Hangtian* project got underway in 1997, though it was approved by Beijing in 1992.

43 This means some 100 transistors per chip.

44 This means some 1,000 transistors per chip.

45 Interview, 8 December 2004, San Jose, CA, USA, and 27 August 2005, Beijing, China.

46 Interview, 30 August 2005, Beijing, China.

47 This periodization is based on an account by a COSTIND official.

48 Both articles appeared in *Junyong Biaozhunhua* (Military Standardization), the bimonthly journal that was initially sponsored by COSTIND's Central Institution for Standardization, but was subsequently sponsored by GAD's Electronics and Information Base Department. The section of the chapter relies heavily on materials from the journal in question whose circulation had been confined to an internal circle within the government between its inception in 1985 and July 1994. In theory, these *neibu* materials would address genuine policy issues because the readership was confined to insiders involved in the work of military standardization. If so, they are more credible and less propagandistic than open-source materials, thereby representing the most authoritative official accounts of the development of China's military standardization system.

49 The Chinese military-industrial complex has developed organizationally and structurally since the 1960s. Major structural reforms took place in the 1960s, 1982, 1988, and 1993. See Liu *et al.* 1986: 653–4; Shambaugh 2004: 230–3.

50 These agencies, for instance, included Beijing No. 2 Radio Factory, Beijing Semiconductor Device No. 3 Factory, Dongguang Radio Factory, Shanghai No. 16 Radio Factory, No. 746 Factory, Huafeng Radio Factory, Xinyun Electronic Apparatus Factory, Hongxing Radio Elements, and Materials Factory, as well as Jiangnan Materials Factory.

51 China began to develop *Dongfeng-5* in 1965 with the goal of developing an ICBM capable of hitting the continental USA and the western Soviet Union. It was China's first missile with an in-flight computer. In 1965, computer engineers at CAS began their one-year research effort to develop the *Dongfeng-5* computer, and produced an instrument assembled from newly designed ICs. However, the Seventh Ministry rejected the computer because it was too complicated and unreliable. In 1974, No. 771 Research Institute succeeded in developing a minicomputer for the missile. The missile did not undergo a full-range test flight until 1980, and it did not enter service until 1981. In the first part of the 1980s, the missile was upgraded through the incorporation of improved guidance and propulsion systems. This gave birth to *Dongfeng-5A* in 1986. See Lewis and Xue 1994: 167–8; Swaine and Runyon 2002: 11–12.

52 Interview, 19 September 2005, Shanghai, China.

53 The authors were members of the 4th Research Institute of the Ministry of Electronics Industry.

54 The establishment of GAD in 1998 aimed to reorganize and rationalize the production and structure of the Chinese defense industrial complex. It was accompanied by a parallel reorganization of COSTIND, which was under the State Council and renamed as State COSTIND, also known as SCOSTIND. Shambaugh argued that the division of labor between GAD and the SCOSTIND was unclear.

55 Liu was the deputy minister of GAD's Electronics and Information Base Department.

56 According to Cheung (2007b), China had issued about 5,700 national military standards by the end of 1998. However, the Chinese military specifications and standards regime has a long way to go to catch up with its more established counterparts in advanced industrial countries. For instance, the Pentagon had an active list of more than 26,000 military specifications and standards in 2001.

57 The top targets for this outer pool include existing and former military entities as well as mainstream civilian firms with advanced technological capabilities in the areas of high military demand.

58 On 10 August 1999, Jiang Zemin, in his capacity as the CMC chairman, put forward the idea of *Junmin Jiehe* and *Yujun Yumin* as a guiding principle for the restructuring of the Chinese defense industry. See Yan 2004: 16. However, a comprehensive pronouncement of the defense sector reform appeared in the 10th Five-Year plan as discussed here.

59 For an analysis of Deng's dictum that has significantly shaped the development of the Chinese defense industry over the past decades with mixed results, see Gurtov 1993; Feigenbaum 2003: 98–104; Commission on Science 2004: 91, 311, 353; Shambaugh 2004: 251–2; Medeiros *et al.* 2005; Cheung 2009. Deng's successor Jiang Zemin (2001: 20) has repeated his dictum on numerous occasions. For an analysis of the 863 program, see Feigenbaum 1999; 2003: 141–3, 157–8, 161–2, 248–58; Gabriele 2002; Mulvenon 2005. For an official account of the program, see Zheng 2007.

60 These regulations are as follows. In October 2002, the CMC promulgated the Regulations on the Armaments Procurement of the PLA. In December 2003, the GAD issued five new provisions to further elucidate various aspects of the above regulations. They include the Provisions on the Management of Armaments Procurement Plans, the Provisions on the Management of Armaments Procurement Contracts, the Provisions on the Management of Armaments Procurement Modes and Procedures, the Provisions on the Management of the Examination of the Qualifications of Armaments Manufacturing Units, and the Provisions on the Management of the Centralized Procurement of Armaments of the Same Kind. These rules constituted a new statutory system for the procurement of armaments. See Medeiros *et al.* 2005: 38–9.

61 On 5 December 2002, COSTIND issued interim procedures regulating the application for a permit to research and to produce specific weaponry. As of Spring 2004, COSTIND had issued 318 permits to defense industry units and private firms in a trial project to implement the interim rules. According to Yu Zhonglin, vice-chairman of COSTIND, COSTIND planned to apply the regulations to every aspect of the Chinese economy as soon as possible. The issuing of permits would be open to any qualified enterprise, regardless of its ownership type. On 16 September 2005, COSTIND issued formal regulations governing the permit system. For the interim rules, see The Commission of Science 2003. For Yu's remark, see Yu 2004. For the formal regulations, see Commission of Science Technology and Industry for National Defense 2005.

62 Yu, Qian and Liu argued that civilian firms in China's high-tech industries would be capable of supporting the Chinese defense industry.

63 Interview, 19 August 2005, Hsinchu, Taiwan.

64 Interview, 17 September 2005, Shanghai, China.

65 Interview with a former member of Taiwan's Industrial Technology Research Institute (ITRI), 24 March 2005, Shanghai, China. Established in 1973, ITRI is Taiwan's premier industrial research institute, which has helped revolutionize Taiwan's semiconductor industry.

66 Interview, 17 September 2005, Shanghai, China.

67 Interview, 24 January 2005, Washington, DC, USA.

68 For an in-depth analysis of the defense conversion initiatives, see Ye 2003: 35; Medeiros *et al.* 2005: 6–8. By 2000, over 80 per cent of the aggregate output of Chinese defense enterprises was civilian goods, but few firms were profitable. This was chiefly because the civilian goods produced were of low quality and uncompetitive. According to a COSTIND official cited in the study by Medeiros *et al.*, China's defense industry ran a net loss in aggregate terms from 1993 to 2001.

69 It was also known as Jiangnan Radio Equipment Factory.

70 Interview, 17 September 2005, Shanghai, China.

71 A major joint venture during the 7th Five-Year Plan period was the establishment of Shanghai Belling in 1988. With its ownership shared by Alcatel, Shanghai Bell, and the Shanghai government, the firm was not commercially successful. See China Semiconductor Industry Association 2002: 153.

72 In 1989, the State 742nd Factory was merged with 24th Research Institute, Wuxi Branch, to form China Huajing Electronics Group Co, which later became the undertaker of Project 908. See China Semiconductor Industry Association IC Design Branch 2005.

73 Interview with Wang Qinsheng, chair of HED, and chair of CIDC, 30 August 2005, Beijing, China.

74 Interview, 19 September 2005, Jiangyin, Jiangsu, China. Other semiconductor joint ventures established during the 8th Five-Year Plan period included Shougang NEC and Advanced Semiconductor Manufacturing Corp (ASMC).

75 This refers to an 8-inch line operating on 0.35 micron process technology.

76 See also interview, 17 September 2005, Shanghai, China.

77 Interview, 19 September 2005, Jiangyin, Jiangsu, China. At the time of the interview, he was General Manager of Jiangsu Changjiang Electronics Technology, Secretary General of Jiangsu Semiconductor Industry Association, as well as Deputy Secretary General of China Semiconductor Industry Association.

78 Interview, 24 August 2005, Beijing, China. Wang became the CEO of Shanghai's Huahong Group in September 2005 and then joined SMIC as its CEO in 2009.

79 Interview with Peter Cheng-yu Chen, chairman of CSMC, 25 September 2005, Shanghai, China. See also interview with Robert N. Lee, president of CSMC, 31 August 2005, Beijing, China.

80 Interview, 27 September 2005, Shanghai, China. Zou left his post at Huahong NEC in April 2004. At the time of the interview, he was chairman of GSMC.

81 Interview with a former top executive of the firm who wished to remain anonymous, 30 September 2005, Shanghai, China.

82 Interview with a former high-ranking executive of the firm who wished to remain anonymous, 10 October 2005, Shenzhen, China; interview with a former top executive of the firm who wished to remain anonymous, 30 September 2005, Shanghai, China.

83 Interview with a former high-ranking executive of the firm who wished to remain anonymous, 10 October 2005, Shenzhen, China.

84 The eight design houses and institutes include Shenzhen-based State Microelectronics, Huahong NEC's IC design house, Huada, among others. See China Semiconductor Industry Association IC Design Branch 2005.

85 To develop its potential IC design customers, Huahong NEC thus engaged in venture capital business with partial funding from the central government, targeting technologically superior emerging IC design teams overseas. See interview with a former top executive of the firm who wished to remain anonymous, 30 September 2005, Shanghai, China.

86 This section will not analyze China-based players in the most value-added segments of the global chip value chain, which include semiconductor equipment, materials, EDA and semiconductor IP. This is because chip firms operating in China are primarily users, licensees, and buyers of products in these segments rather than producers and licensors. By the end of 2004, there were only 40 semiconductor equipment research and manufacturer units in the country generating sales value of 700 million RMB. In the same year, the sales value of silicon material reached 1.63 billion RMB. See China Semiconductor Industry Association and China Center of Information Industry Development 2005: 27–8; PricewaterhouseCoopers 2007: 44, 47–9.

87 This comprises discrete design, manufacturing, packaging, and testing.

88 The firm was subsequently renamed Huada.

89 However, foreign-related operations may not be recognized in the reported revenue totals for China's IC design sector. See PricewaterhouseCoopers 2007: 33–5.

90 They include SOEs, joint ventures and privately owned enterprises.

91 These include, for example, HiSilicon and ZTE Microelectronics Technology.

92 Primary examples include Spectrum Communications and Vimicro.

93 These include, for instance, Datang, Huahong NEC, and Fudan Microelectronics.

94 TD-SCDMA is a 3G format of choice for the national standard of 3G mobile telecommunication in China.

95 Representative cases include Spectrum Communications and Chipnuts Technology.

96 This company case study will be discussed in detail in Chapter 5 because of its connection to Taiwan.

97 NASDAQ originally stood for National Association of Securities Dealers Automated Quotations.
98 RISC stands for reduced instruction set computing, a CPU design strategy with relatively simple instructions.
99 Interview, 30 September 2005, Shanghai, China. See also interview with Wang Qinsheng, Chair of HED, and Chair of CIDC, 30 August 2005, Beijing, China.
100 The original data came from CCID, IEK, MediaTek Inc. in 2004 as well as *IC Insights* in 2005. An exchange rate of US$1 = NT$31.5 was used for the calculations in IEK, MediaTek Inc. data. An exchange rate of US$1 = RMB$8.27 was used for the calculations in the CCID data.
101 Interview with SMIC staff members, 30 September 2005, Shanghai, China.
102 During 2009, however, three MNC IDMs discontinued the operation of their facilities in China because of the worldwide economic recession. They were International Rectifier, National Semiconductor, and Qimonda.
103 Interview with the PRC engineer, 28 September 2005, Shanghai, China; interview with president of a foreign subsidiary, 27 September 2005, Shanghai, China.
104 One of Applied Materials' major expansions in China was the inauguration of its first product development center in Xian in March 2007. See Yan 2007.
105 Presentation by Yu Zhongyu, president of CSIA, at Stanford University, CA, on 6 November 2005.
106 The acronym ARM, first used in 1983, stood for "Acorn RISC Machine." In 1990, when the company was incorporated, the acronym was changed to stand for "Advanced RISC Machines." In 1998, the company name was changed to "ARM Holdings" at the time of the IPO.
107 The quote was attributable to David N.K. Wang, interview, 24 August 2005, Shanghai, China.

5 The migration of the Taiwanese semiconductor industry to China

I don't think there would be much of a semiconductor industry in China today had it not been for Taiwan. So they really should be grateful . . . Taiwan essentially went across the Strait and started to participate in the Chinese semiconductor industry. They took money over there. They took people and management skills.

(Klaus Wiemer, former president of TSMC, and former CEO of Chartered Semiconductor[1])

Through various forms of "internationalization," caliber and capital from Taiwan have entered mainland China and played important roles. GSMC, SMIC, TSMC and HeJian, for example, cannot shake off their links to Taiwan . . . [Taiwan president] Chen Shui-bian is unable to control the trend. Taiwan has already exerted its impact here [China].

(Wang Qinsheng, Chair of HED, and Chair of CIDC[2])

Introduction

The migration of the Taiwanese semiconductor sector to China is a constitutive component of the production globalization of the semiconductor industry, which has been discussed in detail in Chapter 3. This chapter summarizes my empirical findings concerning this understudied case of globalization.

Why is it important to study the migration in this geographical region? The answer to this question must begin with an assessment of Taiwan's position in the global chip supply chain. Taiwan has become one of the world's chip manufacturing powerhouses because of state-led industrialization efforts, technology transfers from foreign firms, and the return of US-educated and trained engineers and technology entrepreneurs since the 1970s. This development has allowed the country to be labeled the "Silicon Island" in East Asia (Meaney 1994).[3] In 2008, Taiwan's IC revenue totalled US$41.8 billion; it comprised US$11.6 billion in design, US$20.3 billion in manufacturing (which includes pure-play foundry and memory business), US$6.9 billion in packaging, and US$3 billion in testing (Taiwan Semiconductor Industry Association 2009). Today, Silicon Island's IC design capability, led by MediaTek, is ranked second in the world only after the USA. In 2007, its IC design houses contributed to 25 per cent of worldwide revenue (Taiwan Semiconductor Industry Association 2007). Led by TSMC and

UMC, Taiwan's pure-play foundry capacity is the strongest in the world, accounting for 67.2 per cent of global market share in 2008 (Industrial Development Bureau 2009). Measured by monthly 8-inch wafer production, Taiwan's foundry capacity remained the strongest worldwide in July 2009 (IC Insights 2009). In addition, Taiwanese foundries also contribute to the success of almost all the IC design start-ups in Taiwan (Chang and Tsai 2002)[4] and in North America (Macher *et al.* 1998; Taiwan Semiconductor Industry Association 2004; Defense Science Board Task Force 2005: 24; Engardio 2005; Hurtarte *et al.* 2007: 3–5; Taiwan Semiconductor Industry Association 2007).[5] Its packaging and testing sub-sector, led by Advanced Semiconductor Engineering, Inc. (ASE) and Silicon Precision Industries Ltd. (SPIL), is ranked first in the world. In 2006, its packaging firms had a 51.2 per cent of worldwide market share, whereas its testing companies accounted for 62.5 per cent of global business (Chen and Chan 2007). The efficiency and competitiveness of Taiwanese semiconductor firms have made them critical participants in the mainstream commercial chip sector of the global digital economy today. Moreover, some of these firms have engaged in production activities for the military market, both domestic[6] and foreign. For instance, seven of them have supported QML-38535 US firms in supplying semiconductors for American defense applications, thereby becoming part of the globalized US defense chip supply chain (Defense Supply Center – Columbus 2007).[7]

As Taiwan emerges as a key player in the global commercial and defense chip supply chains, the migration of its critical industry to China has raised concern for the state actors involved. This results from the importance of the industry to the Taiwanese economy, and the intimate links between semiconductors and national security. To mitigate what it sees as the potentially negative results of the migration, Taipei has introduced a series of policies since 2002. These regulations have conditionally lifted Taiwan's prohibition on domestic investments in semiconductor manufacturing operations in China with the intention of controlling the pace and scope of the migration (Table 5.1).[8] Under such a regulatory framework, to what extent has the actual process of migration occurred and to what extent can the geographical shift be controlled by Taipei? This chapter aims to address these questions through a study of the migration from empirical field research.

The survey is divided into five sections. Following a brief introduction, the second section analyzes the direction and scope of the migration across the Strait through cross-border technology transfers, investment, and talent flows. The third section examines the forces driving the migration, which include enabling, push and pull factors. The fourth section explores the motif of migration by presenting firm-level case studies in IC design and fabrication sub-sectors.[9] The final section summarizes the findings on the migration.

The key arguments in this chapter are three-fold. First, the scope of the migration has been extensive, the direction complex, and the causes manifold. Second, this process has been characterized, in part, by cross-border economic activities that have circumvented or violated Taipei's policies, eroding the power and capacity of the Taiwanese state. Finally, the migration has helped to develop China's chip industry, particularly the mainstream commercial sub-sector. This

Table 5.1 Taiwanese government regulations on domestic semiconductor investments in China

Segments of industry	Regulations
Industry overall	Without prior approval from the government, it is forbidden for any Taiwanese person* or entity established under the laws of Taiwan to directly or indirectly invest in a PRC entity in the chip industry.
IC design	A PRC entity, which has Taiwanese investment approved by Taipei, is forbidden to engage in IC design work, but is permitted to engage in sales activities and to offer technical support to customers.
IC fabrication	• Taiwanese investment in 6-inch and smaller wafer fabs in China has been permitted. • Taiwanese investment in 12-inch wafer fabs in China has been unconditionally banned, although Taipei in 2008 vowed to lift the ban soon. • Since August 2002, Taiwanese investments in up to three 8-inch fabs in China before 2005 have been permitted under three conditions. They are: (a) Any firm that plans to invest in such fab in China must have been engaged in mass production of 12-inch wafers in Taiwan for at least six months; (b) Any such investor is prohibited from transferring technology to the China plant for feature sizes smaller than 0.25 microns; (c) Any such fab is permitted to install used equipment that has been phased out in Taiwan. • Since December 2006, transfers of 0.18 micron process technology to the aforementioned 8-inch fabs in China have been permitted.
Packaging and testing	• Taiwanese investments in transistor packaging and testing operations in China are permitted. • Taiwanese investments in IC packaging and testing plants were banned until April 2006. • Since then, Taiwanese packaging and testing firms have been permitted to transfer technology to their plants in China for certain less advanced wire bond packaging and testing of ICs.

Sources: Interview with Chin Tan Huang, Executive Secretary of the Investment Commission at MOEA, 18 August 2005, Taipei, Taiwan; Howell *et al.* 2003: 73; Wu and Loy 2004; Investment Commission 2005a; TSMC 2006b; ASE 2007a; 2007b: 15–16; Reuters 2008c.

Notes: *(1) A Taiwanese person is deemed to have made an indirect investment in the PRC if the person invests in a non-Taiwan entity that is set up in a third area which then invests in a PRC entity, if the person has "control and influence" over the given entity. (2) The person is deemed to have "control and influence" over the said non-Taiwan entity under any one of the following three conditions: (a) the person holds more than 5 per cent of the equity of the entity or is the largest shareholder of the entity; (b) the person invests more than US$200,000 in the entity; or (c) the person acts as a director, supervisory director or president of the entity.

impact has been evidenced by the flow of much-needed talent, technology, and investment across the Strait and the resultant rise of start-ups which have become China's leading firms in IC design and fabrication. This has also been evidenced by the build-up of long-term assets to the local industry which the migration has created through talent training programs and MPW services to local IC design customers. The migration has thus contributed to the shift of IC capabilities across the Strait with significant long-term ramifications.

Migration across the Strait: an overview

This section summarizes my research findings on the scope and direction of the migration of the chip industry, involving flows of technology, investment, and human resources.

Scope and direction

The migration, which began in the late 1980s and has gathered momentum since 2000, has expanded across the entire span of the IC industry. It has covered IC design, fabrication, and packaging and testing stages of the supply chain (Figure 5.1). The extensive scope of this migration has constituted the new outsourcing trend in the global industry. It has also overturned some conventional views of the economic phenomena, which suggest that the migration has been confined largely to the IC fabrication sub-sector (Howell *et al.* 2003: 67–76; Chase *et al.* 2004).[10]

Even the earliest cases of migration in this industry were extensive in scope. From 1989 to 1999, there was a steady, albeit small-scale, flow of Taiwanese resources into the three main chip sub-sectors in China. The first known case involved the establishment of a Taiwanese packaging and testing company in Zhuhai in 1989 (Wang 2001a; Hsu and Wang 2002; China Integrated Circuit Yearbook Editing Committee 2005: 281; Nanker Group n.d.).[11] In 1994, Hangzhou Youwang Electronics Co., Ltd., the first known Taiwan-invested IC design

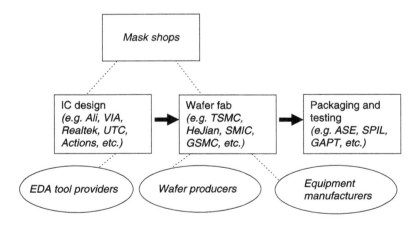

Figure 5.1 The extensive scope of the migration.

Sources: Interviews and secondary data.

Note: The migration has gathered momentum since 2000, although the first known case of migration dates back to the late 1980s.

start-up in China, was incorporated. As will be discussed later, the firm gave rise to two strong IC design powerhouses in China in 2004 and beyond. In 1995, Taiwanese American Tony Liu relocated to Shanghai to lead the joint venture ASMC, marking the first wave of high-tech returnees from the USA to China. In 1996, the first known Taiwan-invested IC manufacturer was incorporated; it was a joint venture which revamped a former state-run 4-inch wafer fab (Fujian Provincial Development and Reform Committee 2007).[12] In 1997, Taiwan-invested CSMC was incorporated. It pioneered Taiwan's successful pure-play foundry model in China, a model which was to be emulated by start-ups on the mainland. In the late 1990s, Ali, a Taiwanese IC design house, began to include China as part of its globalized production base.

This migration has continued to be extensive in scope in recent years. In the IC design sub-sector, Taiwanese firms and investors have continued to build new production capacities in China. By 2004, Taiwan's top ten IC design powerhouses had set up various operations in China.[13] Taiwanese entrepreneurs have also invested in start-ups in China, some of which have subsequently become rising stars in the market. More importantly, the trend toward globalization in this sub-sector in the cross-Strait context is expected to continue. Speaking in 2005, a senior executive at a leading Taiwanese IC design firm noted:

> Certainly on the IC design side, it is starting, and in the next three to five years, and probably earlier in the next two to three years, it will get bigger and bigger and you will see quite a rapid migration across to China in terms of building new capabilities . . . A lot of the companies have already set up design centers. They have already built up software teams, etc. Therefore, that is already happening, and will gather pace.[14]

In August 2009, the same executive reiterated his observation made four years ago.[15] Hence, the IC design production shifts from Taiwan to China have contributed to the growth of offshore design operations in Asia since the mid-1990s, as observed in the studies done by Ernst (2005), Brown and Linden (2005: 298–315).

Of all the segments of China's chip industry on which the migration from Taiwan has exerted an impact, the IC fabrication sub-sector is the most obvious. By the end of 2006, four out of the top five fabrication companies in Taiwan had invested in China, or received approval by Taipei to do so (Taiwan Semiconductor Industry Association 2007). They include TSMC, UMC, Powerchip, and ProMOS.[16] Some senior Taiwanese engineers have initiated various foundry start-ups in China. Others have become top leaders in existing chip-making firms in China. Almost all of China's top chip fabrication companies today, especially in the foundry business, have had significant Taiwanese input. The eight foundry companies in China that were at the top in 2006 benefited from various resource transfers from Taiwan, particularly in human capital.

The extensive scope of this migration of the IC fabrication sub-sector has been noted universally by industry insiders. According to a senior engineer, over 20 proposals for China-based foundry initiatives, which he had read by December

2004, relied on different degrees of Taiwanese involvement. As he put it: "We supply everything: people, money, and know-how. Not officially. But under the table everything involves the Taiwanese people, especially in the foundry business, such as HeJian, SMIC, and GSMC."[17] According to Jann Hwa Hsu, president of HeJian, a start-up established in Suzhou, China, with assistance from UMC, "The Chinese chip fabrication industry did not take off until the expansion of the Taiwanese operations here."[18] His view was echoed by his counterparts in six additional top chip-makers in Taiwan and China,[19] as well as senior US industry players whom I interviewed.

Taiwanese firms and investors have also built up capacities in the packaging and testing sub-sector in China in recent years. Company data on the Taiwan Stock Exchange website indicate that almost all of the major Taiwanese firms have included China as part of their increasingly globalized production bases. By the end of the first quarter of 2009, the accumulated offshore investments in China-based plants remitted by ASE and SPIL exceeded NT$15.2 billion. ASE has become the most influential Taiwanese packaging and testing firm operating in China. Its joint venture with NXP Semiconductors (Wang 2007; ASE 2007b; 2009: 20) and various wholly owned subsidiaries scattered throughout China's greater Shanghai area have made the "ASE Shanghai Development Plan," as envisaged by a company executive in 2004, a reality. In May 2008, it acquired 100 per cent of Weihai Aimhigh Electronic Co. Ltd., now known as ASE (Weihai) Inc., from Aimhigh Global Corp. and TCC Steel (ASE 2009: 20). Accordingly, a growing proportion of ASE's revenue is expected to derive from its China-based production hubs. The company estimated that after it acquired Global Advanced Packaging Technology Co. Ltd. (GAPT), another Taiwan-invested firm in China, about 10 per cent of its revenue would result from its production capacity on the mainland (ASE 2007a).[20]

The direction of the migration has also been found to be far more complex than what was formerly thought (Table 5.2). Subsidiaries of existing Taiwanese chip firms or newly founded Taiwanese enterprises in China are not the only destinations for the Taiwanese resource transfers. The migration has also benefited the development of other types of semiconductor operations in China, which form the backbone of the commercial and defense chip sub-fields of the Chinese industrial base. These comprise newly established MNCs with hybrid sources of talent, technology, and capital, and existing or newly founded joint ventures. They also include university-run semiconductor operations, local research institutes, and incumbent or former state-owned enterprises, some of which have long been associated with Chinese defense microelectronics production activities. The concrete ways in which the migration has contributed to the Chinese defense chip industrial base will be discussed in detail in Chapter 6 where the role of Taiwan in Chinese defense microelectronics activities is analyzed.

Summary of technology, investment and talent flows

The migration can be further analyzed in terms of cross-border flows of technology, investment, and talent. It is argued that while the technological and

Table 5.2 The complex direction of the sectoral migration

Destinations of sectoral migration in China		Primary examples
Chinese commercial chip industrial base	Taiwanese subsidiaries or Taiwanese-owned newly founded enterprises	Nanker Group (1989); Hangzhou Youwang Electronics (1994); VIA Technologies China (2000); Ali Shanghai (2000); GAPT (2000); Realsil Microelectronics (2002); NSSI (2002); TSMC Shanghai (2003); ProMOS Chongqing (2007)
	Newly founded MNCs	GSMC (2000), SMIC (2000), HeJian (2001), Actions (2001), Spreadtrum (2001)
	Existing and newly founded joint ventures	ASMC (1995–2007); Fujian Fushun Microelectronics (1996); Silan Microelectronics (1997); Shanghai SIM–BCD (2000); CAS–CSMC joint venture (2005); Shanghai Huahong NEC (2005–07); ASEN Semiconductor (2007); HeJian–Elpida joint venture (2008)
Chinese defense chip industrial base	Incumbent or former state-owned enterprises	Huajing (1998); Jiangsu Changdian Electronics Technology Co. Ltd., and Tian-shu-hua-tian Microelectronics Co. Ltd. (2005)
	University semiconductor operations	Tsinghua University (1999–2003)
	Research academies and institutes	Huajing (1998); Shanghai SIM–BCD (2000); CAS–CSMC joint venture (2005)

Sources: Interviews and secondary data.

Notes: (1) The line between the Chinese commercial and defense chip sub-sectors has become blurred. Some of the semiconductor operations, such as Huajing, have engaged in both commercial and defense chip production activities. Hence, the division made between the two sub-sectors in the table is arbitrary. The cases of the Shanghai SIM–BCD and CAS–CSMC joint venture are listed under both categories of commercial and defense chip sub-sectors. Both SIM and CAS have long been associated with Chinese defense microelectronics production activities. Moreover, the new joint ventures with Taiwanese input mainly make chips for the commercial market. (2) The number identified in the bracket on the right-hand side of the table shows either the date when a given entity was incorporated or the period Taiwanese input (in terms of technology, talent, and/or investment) was identified.

investment flows have been key to the migration, human resource transfer has been the most important dimension of the geographical shifts.

Technology

The flow of Taiwanese semiconductor technologies to China has followed different paths in the three sub-sectors involved. Four patterns of technology flow have emerged in IC fabrication and/or packaging and testing sub-sectors, whereas three have become apparent in IC design. In the IC fabrication and/or packaging and testing sub-sectors, the first mode of technology flow, and the most common among the firms studied, involves the transfer of mature and low-end technologies from Taiwanese companies to their subsidiaries or local outsourcing partners in China. David N. K. Wang, executive vice-president of Applied Materials, observed in 2005 that the majority of the semiconductor process technologies that had been transferred from Taiwan to China were "not very important" because they were not global state-of-the-art.[21]

In the IC manufacturing sub-sector, most of the firms have transferred process technologies for 6-inch and/or 8-inch wafer fabrication instead of leading-edge 12-inch wafer production expertise. Primary examples include TSMC and ProMOS. In 2002, TSMC shipped part of its second-hand equipment for 6-inch wafer fabrication to Ningbo Sinomos Semiconductor Incorporation (NSSI), a Taiwan-invested foundry start-up in China. TSMC also transferred related process technologies to NSSI, and about 40 former TSMC process engineers joined the start-up. Dubbing NSSI TSMC's outsourcing partner, the chairman of the start-up said: "The best way is to ask the Taiwanese to transfer mature technologies here so that we don't need to start from scratch."[22] Moreover, TSMC (2006b) has also transferred mature process technologies (0.35–0.18 microns) for 8-inch wafer fabrication to its wholly owned subsidiary in China, which was incorporated in 2003. In contrast, the company has retained its state-of-the-art technologies at home, invested heavily in R&D, and continued to expand its high-end production in Taiwan. By December 2007, it had shipped its one-millionth 12-inch 90 nm wafer in just 53 months, and its R&D team at company headquarters had reported manufacturing gains in 32 nm low-power foundry technology (Mokhoff 2007; TSMC 2007e). In addition, Taiwanese memory makers ProMOS and Powerchip were approved by Taipei in 2006 for their China-bound investments.[23] In 2007, ProMOS opened a wafer fab in Chongqing (Zhang 2008).[24] It has transferred matured process technologies for 8-inch wafer fabrication to its subsidiary in China, while retaining high-end ones at home. The first pattern of technology flow has also dominated the packaging and testing sub-sector. ASE and SPIL, for instance, have transferred their low-end technologies to their subsidiaries in China while retaining high-end ones in Taiwan.[25]

In the second pattern of technology flow, manufacturing operations in China which have relied on Taiwan for talent and investment have acquired second-hand equipment from the international market for their production activities that use mature technologies. The technological know-how, nevertheless, has transferred from senior Taiwanese engineers to their junior local counterparts. A vast majority of other Taiwanese-invested or -run IC fabrication and/or packaging and testing operations in China have followed this path. For instance, CSMC, in 2004, purchased Chartered Semiconductor's second-hand equipment for 6-inch fabrication and signed related technological licensing agreements with the Singapore-based foundry.[26]

In the third pattern, start-ups in China's IC fabrication sub-sector, with apparent input from Taiwan in terms of talent, investment and even technology, have further expanded their sources of technologies in order to move towards high-end production. These technology transfers have originated from global chip giants, not limited to those headquartered in Taiwan, and from advanced semiconductor equipment suppliers. Consequently, they have obtained high-end and even near state-of-the-art technologies which are beyond the confines of Taipei's policies. SMIC and HeJian are two prime examples. Both have benefited from transfers of technological know-how from Taiwan at least during the initial stages of their development, as will be detailed in the section on case studies. By mid-2008, however, both firms had advanced their technologies beyond the confines of Taipei's rules largely by involving other foreign technology partners. By March 2008, SMIC was engaged in 12-inch wafer fabrication activities, and was ready to provide an IC manufacturing service for 0.35 micron to 65 nm and finer line technologies (Lapedus 2008; SMIC 2008a). In addition to its continuous reliance on UMC's technologies, HeJian announced in March 2008 that it had decided to work with Japan's Elpida to build a 12-inch memory fab in China (Sung and Chan 2008; Reuters 2008a).

In the fourth pattern, Taiwanese companies, in spite of initial transfers of low-end technologies to their China-based subsidiaries approved by Taipei, have subsequently transferred high-end ones to the subsidiaries engaging in technology transfer which are not approved by Taipei. This has closed the technology gap between company headquarters and offshore operations. For example, a Taiwanese packaging and testing company set up a subsidiary in Shanghai in 2000 with prior approval from Taipei. The plant initially focused on low-end production activities for small signal transistors, which were permitted by Taipei. By September 2005, however, it had engaged in higher-end production activities that were forbidden by the Taiwanese authorities. As the head of the subsidiary admitted, "In terms of technology, what we are doing here is similar to what we are doing in Taiwan."[27]

In IC design, however, the technology flow has occurred in three major patterns. The first and second dictate the relationships between Taiwanese firms and their subsidiaries in China. The third applies to IC design start-ups in China, which are established as independent companies under the helm of senior Taiwanese engineers.

The first pattern involves a limited flow of IC design expertise. This pattern has been followed when Taiwanese subsidiaries in China are engaged in business activities unrelated to actual IC design, which include sales activities or field applications engineering tasks.[28] At least two of the top ten IC design houses in Taiwan, whose senior executives I interviewed, have followed such a pattern.

In the second pattern, which is the most common one followed by the firms that were studied, a substantial flow of IC design know-how, focusing on low-end segments of the design flow, has taken place. This occurred when subsidiaries in China undertook low-end and fragmentary tasks, while firm headquarters in Taiwan performed high-end and integration tasks. Such a transfer of IC design technological know-how, a key component of which involves talent training programs, enables subsidiaries to accomplish their IC design tasks. Consequently a division of labor between the firm's headquarters and its subsidiaries is established. The dominance of this practice among Taiwanese IC design houses results from the lack of sufficient IC design expertise in China, a desire to retain the firm's competitiveness, and poor IP protection in China.[29]

Examples of this practice abound. In one of Taiwan's top ten IC design houses, which has run more than four subsidiaries in China, the firm's headquarters in Taipei dictates the way any given IC design project is divided among the company's geographically dispersed production sites. "Taipei dominates everything. Most of our work here involves software, whereas core and high-end design and R&D are carried out in Taiwan," revealed the head of the firm's major subsidiary in China.[30] In another Taiwanese IC design house, which has operated a subsidiary in Suzhou, China, the firm's headquarters is responsible for core design tasks.[31]

The IC design tasks that these subsidiaries undertake include software work, labor-intensive and low-end tasks in any given design flow (e.g. routing, placements and testing), and small modules of design assignments. Software outsourcing, which is common among some Taiwanese firms, has been driven by the availability of local software engineers in China. As the president of one IC design house put it, "Our investment in China chiefly focuses on software because the human capital in China, in the area of software engineering, is good." His company, as of October 2005, had hired more than 200 engineers in China.[32] The most common practice among Taiwanese firms has been to "dump" the low-end placement/routing and testing tasks of any given design flow on their subsidiaries in China, and to keep high-end specification, architecture (layout) and synthesis work in Taiwan.[33]

Other tasks that subsidiaries in China have often undertaken include the small modules of a given design project. As a general observation, the president of an IC design house said, "The Taiwanese are careful in operating their subsidiaries in China. They divide a single design project into several small modules, outsource some of the modules to their subsidiaries in China, and retain integration work at the headquarters in Taiwan."[34] At least two of Taiwan's top ten IC design houses I interviewed have retained integration tasks at the firm's headquarters. One firm's subsidiary in Shanghai normally undertakes entry-level tasks for a certain segment of the design flow. Once these assignments are completed in China, the firm's headquarters in Hsinchu, Taiwan, will undertake related integration tasks. The company has not requested its subsidiary to develop any core IP, and gives two main reasons for this: the lack of sufficient IC design expertise among its R&D team in China, and a desire to retain the firm's competitiveness.[35]

Taiwanese IC design houses which have followed this second pattern of technology flow have been compelled to transfer necessary IC design expertise to their production hubs on the mainland in order to empower their Chinese employees to accomplish the rudimentary IC design tasks. Given the magnitude of such a practice in the sub-sector, a substantial flow of IC design technological know-how has occurred in spite of the fact that the related know-how transfers have tended to be in areas away from the core competencies of the Taiwanese firms. This finding echoes conventional views on the globalization of IC design production activities, which hold that firms often prefer to keep core, value-added R&D activities at home while globalizing their production operations (Henderson 1989: 45–8, 56–8; Brown and Linden 2005: 305–6; Electronic Engineering Times 2006; Macher *et al.* 2007).

The third pattern of technological know-how flow in the IC design sub-sector applies to start-ups in China that are established as independent companies by senior Taiwanese engineers. In these firms, technological expertise transfers have occurred primarily because of human capital flow. These highly experienced Taiwanese engineers have turned their China-based companies into training grounds for their educated but inexperienced local counterparts, thereby resulting in the transfers of IC design know-how.

In sum, various types of technology transfer have occurred as part of the migration that is under examination here. In the IC fabrication and packaging and testing sub-sectors, the transfer of mature technologies has been the most common practice.[36] Nevertheless, there are important exceptions. Some China-based start-ups, despite their initial dependence on technology transfers from Taiwan, have moved towards the global state-of-the-art by relying on technology partners other than those in Taiwan. There are also instances in which Taiwanese subsidiaries in China have obtained high-end technologies from the firm's headquarters, closing the gap between onshore and offshore operations. In the IC design sub-sector, the most common practice among the firms that I have studied is that the firm headquarters in Taiwan control the most value-added parts of the production flow, whereas subsidiaries in China perform low-end tasks. This leads to a substantial transfer of IC design knowledge across the Taiwan Strait at the infra-firm level, mainly focused on the low-end segments of any IC design flow. However, there are also exceptions. In China-based start-ups that were established by experienced Taiwanese engineers, technological transfer has occurred because of human capital flow. The migration of technology has thus involved a complex array of possibilities, almost all of which are assets to China's nascent IC industry.

Investment

The influx of Taiwanese capital into China's semiconductor industry has been another important part of this migration. The actual amounts concerned may have been underestimated in the official statistics because of the tortuous paths the investments followed in order to circumvent Taipei's scrutiny, according to the industry insiders I spoke with.

According to CCID, Taiwanese capital represented the largest bulk of total foreign investment in China's chip industry in 2005. It was followed by US, South Korean, European, and Japanese capital (Song 2007). A survey (Fu 2005: 49–51), conducted by the Shanghai Municipal Government, further revealed the role of Taiwanese capital in China's IC sector. Among the 124 IC design houses operating in Shanghai in 2004, which generated 15.5 per cent of the Chinese subsector's sales revenue in the same year, 22 per cent had been partly backed by Taiwanese investment, 17 per cent were wholly Taiwanese-owned operations, and 5 per cent were Sino-foreign joint ventures. Thus, partially or wholly Taiwanese-invested IC design firms in Shanghai generated 6.1 per cent of the design sales revenue in China in the same year.

Taiwan's MOEA data (Investment Commission 2007; 2008) indicate a steady growth in approved indirect Taiwanese investment in China in the electronic parts and components manufacturing industry, which includes semiconductor manufacturing (Figure 5.2). Between 1991 and 2007, Taipei approved 2,046 applications, which amounted to US$10.36 billion of indirect investment in the sector. By the end of 2007, the industry surpassed the computer, electronic, and optical product manufacturing sector, to emerge as the largest area for approved indirect Taiwanese investment in China. A substantial part of the investment has targeted semiconductor production since 2002, the year Taipei began to conditionally lift its ban on Taiwanese semiconductor investment on the mainland.

Industry insiders have estimated that the actual amount of Taiwanese investment in China's semiconductor industry may have been higher than is recorded in

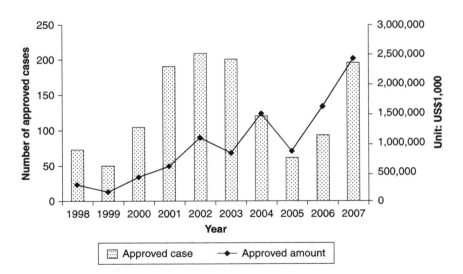

Figure 5.2 Approved indirect Taiwanese investment in China in the electronic parts and components manufacturing industry, 1998–2007.

Sources: Investment Commission 2007; 2008.

the official statistics.[37] This is because the pertinent capital has often followed a convoluted path in order to circumvent regulations at home. As the head of a Taiwanese subsidiary in Shanghai expressed it, "Loads of Taiwanese investment has entered China through many hands and places to skirt Taipei's regulations."[38]

Almost all the senior industry insiders on both sides of the Strait whom I interviewed have underscored the importance of Taiwanese investment in China's semiconductor industry. Zou Shichang, chairman of GSMC, expressed it bluntly: "SMIC, GSMC, and HeJian are from Taiwan. Basically they are established based on investment from Taiwan."[39] In brief, Taiwanese capital has indeed, legally and illegally, supported various types of semiconductor operations in China over the course of this high-tech industry migration.

Talent

Compared to technology and investment flows, talent transfer from Taiwan's competitive semiconductor industry to China has been regarded by industry insiders as the most important component of the migration.[40]

More than 3,000 semiconductor engineers reportedly left Taiwan in the second half of 2001 to work in China when foreign investment accelerated in the sector (Geppert 2005).[41] As of November 2005, at least 1,200 Taiwanese engineers and managers were working in various China-based semiconductor firms whose senior executives I interviewed.[42] A large proportion of them were concentrated in the IC manufacturing sub-sector. SMIC alone housed about 650 Taiwanese employees, and this company was followed by HeJian (200), TSMC (Shanghai) (100), GSMC (100), NSSI (50), and CSMC (20). As senior engineers and managers, these Taiwanese professionals helped train about 13,000 local engineers and technicians in the six companies under study.[43] Concurrently, a small proportion of them were with IC design and packaging and testing firms. Two of China's indigenous packaging and testing companies even absorbed a dozen senior managers from Taiwan.[44] Follow-up field research conducted in 2009 indicated that the number of Taiwanese professionals operating in China's semiconductor sector has increased since late 2005, particularly in the IC design sub-sector.[45] The majority of these Taiwanese professionals, especially the highly skilled managers, had migrated to China because of the key institutional drive behind them, namely the globalization of production spearheaded by these companies either through direct control and ownership or through regularized contractual relationships (Salt 1997: 9–10, 16–18; Ietto-Gillies 2003: 140–4).

Industry insiders have viewed the aforementioned talent transfer as the most critical aspect of the migration for the following reasons. This development has engendered the transfer of important technological and managerial know-how and business connections to the mainland, advancing the development of the Chinese industry. Taiwanese talent transfer has also helped to quench China's thirst for much-needed semiconductor know-how and human resources, the lack of which has long inhibited the country's pursuit of strong indigenous semiconductor capability (Hu 2001: 382). These Taiwanese professionals have also helped to train a large number of local engineers and managers. In the eyes of Xia Zhong Rui, former CEO of Huahong NEC, "The development of China's semiconductor industry in recent years cannot be separated from a move by Taiwanese entrepreneurs to pursue careers in China."[46] In 2006, China's top eight foundry companies were under the control of Taiwanese professionals (Table 5.3), as was also true of China's No. 1 IC design house, Actions.

Given the security tension across the Strait and the centrality of semiconductors to national security, it is nonetheless ironic to see that Taiwanese high-tech personnel have been leading China's endeavors to upgrade its indigenous

Table 5.3 Taiwanese high-tech professionals in China's major pure-play foundries

Ranking in 2006	Foundries	Sales revenues in 2006 (100 million RMB)	Key Taiwanese executives	Total workforce (Taiwanese)
1	SMIC	113.5	Richard Chang, CEO (2000–09); David N.K. Wang, CEO (2009–)	8,400 (650)
2	Huahong NEC	28.48	David N.K. Wang, CEO (2005–07); Tzu Yin Chiu, CEO (2009–)	–
3	HeJian	23.5	Jann Hwa Hsu, President (incumbent)	1,000 (200)
4	ASMC	14.36	Tony Liu, President (2000–07); Hsueh Cheng Lu, President (incumbent)	–
5	TSMC Shanghai	12.87	Y.C. Chao, President (2003–09); J. Chen (2009–)	1,000 (100)
6	GSMC	12.22	Winston Wang, Co-founder and CEO (2000–04); Shi-chung Sun, Chief Technology Officer (2005–08)	1,300 (100)
7	CSMC	9.55	Peter Chen-yu Chen, Chairman and CEO (incumbent); Robert N. Lee, President (1998–2007)	20 (1,600)
8	SIM–BCD	5.20	Tung-Yi Chan, CEO (incumbent)	–

Sources: CCID revenue data cited in Xinhua-PRNewswire 2007; company reports; workforce data from interviews.

semiconductor capabilities in recent years. As Robert Haavind (2003), editor-in-chief of *Solid State Technology*, put it, "It is ironic that the Taiwanese, the people most potentially threatened if hard-line factions gain ascendancy in China, are leaders in upgrading capabilities there."

In a nutshell, the migration has been extensive in scope and complex in direction. The technology flow has involved the transfer of IC manufacturing and design knowledge, the majority of which has been concentrated in mature technologies and low-end design expertise. This flow has contributed to a division of labor in production at the intra-firm level. Nevertheless, there are exceptions to this trend. In some cases, firms in Taiwan have transferred core technologies to their subsidiaries in China. In others, start-ups in China which were established with Taiwanese input have continued to pursue global state-of-the-art technology by relying primarily on foreign partners other than those headquartered in Taiwan. While Taiwanese investment has been an important part of the migration, human resource transfer has been the most critical dimension of the geographical shifts.

Migration across the Strait: driving forces

From the standpoint of the industry players, what are the driving forces behind the migration of this sector to China? Although motivations are always hard to pin down, the weight of evidence suggests that they are a confluence of enabling, pull and push factors. The enabling factors, identified in Chapter 3, have been an important backdrop for the migration. The two primary pull factors have involved a desire to gain access to China's market and an aspiration to exploit China's location-specific resources. The three secondary pull factors have included the desire to reduce costs, policy incentives in China, and political considerations. The major push factor results from concerns over the long-term shortage of engineering manpower in Taiwan. Below I shall elaborate on the manifold forces at play.

Access to China's market

The top driver behind the migration is the desire to penetrate a lucrative market, rather than the desire to lower production costs. In the IC design sub-sector, many firms' executives have identified a desire to gain access to China's rising market as the major driving force behind their decision to set up production operations on the mainland.[47] The size of China's IC market reached US$40.8 billion in 2005 in terms of overall consumption, making China the world's largest IC market for the first time. In 2009, its share of the US$226.3 billion global semiconductor market increased to 41 per cent (PricewaterhouseCoopers 2010: 9). A senior executive at one of Taiwan's leading IC design houses suggested in 2005 that the desire to penetrate the China market had pushed his company to start investing on the mainland in 2000.[48] The head of the company's subsidiary in Beijing further identified the market magnet as China's insatiable demand for chip components in recent years. Such a demand was buoyed by the country's dominance in electronic system production for both domestic and export markets and its fast-growing semiconductor end-use markets, as detailed in Chapter 4. Regarding the huge gap between China's surging IC demand and its limited domestic IC supply, as discussed in Chapter 4, as a "good market opportunity," the aforementioned firm decided to begin offshore operations on the mainland in order to penetrate the local market.[49]

Another Taiwanese IC design house set up a subsidiary in Suzhou, China, for similar reasons. In 2005, the company chairman argued: "China's market represents tremendous opportunities. One cannot grasp these opportunities without a physical presence here. This is in spite of the fact that your presence here may or may not guarantee your success."[50] As the company's president put it, "Because everybody pins high hopes on China's booming market, we cannot afford to be excluded from this market. We want to grab these opportunities by starting to deploy our workforce in China."[51]

Related to these IC design firms' aspiration to penetrate China's local IC market is their desire to be close to their customers in China, which include Taiwanese electronic systems houses that have already shifted their production operations to the country.[52] In one instance, a leading Taiwanese IC design house decided to set up subsidiaries in China in order to support its Taiwanese customers on the mainland, such as motherboard and notebook makers.[53] The close-to-the-customer consideration also drove another Taiwanese IC design house to run a subsidiary in Shanghai. According to the vice-president of the consumer IC firm, it was "Because many of our Taiwanese customers have already moved there, that we need to make sure we will be able to communicate with them efficiently in order to offer them customer services through our presence there."[54]

In the IC fabrication sub-sector, the majority of senior executives I interviewed identified China's market as the main driving force behind their strategic move to extend their geographical reach across the Strait.[55] Insiders have identified UMC's desire to gain access to China's market as a major reason for the company's move to establish HeJian in China. According to Jackson Hu, CEO of UMC, "We've decided to help to establish HeJian because we want to understand China's market and to get close to China's customers through HeJian's operations."[56] The president of HeJian echoed Hu's view. He did so by identifying China's market and its engineering talent as the two major pulling forces behind HeJian's birth: "We came chiefly because of the market and the talent." He referred to China's market as the country's insatiable thirst for ICs. More importantly, UMC feared that Beijing might change its policy one day to request local IC users to use China-based foundries alone. In view of China's large IC market, UMC could not afford to lose the market because of a possible policy change. This fear of market loss reinforced UMC's resolve to set up HeJian. As the president of HeJian admitted:

> Mainland China is a society based on rule by the people. What if one day the government here requires IC users to purchase locally fabricated semiconductors? To deal with such a possible policy change, which would benefit fabs in China rather than those outside, we decided to set up a production base here. The decision would entitle us to be viewed favorably in a market this size if the policy change occurs.[57]

Senior executives involved in TSMC's China expansion project have argued that the mainland's potential market was the primary reason for the firm's decision to build a fab in China. According to F.C. Tseng, vice-chairman of TSMC and chairman of TSMC Shanghai, "[We went there] in order to get our market share in mainland China's domestic market."[58] The head of the company's China subsidiary said that TSMC built the fab in Shanghai because of China's "future market." He construed this as the ability of China's local IC design houses and IDMs to place orders with TSMC, which would give the company access to a market segment that was different from China's IC consumption market.[59]

Although market considerations have driven TSMC to build a fab in China, its operation in Shanghai, as of mid-2005, had not received many orders from local customers. According to Tseng:

> China's local market seems enormous, but it may not be the case in reality. Such a development may be uncertain ... We went there because of the market, but so far the market is not yet in sight. Its rise will depend on the development of local IC design houses.

At the time of the interview, the fab was producing 8-inch wafers using 0.35 and 0.25-micron processes, which would be more advanced than those in demand by a majority of local design houses.

However, Tseng stressed that a large foundry market in China could arise, due to the combined forces of China's enormous demand for IC consumption and the continuous efforts made by local IC design houses to strengthen their capabilities. Once China's foundry market expanded and the demand for high-end foundry services increased accordingly, Tseng said, it would be highly likely that some of the local IC designers would prefer working with fabs inside rather than outside China. In view of this possibility, his company decided to go ahead with its China project. "Now we are based in Shanghai, observing the development of local IC design houses while engaging in some production activities," he concluded.

Taiwanese foundries like TSMC have started production in China, using more advanced technologies than those demanded by the majority of local design houses and IDMs. Their aim is to penetrate the local market in the long run rather than in the short run. As a director at a US IDM subsidiary in China observed in 2005:

> The "close to the market" argument is wrong as a short-term driver behind the decision by the Taiwanese foundries to set up operations in China using relatively advanced process technologies. But it is perhaps correct in the long run. Currently, 6-inch wafer fabs using 1-micron or even 1.5-micron process technology would suffice to meet a majority of the domestic IC demand.

He concluded that in the short run, these foundries would churn out semiconductors mainly for the international rather than the domestic market.[60] His observation has been substantiated by my interviews with the senior executives of four major Taiwanese-owned or -run foundries that used process technologies for 8-inch wafer fabrication in China in 2005. These firms exported more than 80 per cent of their chips to the international market at that time. They included SMIC,[61] GSMC,[62] HeJian[63] and TSMC Shanghai.[64] By contrast, Taiwan-related foundries that were engaged in 6-inch wafer fabrication in China largely served China's domestic market. The case of CSMC was illustrative. From September 2004 to September 2005, over 70 per cent of CSMC fabricated chips in its 6-inch wafer fab were for domestic consumption, with customers including six of China's top ten fabless design houses.[65] Table 5.4 indicates the varied degree of dependence that these foundries had on China's domestic market for their revenue.

Aside from the cases of UMC and TSMC, other critical Taiwanese chip-makers involved in establishing foundry operations in China have also emphasized the market driver behind their decisions. They include the CEO of SMIC and the Chairman of CSMC.[66] Even in the packaging and testing sub-sector, Taiwanese firms have singled out the desire to gain access to China's market as one of the major explanations for the extension of their geographical reach across the Strait. One of them set up a subsidiary in Shanghai in 2000 in order to increase its local market share. Prior to that, relatively low labor costs in China had driven the company to outsource part of its production to its local partners on the mainland for a few years. By 2005, the Shanghai subsidiary enabled the firm to increase its local market share. However, for that subsidiary, local market access would have been problematic as local firms catch up fast and foreign competitors have established production operations in China to increase their market share.[67] In another case, the expectation of an expanding Chinese market and a desire to have a fair

Table 5.4 Fabricating chips for export? A comparison of representative first-tier and second-tier Taiwanese foundries in China in 2005

Representative Taiwanese-owned or -run chip makers in China		*Sales revenue distribution*	
		International (%)	*Domestic (%)*
First-tier players (8″ and/or 12″)	SMIC	90	10
	GSMC	> 80	< 20
	HeJian	80	20 *
	TSMC Shanghai	90 **	10
Second-tier players (6″)	CSMC	30	70

Sources: Interviews with the following top players: a public relations manager at SMIC, 30 September 2005, Shanghai, China; Zou Shichang, the chairman of GSMC, 27 September 2005, Shanghai, China; the president of HeJian, 21 September 2005, Suzhou, China; F.C. Tseng, the vice-chairman of TSMC and the chairman of TSMC Shanghai, 30 June 2005, Hsinchu, Taiwan; and Peter Cheng-yu Chen, the chairman of CSMC, 25 September 2005, Shanghai, China.

Note: Those marked with (*) include local IC design houses and foreign design houses which require local chip delivery in China. Those marked with (**) include foreign customers who require local chip delivery in China.

share of this market drove a heavyweight Taiwanese businesswoman to invest in a packaging and testing company in Shanghai in her private capacity in 2000.[68]

More importantly, the desire to help China-based customers deal with the issues of inventory and time to market, which is related to market considerations, has also enticed some Taiwanese chip firms to set up operations in China. Generally, electronic systems houses, as IC users, regard IC inventory control and time to market issues as crucial to their competitiveness. The fewer ICs that are stored in warehouses next to their manufacturing hubs, the easier it is to minimize related costs.[69]

Similar inventory and cost considerations have driven many of HeJian's customers, who sell their ICs to electronic systems manufacturers in China, to prefer using fabs on the mainland.[70] This is because it will take a minimum of three to four days to deliver ICs that are fabricated in Taiwan to systems manufacturers in China, partly because, prior to the opening of direct air links across the Strait in 2008, they have to go via Hong Kong. The transportation of these components often involves time-consuming customs inspections in Hong Kong. To ensure that their manufacturing activities in China are on track, electronic systems manufacturers need to stockpile a sufficient number of ICs in their production hubs on the mainland and to shoulder the resultant costs. As a manager at HeJian put it, "Each additional day is a risk for our customers because their ultimate goal is to minimize the number of chips in stock."[71] The IC users I consulted dislike this scenario, particularly when production capacity runs high and they are compelled to stock large numbers of ICs in their warehouses. Moreover, inventory problems may be exacerbated because of the disruption to international transportation in the region in the event of Strait contingencies.

By contrast, the problem of large inventories can be eliminated by using a comparable foundry in China. Having such a foundry there would allow timely local delivery of the ICs to the given systems houses, helping them to ensure time to market for their products. For instance, HeJian's president indicated that it would take HeJian one day to ship its components to a local systems house. Because its packaging and testing partners, also from Taiwan, are located just 500 meters away from its fab facility in Suzhou, it would take HeJian less than ten minutes to send chips churned out by its fab to its next-door partners for backend services before these chips make their way to their local customers.[72] Hence, some of the China-based foundry customers, such as those of HeJian, have urged their Taiwanese foundry partners to set up fabs in China.[73]

To conclude this section, the desire to gain market access has been a powerful driver behind the migration. Related considerations, including an aspiration to be close to the customers, to help customers solve inventory problems, and to ensure an efficient time to market for their products, have also been important motivating factors.

Access to location-specific resources

The second major pull factor behind the migration has involved a desire to gain access to location-specific resources in China (particularly engineers and technicians). This has often been accompanied by a related push factor, namely the shortage of engineering talent in Taiwan.

This consideration, together with an enabling factor concerning the increasing complexity in IC design, has driven numerous Taiwanese IC design houses to set up operations in China.[74] As discussed in Chapter 3, the increasing complexity of IC design has been one of the enabling factors behind the globalization of IC design production activities. The same consideration applies to the Taiwanese case. Various Taiwanese IC design houses have contended that as IC design technological complexity increases with the emergence of SOC as the dominant trend in the global chip industry, any Taiwanese IC design house focusing on SOC needs to employ some of the growing number of hardware and software engineers in order to fulfil the demands of the IC design flow. The need for more designers than are available in Taiwan has thus fostered their offshore IC design activities.[75] Moreover, the ratio between hardware and software expertise in any IC design firm that focuses on SOC has been changed from 3:1 or 1:1 to 1:2 or 1:3 in order to ensure firm competitiveness. Software expertise has become increasingly important for competitiveness.[76] The growing demand for software expertise and the availability of software engineers in China have driven numerous Taiwanese firms to outsource software tasks to their subsidiaries on the mainland.[77]

The availability of entry-level engineers in China has also propelled some of these Taiwanese companies to open operation centers there in order to outsource low-end design tasks to their China hubs.[78] As a vice-president of a Taiwanese firm admitted, "China allows us to gain access to a large quantity of entry-level engineers, which helps us to mitigate our R&D burden."[79]

Although China's IC design sub-sector, which remained backward through the 1990s, has deprived local engineers of the professional experience that their Taiwanese counterparts have had, the quantity and potential of Chinese engineering graduates are highly regarded by numerous Taiwanese firms. These companies have identified China's engineers and technicians as the most important location-specific resource that has driven them to set up operations there. This consideration is compounded with a related push factor, namely these firms' anxiety over the shortage of outstanding IC designers on the Silicon Island. In the words of the head of a Taiwanese subsidiary in Beijing:

> Although university graduates in microelectronics in China are not yet ready to contribute to our daily IC design tasks, many of them have a solid foundation in the theory of microelectronics and are very intelligent. Therefore, we decided to enter China early in order to train this group of engineers. Besides, there is just not enough outstanding IC design talent in Taiwan.[80]

Similar push and pull factors also explain why another IC design house has emerged as a forerunner outsourcing its IC design tasks to its subsidiaries in China. As the company's president admitted, "We need to leverage the resources that are available in China, including talent. After all, Taiwan's talent resources are limited."[81] Another firm has established a subsidiary in China for similar reasons. On-the-job training programs at its Shanghai hub have enabled local engineers to undertake various IC design tasks.[82]

In the IC fabrication sub-sector, UMC and SMIC have also expressed their desire to gain access to China's location-specific resources as a primary explanation for their decision to begin making chips on the mainland. The foreseeable shortage of talent in Taiwan in the future has further strengthened UMC's resolve to establish HeJian.[83] As for the IC packaging and testing sector, presidents of two Taiwanese subsidiaries in China have identified the relatively cheap labor in China as a principal driver for the production shifts.[84]

While China's market and location-specific resources mainly account for the migration, they are often combined with other forces, including policy incentives in China, a desire to reduce cost, and political factors.

PRC government policies

Industry insiders regard Beijing's liberalized policies, which are designed to attract foreign semiconductor investment, as catalysts for the migration. As discussed in Chapter 4, these new policies, introduced since 2000 as part of the 10th Five-Year Plan, include offering subsidies and tax incentives, promoting new businesses, and improving infrastructure arrangements involving land, water, and electricity supplies.

The Taiwanese semiconductor actors say that the policy incentive that was most relevant for their decision to migrate was a set of promotional measures specified in State Council Circular 18, which was introduced in 2000. Because these measures apply to eligible foreign and domestically owned semiconductor firms operating in China, the Circular was a powerful tool used by Beijing to lure foreign investment into advanced IC design and fabrication segments of the local sector. The key provisions in the Circular include refunding VAT, allowing taxation holidays/reductions and permitting foreign currency retention. In particular, the VAT policy would entitle qualifying IC manufacturers and design houses operating in China to VAT refunds (Howell *et al.* 2003). Beijing also shifted from depending on government bank loans to obtaining capital investment from foreign or local financial markets.

In the IC fabrication sub-sector, the CEO of SMIC argued in 2005 that Beijing's pertinent new policy measures, particularly preferential tax treatments and the availability of preferential infrastructure arrangements concerning land, water and electricity supply, had attracted foreign investment in the sector.[85] When asked if Circular 18 was relevant to UMC's decision to help establish HeJian, the president of HeJian described the policy as a "catalyst." The policy incentives alone did not determine the decision because the primary drivers behind the initiative were China's market and engineering talent.[86] TSMC's China-bound investment was not primarily driven by a desire to cash in on China's pertinent preferential tax treatments, according to the company's vice-chairman.[87] These policy incentives were secondary considerations behind the firm's decision to build a fab in China.[88]

As for the IC design sub-sector, the vice-president of a Taiwanese firm admitted that tax incentives, such as those embodied in the VAT policy of Circular 18, partly accounted for his firm's decision to open a subsidiary in China.[89] As such, when China decided to scrap its original VAT policy specified in the Circular in 2004, it disappointed Taiwanese firms which had set up production centers in China partly to take advantage of the VAT incentive.[90]

Cost reductions

My interview data indicate that a desire to reduce costs is a secondary rather than a primary explanation for the migration. In the IC design sub-sector, some of the senior executives admitted that the prospect of reducing costs by paying IC designers in China relatively low salaries partly motivated their firms' decision to establish subsidiaries in China.[91] One Taiwanese firm decided to set up a subsidiary in China in order to reduce production and logistics costs. The vice-president of the company predicted that these costs would be reduced because of the subsidiary's access to a strong local logistics support system comprising nearby fabs (e.g. HeJian and TSMC) in China, as well as the tax benefits promised by Beijing.[92]

In the IC fabrication sub-sector, while the CEO of SMIC argued that lower costs attract foreign firms and investors to set up foundries in China,[93] others disagree. A desire to reduce production costs was not the primary driver behind TSMC's decision to set up a fab in China. As a company senior director put it, "We are not there for the low costs."[94] A SIA study (Howell *et al.* 2003) also failed to support the hypothesis that lower construction, manufacturing or operating costs in China accounted for an increasing concentration of foundry investment in the country by external firms (including those from Taiwan).

Executives of several Taiwanese firms in the packaging and testing sub-sector have admitted setting up subsidiaries in China partly to exploit low-cost labor there. The president of a Shanghai-based subsidiary estimated that since labor costs in the country would be about 10 per cent lower than in Taiwan, the desire to reduce labor costs partly accounted for his firm's decision to set up an operation in Shanghai in 2000.[95]

Since cost reduction has not been viewed by industry insiders as a major driver behind the migration, this finding challenges the conventional understanding of the globalization of semiconductor production activities, which is that cost reduction primarily drives the globalization of this sector.[96]

Political factors

So far, we have identified several economic drivers behind the migration. The weight of evidence indicates that at least two types of political factors also help account for the migration, namely the fear of cross-strait political instability and a desire to contribute to the rise of China. A Taiwanese scholar-turned industry player admitted that it was the 1995–96 missile crises and his fear of any future

Strait contingency that had driven him to give up his academic career in Taiwan and to establish a foundry company in China. As he recalled:

> The core driver behind my decision to relocate to China was the missile crises in 1995 and 1996. I was nervous about the crises. I feared that Taiwan would be destroyed, and I kept thinking what the future would hold for Taiwan. Subsequently, I decided to help strengthen cross-Strait exchanges, to leave Taiwan, and to give up academia.[97]

Moreover, a "Greater China mentality" has driven other Taiwanese to move to China to work in the country's IC sector. They subscribe to the belief that they can enhance China's capability in the strategic industry that is important to China's overall power. As a senior Taiwanese industry player admitted:

> One mentality that has driven the migration at the individual level is the view that "I must go to China to help with the country's rise." Such a mentality reflects the identification of China as the motherland. It has motivated some of my university classmates [at National Taiwan University] to give up successful careers in the US in order to work in China.[98]

In sum, against the backdrop of the enabling factors that characterize the semiconductor industry, a combination of the push and pull factors identified above has driven the migration to date. If these forces continue to combine in the years ahead, Taiwan's chip production activities will continue to shift to the mainland.

Migration across the Strait: case studies

To deepen our understanding of the understudied case of sectoral migration, the following section summarizes eight firm-level case studies in IC design and fabrication sub-sectors.[99] In each of the studies, the following dimensions are analyzed: the role of Taiwan in terms of talent, technology and investment; the evolution of business operations; and the connections with Chinese domestic semiconductor players.

Four cases in the IC design sub-sector

An analysis of four Taiwanese IC design firms follows. These companies have been chosen for the following reasons. Three of them were among Taiwan's top ten companies in 2004 in terms of sales revenue: VIA Technologies Inc. (VIA), Realtek Semiconductor Corporation, and Ali Corporation. Today, VIA's operation in China remains one of the largest among Taiwanese IC design houses. Realtek has followed a unique dual-track development strategy on the mainland which has helped Actions to emerge as China's largest fabless design firm by 2005. Ali has pioneered the migration of Taiwan's IC design activities across the

Strait since the late 1990s. And one of them, Unisonic Technologies Co., Ltd. (UTC), is a small-sized pioneering firm whose expansion into China has enabled two start-ups to emerge among the mainland's top ten IC design houses by the end of 2004.

VIA

The first of these companies is VIA, a Taiwanese design house specializing in microprocessors, graphics ICs, and chipsets.[100] In 2001, it was Taiwan's biggest IC design company and, after Intel, the world's second largest supplier of chipsets (Business Week 2001). In 2004, it generated US\$579 million in revenue, becoming Taiwan's second largest and the world's eleventh largest fabless design house.[101]

This company opened its first subsidiary in China in 2000.[102] "VIA began to set up offshore operations in China when it was at its peak," an industry player recalled in 2005.[103] With its then budding ambition to break new ground in microprocessor design, VIA began to expand its production bases on the mainland by capitalizing upon China's long-standing need for indigenously designed and manufactured state-of-the-art microprocessors. The company built its public image in China by portraying itself as a leading semiconductor design house capable of designing "China Chip" (*zhongguoxin*). Its "China Chip" advertisements often caught travelers' attention on their way from Beijing airport to the downtown area.[104]

Since then, VIA's offshore operations in China have expanded to run one of the largest operations on the mainland among Taiwanese IC design firms. Between 2000 and 2003, Taipei approved its investments in five high-tech subsidiaries located in four Chinese cities.[105] From 2001 to 2007, the paid-in capital of its subsidiary in Beijing grew ten times, from NT\$65.3 million at the end of 2001 to NT\$653.6 million by the third quarter of 2007.[106] The Beijing subsidiary had about 400 employees in September 2005, with a 1:30 ratio between Taiwanese and Chinese staff.[107] While its employees in China were about 800 in 2004 (Lee 2004), it began to hire more engineers in China than in Taiwan in 2008. As of August 2009, it had hired more than 2,000 engineers on the mainland.[108]

The main business scope of these subsidiaries, as disclosed in the MOEA data, involves sales operations and post-sales technical services to customers. These activities are permitted by Taipei and these operations have engaged in their officially stated businesses. As a company executive stressed in 2005, these subsidiaries provided their customers with timely technical support on the ground. He expressed it this way:

> A lot of the Taiwan customers have moved there. For example, there is little motherboard production done now in Taiwan. Most of the production is done in China . . . So you need to support your customers there. That generally means simple engineering. But the stronger the customers become there, the more R&D they start doing there, and the more R&D you need there to support them.[109]

However, these offshore hubs have also engaged in IC design work, which is still forbidden by Taipei. A company senior executive has revealed that a hierarchical division of labor has been established at the firm's operations in China, Taiwan, and the USA. Design capabilities that the company has developed in its subsidiaries in China[110] have enabled these offshore hubs to engage in low-level tasks in the design flow. By contrast, operation centers in Taiwan and the USA have undertaken most of the high-end design work (i.e. architecture and product definition). While CPUs and graphics ICs are primarily designed in the USA and chipsets are designed in Taiwan, support engineering (i.e. validation and software work) is completed by subsidiaries in China.[111] The subsidiary in Beijing has dealt with CPU work, whereas the subsidiary in Shanghai has focused on graphics ICs.[112]

To strengthen its ties with local semiconductor players in China, VIA established IC design R&D labs in several universities. It has established one lab for CPU development at Tsinghua University and another for communications ICs at Beijing University of Posts and Telecommunications. Both programs were designed to make VIA a familiar brand name to participating students and to enable them to become familiar with the design tools VIA used, in order to entice them to join the company in the future. However, the students involved in these labs were unable to deliver, forcing the company to discontinue both schemes earlier than expected. As the firm's China chief put it, "Initially we were hoping for some substantial cooperation, but we realized that it was a difficult task for them."[113]

Today, VIA's operation in China remains one of the largest among Taiwanese IC design houses, and is an important illustration of the migration of Silicon Island's IC design sector across the Strait.

Realtek (and Actions)

The second case involves Realtek. Established in 1987, the company has been one of Taiwan's top ten fabless design houses specializing in communications ICs. In 2004, its sales revenue reached US$278 million; it became Taiwan's seventh largest and the world's no. 22 fabless IC supplier.

Realtek's expansion in China has involved offshore investments in fully-owned subsidiaries and informal assistance with the establishment of a start-up. On the one hand, its subsidiaries in China include Realsil Microelectronics Inc. in Suzhou, and Realtek Semiconductor (Shenzhen) Ltd. in Shenzhen. The amount of the initial investment in Realsil, which was approved by Taipei in April 2002, was US$8.8 million. It amounted to US$13.3 million by the end of 2005 (Realtek Semiconductor Corp. 2006: 48; Ministry of Economic Affairs 2007). As of August 2005, Realsil employed about 100 engineers. Three to four of them were Taiwanese, while the rest were Chinese.[114] The original amount of investment in the Shenzhen subsidiary reached US$200,000 by the end of 2005 (Realtek Semiconductor Corp. 2006: 48). According to Realtek's company report in 2005, the stated main business scope of these subsidiaries is as follows. The subsidiary in Shenzhen is involved in "R&D and information service," whereas the Suzhou hub focuses on semiconductor "design, research, development, selling, and marketing." However, the actual business scope of these subsidiaries is hard to pin down. Several senior executives at the company headquarters where I interviewed were tight-lipped over the issue. A former chief of the subsidiary in Suzhou merely praised his local engineers as "extremely outstanding" in quality. This contrasted with what the company told the government when it sought Taipei's approval of its investments in China. It said its subsidiaries on the mainland would only focus on IC sales "because IC design engineering talent in China is not very good."[115]

On the other hand, Realtek has informally shifted its resources to China's IC design sub-sector to help set up a new IC design house headquartered in Zhuhai, China. Founded in 2001, the start-up was initially known as Cristo Capital Inc. In 2005, it was renamed as ASC, a wholly owned subsidiary of Actions Semiconductor (Mauritius). The investment and talent flows from Realtek to Actions have been substantiated by my interviews with industry insiders and secondary data. As a senior Taiwanese IC design engineer observed, "Realtek has a design house in Zhuhai. Its name is Actions. It's a product of indirect investment by Realtek."[116] Another industry interviewee described Actions as a "Taiwanese company."[117] When asked to clarify the alleged linkage between Realtek and Actions, a vice-president at Realtek downplayed the inter-firm tie: "Actions has nothing to do with our company . . . It's just that Realtek's major shareholders have invested in Actions in their personal capacity." These shareholders, he claimed, withdrew their investment in Actions for fear of a government inquiry after Taiwan prosecutors launched a high-profile investigation into UMC's problematic investment in HeJian in early 2005. "This is because Taipei forbids investment in IC design houses in China," he explained.[118]

Even if Realtek's major shareholders did withdraw their investments in 2005, which I cannot verify independently, the intimate ties between the two companies have continued. In September 2005, the same vice-president of Realtek whom I interviewed traveled to China for a business trip: his first stop was the subsidiary in Suzhou, and his second was Actions in Zhuhai.[119] In early 2007, he reportedly became the president of Actions (Chen 2007).

Furthermore, Actions' 2006 annual report (ASC 2007: 50–1) mentioned the linkage between Actions and Realtek. Five of Actions' eight board members are from Taiwan, including the CEO and the chief financial officer (CFO). Since August 2005, the company's CEO has been a co-founder and former vice-president of Realtek.[120] More importantly, the report has regarded its Taiwan link as a risk factor that is potentially detrimental to the company's future operations. In view of Taiwanese regulations on IC design investments in China, the report (ibid.: 15–16) made the following statement:

> Certain of our employees and management, including our chief executive officer, are Taiwanese Persons. The Taiwan government may interpret its current regulations and policies . . . such that it considers one or more of our

employees and management as having inappropriate control and influence. Such employees or managers, including our chief executive officer, may be required to leave the company or be subject to sanctions by the Taiwanese Government.

Despite Actions' fear that its Taiwan connection might endanger its operations, the start-up has grown dramatically in China's IC design sector, due partly to resource transfers from one of Taiwan's most powerful fabless design houses. In 2005, its revenue reached US$155 million. It became China's first IC design firm to break the US$100 million revenue mark, and emerged as China's largest fabless design house. It was also one of the first two Chinese IC design companies to have completed successful NASDAQ IPO. In its focused market niches, it gained a 30 per cent share of the worldwide market for non-iPod MP3 player SOCs (PricewaterhouseCoopers 2007: 14, 35).

To sum up, Realtek has crossed the Strait both to set up subsidiaries that are approved by Taipei and to help establish a start-up, namely Actions through investment and talent input which may have broken some Taiwanese regulations. Actions, in 2005, emerged as China's largest fabless design house. Given Realtek's continuous growth at home and Actions' stunning performance in China to date, Realtek's dual-track migration strategy has succeeded – at least from the company's standpoint.

Ali Corporation

The third case involves Ali Corporation. The company, incorporated in 1987, has been one of the world's major suppliers of ICs for the PC, peripherals, and multimedia markets. Between 1987 and 1992, it was the world's third largest chipset supplier. With sales revenue of US$182 million in 2004, it became Taiwan's tenth largest and the world's No. 39 fabless design house. In August 2004, MediaTek, Taiwan's largest IC design house, gained 31.28 per cent of ownership of Ali (Mediatek Inc. 2005: 25; 2006: 99).

Since the late 1990s, Ali has pioneered the migration of Taiwan's IC design activities to China. "We began to cultivate talent in China in the 1990s," recalled a co-founder of the company. He had led Ali's ambitious expansion in China until MediaTek became the largest shareholder of the firm. Ali acquired a local IC design house in Zhuhai, subsequently set up a wholly owned subsidiary in the same city, and then merged the two companies into one.[121] Its first few investments in China may have been made without prior government approval as Taiwanese official data indicate that its first approved investment in China dated back to 2000 (Ministry of Economic Affairs 2007). By August 2004, it was running one of the largest IC design production operations in China among Taiwan's IC design firms. These bases were located in Beijing, Shanghai, Xi'an, Shenzhen, and Zhuhai, according to the co-founder.[122] Its headcounts in China at the time accounted for about 39 per cent of its workforce (i.e. 500 in China, 800 in Taiwan and another 50 in the USA).[123]

MediaTek's acquisition of Ali led to the downsizing of Ali's operations in China because their post-acquisition assessment concluded that the quality of the local engineers at Ali's hubs in China was unimpressive.[124] By October 2005, Ali became a company of 400 employees. Its operations in China shrank accordingly, hiring only about 200 local engineers and five Taiwanese ones. By the end of 2005, only subsidiaries in Shanghai, Beijing, and Zhuhai remained (Mediatek Inc. 2006: 102, 175, 180). The co-founder viewed such a development as a blow to his vision of cultivating local engineers through the firm's extensive China hubs in order to benefit the company in the long run. "Now that MediaTek has cut Ali in half, Ali will be unable to reap returns on their investment in local Chinese engineers, who had accumulated precious engineering experience over the past few years," he lamented.

As for the main business scope of these China-based subsidiaries, there has been a gap between what has been reported to the Taiwanese authorities and what has happened on the ground. According to the MOEA data (Ministry of Economic Affairs 2007), these subsidiaries engage in "sales of IC products and post-sale services." However, the co-founder admitted that these operations, prior to MediaTek's downsizing endeavors, had also engaged in IC design. "When we applied [for offshore investments in China], we stated that the subsidiaries in question would offer our customers 'technical support.' But once one arrived there, one did whatever one thought appropriate. We did some IC design projects there," he admitted. Since 2005, major R&D work has been primarily carried out by subsidiaries in Shanghai and Zhuhai, involving mostly software-related tasks. As a senior executive at the company revealed, "Our major investment in China is in software. The country's human capital in software engineering is good, but not in hardware engineering."[125]

Ali's pioneering move to shift Taiwanese IC design production activities to China has been controversial. Some industry interviewees have regarded Ali's move to China in the late 1990s as premature because of poor IP protection, the lack of loyalty to a single firm among local engineers, and the lack of a sufficient number of experienced local engineers.[126] Others have justified the production shift as a rational move to cultivate local talent ahead of its competitors in order to benefit the company in the long run. One thing nonetheless is certain about Ali's China operations: their business scope has included IC design production activities, in defiance of Taiwanese regulations.

UTC (Hangzhou Youwang, and Hangzhou Silan)

The fourth case involves UTC, a small-scale company established by K. H. Kao in Taipei in 1990. The most important outcome of the company's expansion into China involves the rise of two start-ups, Hangzhou Youwang Electronics Co., Ltd. and Hangzhou Silan Microelectronics Co. Ltd., which were among China's top ten IC design houses by the end of 2004.

In 1994, Kao set up Hangzhou Youwang as a wholly owned subsidiary of UTC. To absorb local talent, Kao promised a seven-member PRC team, led by Chen

Xiangdong, that he would transfer 40 per cent interest in the company to them in the future. Prior to their switch to the commercial chip business, almost all of Kao's local partners in the team had worked at state-run semiconductor operations including State 871st Plant, a QPL chip supplier to the Chinese military,[127] and Hua Yue Microelectronics Co., Ltd. In 1997, they founded Hangzhou Silan Electronics Co., and Kao initiated a 20-year equity interest transfer agreement between Hangzhou Youwang and the Chinese start-up. The deal transformed Hangzhou Youwang into a joint venture by entitling the Chinese start-up to 40 per cent interest in the company.

More importantly, the agreement supplied the start-up, which was subsequently renamed Hangzhou Silan Microelectronics Co. Ltd., with much-needed capital. According to its prospectus in 2003 (Shanghai Stock Exchange Newspaper 2003): "The equity interest transfer agreement increased company assets, and the subsequent investment income out of Hangzhou Youwang Electronics Co., Ltd. had a positive impact on company profits."[128]

By 2004, Hangzhou Youwang generated annual sales revenue of 247 million RMB; it became China's seventh largest IC design house and the country's 16th largest semiconductor company. Hangzhou Silan did even better. In 2004, it generated 509 million RMB in sales revenue, emerging as China's second largest IC design house and the third largest Chinese chip firm (PricewaterhouseCoopers 2006: 12, 18).[129]

Although UTC has never been a top-notch performer in Taiwan's IC sector, it has fostered the rise of two Chinese start-ups over the course of its investment shift across the Strait, and these were among China's top ten fabless design houses by 2004.

In sum, we should first note that the four cases presented above indicate the various degrees in which the migration of Taiwan's IC design sector to China has taken place. The case of UTC shows that the migration began as early as 1994, and has expanded steadily. Second, some Taiwanese IC design houses and engineers have operated in China by circumventing Taiwanese regulations. Although Taipei has attempted to control investments in PRC entities in the IC design subsector by Taiwanese persons or entities established under the laws of Taiwan, most of the cases above indicate that these regulations have been violated or circumvented in four major ways. First, senior Taiwanese executives have invested in China's IC design houses in their personal capacity, without prior government approval. A key example is the investment in Actions by Realtek's main shareholders.[130] The second way was for Taiwanese IC design firms to set up subsidiaries in China without prior approval from Taipei, as evidenced in the cases of UTC and Ali. The third way was for a Taiwanese firm to help set up a Chinese start-up through tortuous investment and low-key talent support, even though the two companies were seemingly unrelated. The linkage between Realtek and Actions is a case in point. The fourth way was to engage in IC design production activities at China-based subsidiaries, as evidenced in the cases of VIA and Ali.

Third, several China-based IC design start-ups (e.g. Hangzhou Youwang, Silan and Actions) have benefited from the sub-sector migration, thereby entering China's top ten list by 2004 or 2005.[131] Hence, Taiwan's IC design capability, which is only second to that of the USA, has been gradually shifting to China primarily at the infra-firm level and secondarily at the inter-firm level, despite Taipei's efforts to prevent this from happening.

Four cases in the IC fabrication sub-sector

What follows is an analysis of four case studies involving IC fabrication, the segment of China's semiconductor supply chain in which Taiwanese input has exerted its strongest impact over the course of the migration. They include SMIC, GSMC, HeJian, and TSMC Shanghai Co. Ltd. They have been chosen for the following reasons. SMIC has emerged as the most successful foundry company in China, housed with the largest number of Taiwanese semiconductor professionals because of the migration. GSMC is known for its high-profile connections in Taiwan, China, and the USA. HeJian is one of the most controversial foundry companies, spearheading the shifts of chip production activities across the Strait. TSMC Shanghai is the first Taiwanese chip-maker that has invested in 8-inch wafer fabrication activities on the mainland with prior approval from Taipei.

SMIC

Incorporated in April 2000 in the Cayman Islands, SMIC is headquartered in Shanghai. It was established by Richard Chang, an experienced Taiwanese American, with the backing of a multinational array of managers and engineers, capital and technologies, including those from Taiwan. Unlike state-owned enterprises or the 50–50 Sino-foreign joint ventures established in the 1990s, SMIC represents a new business model which has emerged in China's IC manufacturing landscape since 2000. Its emergence as China's flagship chip-maker cannot be separated from the role of Taiwan. In 2005, it became the third largest pure-play foundry in the world, and its China foundry market share was over 50 per cent (SMIC 2006b: 13). With a total of 10,048 employees as of the end of 2006, it remains the largest and most advanced foundry player in China today (ibid.: 43; SMIC 2007a: 57).

To begin with, the nature of SMIC as an MNC is evident in the multinational nature of its talent, technology, and investment. Its employees have been multinational. In December 2001, its headcount reached 1,300, including 400 engineers from the USA, Taiwan, Singapore, Japan, and South Korea (SMIC 2001a; 2001b). Soon after the firm was established, George Scalise, president of SIA, was told by the company that 25–30 per cent of its workforce was expatriates. "This is unheard of in my experience," he recalled.[132] By December 2004, more than 1,000 foreigners worked at SMIC.[133] In September 2005, 1,100 out of its 8,400 employees were expatriates; they came from 26 countries, including the USA, Taiwan, Singapore, Japan, South Korea, Malaysia, and Italy.[134]

Since 2001, SMIC has relied on several semiconductor giants for the transfer of manufacturing technologies in exchange for either equity stakes or access to production capacity at SMIC. These MNCs include mainly Germany's Infineon (SMIC 2002; 2003b; 2006a) and Qimonda (SMIC 2007b), Japan's Toshiba (SMIC 2001c; 2003a), Singapore's Chartered Semiconductor, and the USA's TI, IBM and Spansion. For instance, in December 2001, Chartered decided to transfer its 0.18 micron baseline logic process technology and to grant patent license rights to SMIC in exchange for an equity stake and access to capacity at SMIC (SMIC 2001d). In December 2007, IBM decided to license its 45nm bulk CMOS technology to SMIC in exchange for 12-inch wafer foundry service (SMIC 2007d; Lapedus 2008).

SMIC has also relied on multinational sources of capital for its development. Its first-phase investment amounted to US$1.48 billion by November 2001, which included about US$1 billion in equity financing and US$480 million in loans from four Chinese government banks (SMIC 2001a; 2004a). Related equity investors included Taiwanese and American private investors and Chinese and Singaporean quasi-public investment organizations (Howell *et al.* 2003: 93). In September 2003, SMIC raised an additional US$630 million in private placement of shares (SMIC 2003c). In January 2004, several Chinese banks financed SMIC for the second time, which involved a US$285 million loan agreement with the company (SMIC 2004a). In May 2005, a group of Chinese banks agreed to sign a US$600 million loan pact with SMIC's subsidiary in Beijing to help expand the capacity at the company's three 12-inch wafer fabs in the city (Clendenin 2005; Engardio and Einhorn 2005; SMIC 2005c). In March 2004, SMIC went public in Hong Kong and New York, raising more than US$1.7 billion (SMIC 2004b). Since then, the performance of its shares has disappointed investors. By March 2006, its shares had lost 58 per cent of their value since it went IPO (Brooker 2006). By the end of 2006, Shanghai Industrial Holdings Ltd. remained the single largest shareholder of SMIC, holding 15 per cent of the shares in SMIC. The shareholder is believed to be the "investment arm" and commercial arm of the Shanghai municipal government (Shanghai Industrial Holdings Limited 2005: 18; SMIC 2005b: 31; 2006b: 37; 2007a: 51).[135] In November 2008, however, Datang Telecom Technology & Industry Holdings Co. Ltd., which is closely linked to the Chinese government, agreed to invest US$172 million for a 16.6 per cent stake in SMIC, thereby becoming SMIC's largest shareholder (Electronic Engineering Times Asia 2008).[136]

Based on the above analysis, SMIC is indeed an MNC. Rejecting any attempt to label itself as a Taiwanese firm, SMIC has been eager to categorize itself as an MNC by highlighting the multinational sources of its capital and technologies. Despite this, the majority of my industry interviewees and secondary sources have either described SMIC as a Taiwanese firm or at least cited it as a primary example of the contribution that Taiwan has made to help China catch up in the global semiconductor race.[137] This prevailing view has emphasized the large proportion of Taiwanese employees in the company, its acquisition of proprietary technologies from Taiwan's TSMC, and its absorption of Taiwanese investment (Howell *et al.* 2003: 93–5; Jones 2005).[138]

The human resource flow from Taiwan to SMIC has been obvious from the outset. Over a two-year period starting from mid-2001, TSMC alone lost more than 100 employees to SMIC, and many of them were top semiconductor engineers and senior managers (TSMC 2003; Jones 2005; ASC 2007: 50).[139] By September 2005, about 650 Taiwanese worked in SMIC, accounting for 59 per cent of the company's expatriate workforce.[140] As of 2005, the majority of the vice-presidents and fab directors at SMIC were either Taiwanese or American.[141]

Richard Chang is the single most important figure from Taiwan who epitomizes the talent exodus to SMIC. Born in China, Chang completed university in Taiwan before heading to the USA for post-graduate studies. Over the course of 20 years at TI, he helped build and manage about 10 semiconductor fabs in the USA, Japan, Taiwan, Italy, and Singapore. In 1997, he returned to Taiwan to join Worldwide Semiconductor Manufacturing Corporation (WSMC), and led the company in his capacity as president from 1998 to 1999 (Einhorn 2002; SMIC 2007a: 36). Following TSMC's acquisition of WSMC in 1999, Chang and his potential venture capital partners were in Hong Kong for talks on launching a Silicon Harbor project there, but to no avail.[142] In 2000, he landed in Shanghai to launch SMIC.

Since then, Chang has led SMIC to become China's leading foundry amid controversy. On the one hand, his contribution to China's chip sector has been widely recognized. As a mainland Chinese engineer with more than 37 years' experience in the industry put it, "He is very important. The world microelectronics industry might have looked at China in a different light but for Chang's 12-inch wafer fab."[143] In the eyes of the president of an IC design start-up in Beijing, "Chang came to China in an attempt to help the IC industry in China, and he was assisted by Taiwanese power, meaning people and dollars ... SMIC is changing the face of China's IC industry. He takes the lead and he is the seed."[144] Chang has been crowned with numerous awards in China and abroad in recognition of his contribution to China's semiconductor industry.

On the other hand, the way in which Chang has spearheaded the migration has been controversial. Taiwan's MOEA has viewed his involvement in SMIC as illegal. Despite Chang's emphasis on his US citizenship, the MOEA argued in 2005 that he had household registration in Taiwan as a Taiwan citizen when he, without prior approval from Taipei, founded SMIC. Viewing Chang as having made an illegal indirect investment in China's semiconductor industry, MOEA fined him NT$5 million (US$156,000) in March 2005, and gave him six months to withdraw his investment in SMIC or face further penalties (Nystedt 2005; Investment Commission 2005a).

To refute Taipei's accusation, Chang has taken several measures. Despite his dual US-Taiwanese citizenship, he has repeatedly stressed that he is a US citizen. He has also appealed to the Taiwanese Administrative High Court in order to overturn MOEA's decision. In July 2005, he applied to give up his Taiwanese citizenship, which was rejected by a court ruling in February 2007. Concurrently, MOEA reportedly said it would impose further penalties on Chang if he remained the president of SMIC Shanghai and if the company was to migrate to 65nm process technology as reported by the press. As of that time, Taipei had banned the transfer of process technologies more advanced than those of 0.25 microns to China. In the first half of 2007, the Taiwanese Administrative High Court cancelled the fine imposed by MOEA on Chang, on the grounds that there was insufficient evidence to prove that Chang had actual control over SMIC. In July 2005, a senior MOEA official admitted that if Chang's application to relinquish his Taiwanese citizenship succeeds, his involvement in SMIC would not be subject to Taiwanese regulations (Wang 2005; Hsiang 2007; Tsar & Tsai Lex News 2007).[145] In response to the charges against him and his company, in 2005 Chang said that "SMIC does not belong to Taiwan, and yet the Taiwanese government has made noises about SMIC's operations." He described Taipei's handling of the case as "*shu quan fei ri*," a classical Chinese idiom used to describe uncalled-for fuss out of sheer ignorance.[146]

Several industry interviewees have nonetheless raised their eyebrows at Chang. The president of a Taiwanese subsidiary in China criticized his understatement of the importance of Taiwan to SMIC:

> Chang is a dual US-Taiwanese national. His company was incorporated in the Cayman Islands. A majority of the vice-presidents and fab directors at SMIC are either Taiwanese or American. Yet he has often described his company as an MNC externally, and stressed the Chineseness of his company internally, in China, in order to downplay the company's link with Taiwan.[147]

The president of a fabless design start-up in Beijing highlighted Chang's awkward position: "He is brave enough to take the leadership, to be called a traitor in Taiwan, to come to China to develop China's IC industry. In the eyes of some in the Taiwanese and US governments, he is a bad guy."[148]

Despite the controversy, Chang has been an icon in China's semiconductor industry. With support from more than 650 Taiwanese employees, he, until his resignation in 2009, had led the development of SMIC, which houses the largest number of Taiwanese semiconductor talents on the mainland. As Victoria D. Hadfield, president of SEMI North America, observed in 2005, "Expertise and know-how about how to run a business and how to make it profitable have really helped with SMIC, which is a major player; this expertise has come from Taiwan."[149]

The second key indication of the importance of Taiwan in SMIC's operations involves their acquisition of trade secrets from TSMC. According to a senior executive at a fabless design firm that has used foundry services at SMIC and TSMC, "SMIC's 0.18 micron process technology was copied from TSMC."[150] Moreover, a former SMIC engineer, as cited by TSMC's lawsuit against SMIC in the US federal court in 2004, said that 90 per cent of the process that SMIC had used to make 0.18 micron chips was copied from TSMC. TSMC also examined an SMIC-manufactured chip, bought in the open market, to verify SMIC's use of proprietary technologies. In the same filing, TSMC stated that the device in question contained features similar to those in TSMC-fabricated components.

Furthermore, these features differed from those in Chartered made chips, even though Chartered was SMIC's only legitimate licensor of the 0.18 micron process.

In various court filings in the USA and Taiwan, TSMC identified detailed methods that SMIC had allegedly used to purloin its trade secrets and to infringe on its IPs. A primary method was to lure TSMC employees to join SMIC by stressing that those who came with TSMC's trade secrets would be offered SMIC shares and stock options. Such a tactic seemed to have worked. In an email from a SMIC executive to a then TSMC manager, instructions were given to the manager for obtaining six TSMC process flows, process targets, and equipment types. It ended with the following remark: "Sorry for the long list, but we need a lot of material to set up the new operation" (TSMC 2004; Jones 2005). According to a TSMC senior director, who was formerly the said manager's superior, "She followed the instructions and sent out the materials . . . Then she left TSMC and joined SMIC . . . Several other people were like her. They were all spies."[151] Siding with TSMC's accusations of SMIC, a senior executive at a Taiwanese IC design house described the CEO of SMIC as "a criminal": "He stole the IPs from TSMC."[152]

To refute TSMC's charges, SMIC stressed that it had received technological assistance legitimately by signing technology transfer agreements with major industry players and that it had "a strong and experienced technical team in house that contributed to SMIC's success" (SMIC 2004c).

Despite its defense, SMIC agreed to resolve all pending patent and trade secret litigations with TSMC in January 2005. According to the agreement, SMIC would pay TSMC US$175 million, which would be payable in instalments over six years. Both parties also agreed to cross-license to each other's patent portfolio through to December 2010. The agreement did not grant SMIC a license to use any of TSMC's trade secrets, and it could be terminated upon any breach by SMIC (SMIC 2005a; 2007a: 16–17; TSMC 2005b).

Nevertheless, the dispute has continued. In July 2006, TSMC accused SMIC of having used and retained TSMC's proprietary information on its 0.13 micron and smaller technologies which should have been returned to TSMC under the agreement. It also argued that it would be harmed if SMIC were to disclose or transfer the above information to SMIC's third party partners before the final resolution of the case. In September 2006, SMIC filed a cross-complaint against TSMC, describing TSMC's accusations as "unfair and misleading." In September 2007, the US court requested SMIC to provide advance notice and an opportunity for TSMC to object to any such disclosure or transfer (SMIC 2006c; TSMC 2007c).

The weight of evidence presented above indicates that some of TSMC's technologies did migrate to SMIC without any preceding agreement. This has underscored the importance of TSMC in fostering, albeit unwillingly, the emergence of SMIC as China's top-notch foundry.

The third indication of the key role Taiwan has played in SMIC's operations concerns Taiwanese investment in the start-up (Howell *et al.* 2003: 93–5).[153] At least two Taiwanese venture capital firms attempted to invest in SMIC without prior government approval (Huang 2002; 2003; Asian Private Equity Review

2005).[154] In addition, H&Q Asia Pacific Ltd., one of SMIC's private equity investors, has been closely connected to Taiwan. The company's founder and chairman, raised and educated in Taiwan, was a founding member of the Technology Review Board which advises Taipei on technology matters. Since 2001, he has been a director at SMIC, an appointment he received due to his company's investment in SMIC (Howell *et al.* 2003: 94; SMIC 2007a: 37).[155]

The aforementioned link between SMIC and Taiwan in talent, technology and investment has thus propelled numerous industry insiders to regard SMIC as a primary, albeit controversial, product of the migration.[156]

Since its establishment, SMIC has made tremendous progress in terms of technology, production capacity, and market share. Within 13 months, it had completed the construction of Fab 1, its first 8-inch wafer fab in Shanghai in 2001, using 0.25 micron process technology. By August 2002, the fab became the first in China to implement a 0.18 micron logic process. By March 2008, it was ready to provide IC manufacturing services at 0.35 micron to 65 nm and finer line technologies (SMIC 2008a). An analysis of its annual wafer revenue from 2003 to 2007 by technology (Figure 5.3) indicates its migration towards higher process technologies (SMIC 2006b: 13; 2007a: 13; 2008a).

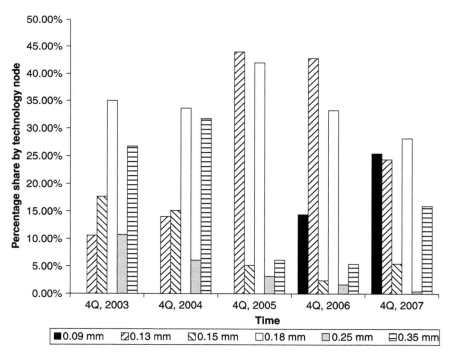

Figure 5.3 Analysis of SMIC's wafer revenue by technology, 2003–07*.

Sources: Company reports.

Note: * This includes logic, DRAM, and copper interconnect only.

Based on his visits to SMIC in 2001 and 2004, Yoshio Nishi, director of the Center for Integrated Systems at Stanford University, said it was "rare" that a start-up like SMIC has been able to execute its technology roadmap as planned during the first few years of its operations. "In May 2001, they showed me their product roadmap. In November 2001, their roadmap was executed . . . They did the same in 2002 . . . In 2004, I visited again. So they are really on track. This is kind of rare," he recalled.[157]

Industry leaders have argued that SMIC's technological progress has largely relied on foreign technology transfers from global giants and advanced semiconductor equipment suppliers.[158] As C. Mark Melliar-Smith, former president and CEO of International SEMATECH, observed in 2005, "SMIC is dependent a lot on its equipment suppliers and a lot on transfer of technology. I don't think it's really totally indigenous."[159] Yu Zhongyu, president of CSIA, also admitted that even though SMIC rose to become China's top foundry firm, its technologies and equipment had largely depended on foreign suppliers, rendering the argument that China is in control of related technologies problematic.[160]

SMIC's fab expansion has been equally impressive. By early 2008, SMIC had one 12-inch wafer fab and three 8-inch wafer fabs in its Shanghai mega-fab, two 12-inch wafer fabs in its Beijing mega-fab, one 8-inch wafer fab in Tianjin, and an in-house joint venture specializing in assembly and testing operations in Chengdu. It also plans to start a new IC production project in Shenzhen, which involves an IC tech R&D center, an 8-inch wafer production line and a 12-inch fab. In addition, it plans to manage and operate at least two additional fabs owned or funded by local governments in Chengdu and Wuhan (Lapedus 2008; SMIC 2008a; 2008b).

The number of 8-inch wafer equivalents shipped by SMIC from 2002 to 2007 increased steadily. While the company shipped 82,486 8-inch wafer equivalents in 2002, it shipped more than 1.8 million in 2007 (SMIC 2007a: 20, 23; SMIC 2008a). At the end of the fourth quarter in 2007, its production capacity increased to 185,250 8-inch-equivalent wafers per month, with a utilization rate of 94 per cent.

From 2002 to 2007, SMIC's annual sales revenue grew as well. Its sales reached US$1.56 billion in 2007, an increase of 5.8 per cent over 2006. With the exception of 2004, however, it had a net loss every year for the rest of the period. In 2007, its net loss reached US$40 million (Figure 5.4). Its share of the worldwide foundry market has also grown over time. In 2004, it became the world's fourth largest pure-play foundry player in terms of sales. In 2005, it surpassed Chartered to become the world's third largest player with a 6.4 per cent global market share, behind only TSMC and UMC (Lapedus 2006a). In the first half of 2007, SMIC remained the world's third largest player (Lapedus 2008).

Since its technological advances have partly relied on semiconductor equipment companies, SMIC has acquired global state-of-the-art fab tools by complying with export control regulations. A core part of their compliance has involved assuring the US regulators that it has not fabricated chips for the Chinese military. The company's CEO said in December 2004 that the Chinese military had never contacted his company to ask for help, for which he felt "grateful." In my

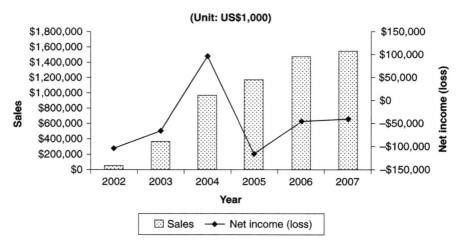

Figure 5.4 SMIC's sales and net (loss) income, 2002–07.

Sources: Company reports.

subsequent interview with him, he shrugged off allegations that his company had supplied chips to the PLA. Claiming that only CPUs could be used as dual-use ICs for military end-uses (a statement which is untrue according to pertinent analysis in Chapter 2), he said that SMIC "is not involved in the fabrication of CPUs, nor does it make advanced chips."[161] SMIC managers have also stressed that their company has never used imported US equipment to manufacture products for military applications. They have identified measures that SMIC has introduced to comply with US regulations and to assuage Washington's misgivings[162] that the company may have fabricated chips for the PLA.[163]

So far, SMIC's efforts have paid off. In 2003, it became a consignee to a special comprehensive license (SCL) granted to Applied Materials by Washington. Since SCL encompasses new and future follow-on orders, the US fab-tool maker with SCL does not need to file any separate license application for each order. This shortens the lead-time needed for the delivery of Applied Materials equipment to SMIC (Lapedus 2003). In October 2007, SMIC became one of the first five companies to be approved by the US Department of Commerce as "validated end-users" (VEUs) of US technology exports to China. Washington will thus remove individual export licensing requirements on shipments of controlled items to the firm. In turn, the company agrees to on-site audits by US officials and stringent record-keeping requirements. SMIC became a VEU based on, among other criteria, its previous compliance with US export controls, its non-military business operations in China, and its record of transparent commercial use of US-controlled technology (Lapedus 2007; Leopold 2007; Bureau of Industry and Security 2007a; SMIC 2007c). Its VEU status helps advance its technological capabilities, as it can now receive and use approved US equipment and materials without the delays and costs associated with traditional export licensing.[164]

As SMIC establishes itself as the strongest foundry player in China, it has also deepened its connections with local semiconductor players. Two major programs are at stake: the MPW service to China's fabless design houses; and local talent training schemes. SMIC, like its competitors in China and elsewhere, has offered the MPW service to its customers on a shared-cost basis for prototyping and verification by cooperating with government-financed incubation centers in China.[165] The company subsidizes its local customers for their participation in any given MPW program by sharing the costs of the exorbitantly expensive masks used, and these subsidies are often supplemented with government ones. The impact of the program is two-fold. First, it enables SMIC to expand its local customer base because it helps upgrade the technological capabilities of the indigenous IC design firms. It does so in a country where most of its local design companies are still technologically backward, trailing far behind global giants in technology and market share (Brown and Linden 2005; China Semiconductor Industry Association IC Design Branch 2005; Electronic Engineering Times Asia 2006; 2007a; PricewaterhouseCoopers 2007; 2008). Second, it helps improve the country's IC design capabilities in due course. To cultivate local semiconductor talents, SMIC has signed agreements with local universities to offer academic degrees in either microelectronics or solid-state circuitry to its technicians. In sum, both schemes have deepened the company's connections with local semiconductor players potentially fostering the long-term development of the indigenous industry.

To conclude, SMIC epitomizes the migration. It has risen in China's semiconductor industry in ways that have stunned many in the global sector, and its ascendancy cannot be separated from the role of Taiwan. The Silicon Island has contributed to the rise of SMIC primarily by supplying more than 650 experienced semiconductor engineers and managers. Moreover, there has been a flow of process technologies and IPs from TSMC to SMIC allegedly because of economic espionage. Taiwanese investment is believed also to have fostered the establishment of the firm. None of the above cross-border flows is without controversy, as revealed by the disputes involving SMIC and its first CEO, TSMC, and the Taiwanese government. Admittedly, the company is faced with numerous challenges in its pursuit of a greater market share, technological advances, and better financial performance. These challenges include the ability to pursue endogenous technological innovation effectively, the viability of its questionable pursuit of price-cutting strategy as a way to enlarge its market share,[166] the viability of its fab expansion project,[167] and the ability to continue to ensure access to advanced fab tools. In a country long troubled with sluggish performance in what it sees as a strategic industry, irrespective of continuous government-led efforts, SMIC has emerged as a top-notch chip-maker ready to compete with global giants, thanks to the Taiwanese migration. The semiconductor landscape in the world's fastest growing economy has altered accordingly, alarming some in other countries who fear that they are losing their edge in an industry vital to their national economy and security.

GSMC

The second case study involves GSMC. GSMC was incorporated in November 2000 in the Cayman Islands. It operates a wholly owned subsidiary, the Shanghai GSMC, or Grace, in China. Grace is a pure-play foundry company with operations in Shanghai. It has become well known for its high-profile connections in Taiwan, China, and the USA. The co-founders of GSMC include Winston Wang, the son of the late Formosa Group Chairman Yung-ching Wang, one of the most powerful businessmen in Taiwan, and Jiang Mianheng, the son of China's former President Jiang Zemin and the vice-president of CAS (Smith 2000; Forney 2001; Balfour *et al.* 2003; Howell *et al.* 2003: 96–7). "Jiang has aspired to foster the take-off of mainland China's IT industry by accelerating the indigenous development of core technologies. That's why he was involved in Grace," recalled GSMC Chairman Zou Shichang.[168] The company also had a high-profile connection with the USA, namely Neil Bush, the brother of the then US President George W. Bush (Balfour *et al.* 2003; Carlson 2003).[169] Like SMIC, GSMC is an MNC, and its birth cannot be separated from the critical input from Taiwan. With more than 1,300 employees, Grace generated US$148 million in revenue in 2004 (Lapedus 2005). It thus became China's fifth largest foundry player accounting for 6 per cent of the total sales revenue.[170]

To begin with, multinational sources of talent, technology, and investment have fostered the development of GSMC. In September 2005, the majority of its 1,300 employees were Chinese, Taiwanese, and US nationals.[171] Of its 150 expatriate employees, 100 were Taiwanese and 30–40[172] were Americans.

Grace, like SMIC, has relied on technology transfers from several foreign companies. They include Silicon Storage Technology, Inc. (SST) and Cypress Semiconductor from the USA, and Sanyo Semiconductor and Oki Electric from Japan (GSMC 2006c). For example, Sanyo started a strategic partnership with Grace in 2003. In 2008, Grace began to produce automotive chips used in car navigation systems for Sanyo, using Sanyo's 0.18 micron Not-OR flash process technology (GSMC 2008). In 2005, Cypress began to transfer its technologies to Grace in exchange for preferential access to Grace's foundry service. In 2006, Cypress announced that it would soon transfer to Grace its 0.13 micron C8 process technology (GSMC 2006b).

The sources of Grace's capital have also been multinational. Its first-phase investment of US$1.63 billion included private equity financing from several countries and Chinese bank loans (Howell *et al.* 2003; GSMC 2004). In 2004, Hong Kong tycoon Li Ka-shing invested US$90 million in Grace (Brooker 2006). In 2002, Grace initially sealed a US$680 million, five-year loan deal with seven local banks. By March 2006, Grace sought to delay payment on the loans, and the Shanghai government-controlled Shanghai Alliance Investment Ltd., chaired by Jiang Mianheng, decided to grant Grace US$190 million in loans convertible into shares. By 2007, Shanghai Alliance Investment Ltd. was the leading shareholder of Grace (GSMC 2007). In contrast to SMIC's successful IPO in 2004, Grace's initial plan to raise US$1.8 billion selling shares in Hong Kong and New York has been deferred more than three times (Brooker 2006). As of early 2008, there was no sign that Grace would go IPO in the near future. By January 2009, Grace was in the process of merging with Huahong NEC.

The development of the company has been fostered by Taiwan. As of October 2002, Taiwan was the largest source of non-Chinese employees for Grace, accounting for 19 per cent of the company's workforce. Many of the Taiwanese engineers at Grace came from TSMC, UMC, and Nanya Technology (Chase *et al.* 2004: 132–3). In 2005, two-thirds of its expatriate employees were Taiwanese. In September 2005, the firm had about 100 Taiwanese employees. Winston Wang is the key Taiwanese figure who epitomizes the migration of Taiwanese talent to Grace. Wang joined Grace to expand his business in China. He recruited several experienced Taiwanese engineers and managers to form the inaugural senior management team of Grace. They included Nasa Tsai as the president of the start-up, and Juei-chen Tsai as the first chairperson of the firm.[173] Although all the aforementioned Taiwanese professionals subsequently left Grace, some of the company's important positions, such as chief technology officers (CTOs) and chief operating officers (COOs), have been held by other Taiwanese.[174]

When Grace was inaugurated, optimists viewed its debut as a symbol of China's likely emergence as a dynamic hub of semiconductor manufacturing. Since then, numerous hurdles have obstructed the company's performance in terms of technology, production capacity, and market share.

In terms of technology, Grace began to construct two fabs based on 12-inch specifications in November 2000. In September 2003, its commercial production began at its first 8-inch fab using 0.25 micron process. In September 2005, it was able to manufacture 8-inch wafers using technologies ranging from 0.25 to 0.13 micron processes, according to the company's chairman. By early 2008, its technologies for production ranged from 0.25 to 0.15 micron processes. However, Grace, like SMIC, has relied on foreign technology transfers rather than in-house R&D efforts for its technological advances.

In terms of production capacity, Grace had one 8-inch fab in full production as of early 2008. In September 2005, it fabricated 25,000 8-inch wafers per month, which rose to 27,000 in 2006. In 2005, it shipped about 300,000 8-inch wafers to its customers. Its wafer shipments grew 108 per cent for 2006; it churned out about 324,000 wafers for its customers (Sung and Chan 2007).[175]

As Grace's revenue reached US$148 million in 2004, it ranked fifth among China's foundry firms, trailing behind SMIC, Shanghai Huahong NEC, HeJian and ASMC. Concurrently, Grace accounted for 6 per cent of the Chinese foundry sales revenue, whereas SMIC had a 42 per cent share (Lapedus 2005). Its revenue for 2006 grew 127 per cent compared to 2005 (Sung and Chan 2007). With revenue of 1.22 billion RMB in 2006, it ranked sixth among China's foundry companies (Xinhua-PRNewswire 2007).

Despite its growths in revenue and wafer shipments, Grace has remained loss-ridden (Brooker 2006). Although it has shown signs of progress in its market performance since former Infineon CEO Ulrich Schumacher became its CEO and President in the latter half of 2007 (GSMC 2007; Tu 2008), it was in the process of merging with Huahong NEC by early 2009.

As for Grace's cooperation with China's domestic semiconductor actors, the company's chairman Zou initiated an important program to deepen the links between the firm and local research institutes in order to cultivate local talent. Zou has been in a position to initiate the scheme because of his capacity as the company's chairman and as a senior semiconductor scientist at the Shanghai Institute of Metallurgy (SIM) of CAS. Zou brought his students from the CAS institute to Grace after they completed a one-year postgraduate training program. Zou explained his rationale as follows:

> Grace should work jointly with research institutes in Shanghai including SIM. Such an initiative has to center around the enterprise, as research institutes do not have advanced semiconductor equipment. Having not yet retired from CAS, I work both in Grace and in SIM. A dozen of my research students are being cultivated in Grace. I bring them to Grace to encourage a marriage between research and production.[176]

Such an initiative, if continued, has important implications for China's IC industry and China's national security. The program helps to accelerate the transfer of technological know-how from the company's experienced engineers, some of whom are from Taiwan, to up-and-coming local engineers at the research institute. These local engineers thereby become major assets for the long-term development of China's IC manufacturing sector. More importantly, given the institute's history as a supplier of semiconductors for military end-uses, including atomic bombs, these trained local engineers at the institute may well use their enhanced know-how, which is refined at Grace, as a commercial fab, to contribute to Chinese defense microelectronics undertakings. The case of Grace and its talent initiative therefore showcases the intricate ties between Taiwanese and Chinese chip actors over the course of the migration, which may have significant long-term security ramifications.

In a nutshell, Taiwan has contributed to the development of Grace though the initial Taiwanese input far from guarantees the company's success. The firm has struggled to cope with numerous challenges including an unusually frequent changing of the guard in the firm's senior management team,[177] its problematic in-house technological innovation capability,[178] and its poor financial performance. In spite of this, the company's initiative to train local graduate students from the CAS research institute in Shanghai, which has a historical link to China's defense chip industrial base, may have an enduring impact on China's security, which we will explore in more detail in Chapter 6.

HeJian Technology (Suzhou) Co., Ltd

The third case involves HeJian Technology (Suzhou) Co., Ltd. HeJian was incorporated in November 2001 in the British Virgin Islands. It then set up a wholly owned subsidiary in Suzhou, China, namely, HeJian Technology (Suzhou) Co., Ltd. through Invest League Holdings Ltd., a vehicle registered in the British territory (Dean 2002; Howell *et al.* 2003: 98; Hille 2005).[179] Headquartered in China, HeJian runs fab operations in Suzhou. It was established by two retired senior executives from UMC with what the UMC Chairman Robert Tsao subsequently described as "assistance" from his company. With its sales revenue of US$239 million in 2004, it ranked third among China's foundry companies, accounting for 10 per cent of the Chinese foundry sales revenue (Lapedus 2005). With its sales of US$310 million in 2006, the company, with about 1,600 employees, remained China's third largest and the world's ninth largest foundry firm (IC Insights 2006).

Like SMIC and Grace, HeJian has positioned itself as a China-based MNC with "abundant foreign capital and state-of-the-art IC technologies." First, its employee force is multinational. About 80 per cent of its 1,600 headcount were locals, and the remaining 20 per cent were expatriates, with the majority being Taiwanese, according to the company's president.[180] The number of Taiwanese employees doubled from about 100 in 2001 to about 200 in early 2005.[181]

As for its technologies, HeJian has not established strategic partnerships with foreign chip companies on a scale comparable to that of SMIC and Grace. Instead, it has primarily depended on UMC for technology transfers, as will be discussed later. Its capital is also multinational. By the end of June 2005, the holding company of HeJian had issued a total of 700 million shares, with the subscription price per share in the last offering being US$1.1. The market value of the holding company was estimated over US$700 million (UMC 2005: 40). By early 2008, HeJian's investment was over US$ 1.6 billion. Although its registered capital was US$400 million, it had invested, as of September 2005, more than US$800 million in its operations.[182] The company's capital relies on multinational equity investment and bank loans. In February 2005, the UMC Chairman identified seven companies and institutions from the USA, Japan, and Singapore as initial investors in HeJian in order to deny charges that UMC had invested in HeJian. Among these investors, ICS is a Taiwanese-owned, US-based chip company, and Xilinx, one of the world's leading suppliers of FPGAs, is UMC's longstanding customer (Xilinx 2004).

However, HeJian's operation as an MNC has relied heavily on Taiwan, primarily through UMC's contribution to the start-up. HeJian's local employees describe HeJian as a Taiwanese company instead of an MNC.[183] The most revealing remark regarding HeJian's Taiwanese linkage came from the president of HeJian. He admitted in 2005 that UMC "made use of the gray areas" in existing Taiwanese regulations in order to establish HeJian. Other industry interviewees have argued that HeJian would not have come into existence without UMC. As a manager of HeJian admitted, "HeJian is just another fab run by UMC."[184] A senior executive at UMC's chief rival put it directly: "HeJian is equal to UMC."[185]

Industry insiders identify the apparent flows of UMC talent and technology to HeJian, while admitting that it is hard to pin down the inter-firm flow of capital even if it does exist.[186] As of March 2005, HeJian was staffed by about 200 Taiwanese engineers and managers, which accounted for nearly 20 per cent of its employees.[187] Numerous former UMC employees have joined HeJian's senior management and R&D teams. The company's president, for one, was formerly a fab director at UMC. As a UMC customer recalled, "Once when I was at UMC, the president of HeJian was there. My UMC partner jokingly introduced the HeJian chap to me as 'UMC's traitor.' "[188]

Industry insiders revealed in 2005 that for a long time, UMC had knowingly transferred its proprietary technologies to HeJian in a low-key fashion in order to escape government scrutiny, while denying such linkages in public. In February 2005, Taiwan prosecutors raided UMC headquarters in order to investigate the inter-firm tie. One month later, UMC Chairman announced that his company would receive a 15 per cent stake in HeJian, worth more than US$110 million. This reward was to compensate for the "past assistance" UMC had offered to HeJian, and was for its future cooperation with HeJian as long as the deal is permitted by Taipei (Tang 2005; UMC 2005: 40–1; Wang 2005; Dean 2006). Following this announcement, a HeJian manager expressed a sigh of relief over what by then had been a secretive inter-firm technology transfer: "Now we can change dark into light: UMC can transfer technologies to us guilelessly in the future so long as we pay for the technologies."[189] In public, however, UMC has reiterated that it has never transferred any technologies to HeJian.

More importantly, the disputes between the Taiwanese authorities and UMC have highlighted the controversy over the HeJian–UMC case. The UMC chairman was subject to official penalties due to UMC's involvement in HeJian. In April 2005, Taiwan's Financial Supervisory Commission fined him NT$3 million, or US$95,600, for breaching securities laws with late disclosure of information relating to what the commission saw as UMC's illegal engagement in HeJian's operations. UMC denied the charge. Robert Tsao and John Hsuan then resigned from their positions as chairman and vice-chairman, respectively, of UMC. In January 2006, both were indicted for their allegedly illegal assistance with HeJian's operations without getting anything in return. Prosecutors stated that both former UMC executives were convicted of "breach of trust" under Taiwan's criminal law by assisting HeJian, a move which was against the interests of UMC and its shareholders (Dean 2006).

In 2007, the dispute saw a U-turn as two court rulings were made in favor of UMC. According to the first ruling, UMC's investment in HeJian could not be classified as an investment because UMC only provided technical assistance to HeJian. The ruling added that the Taiwanese government had failed to prove otherwise. It thus invalidated the fine imposed on UMC. Based on the second ruling, former UMC executives were acquitted as they were found not to be breaking any laws by helping to set up HeJian. Related prosecutors, however, would appeal the verdict (Solid State Technology 2007b).

In reviewing this case, Maw-Kuen Wu, Minister of Taiwan's National Science Council, argued in 2005 that UMC should not have done what it did to help establish HeJian. Nevertheless, the official had a resigned attitude towards Taiwanese chip actors which had used illicit measures to shift their production bases to China. As he put it, "The question is whether we can find a better way to punish or regulate rule breakers."[190]

Despite its disputed ties with UMC, HeJian, since its inception, has performed well in its technology, production capacity, and market share. It began to build fabs in Suzhou in November 2001, and its pilot run and capacity ramp-up commenced in March 2003. In June 2003, it began 8-inch wafer production using 0.5 micron process technology. In September 2003, it completed the set-up of 0.3, 0.25 and 0.18-micron logic processes and a 0.35 micron high-voltage process. In November 2003, it finished a 0.35 micron flash memory process setup. Such swift progress in technological terms is believed to have resulted from UMC's technology transfers. In February 2004, it accomplished a 0.25 micron mixed-mode process setup. In September 2005, its fab was able to mass-produce chips using 0.18 micron process, and to fabricate small-volume semiconductors using 0.15 micron process.[191] In May 2006, it fabricated multi-media player chips using a 0.18 micron process for its customer, Actions. In March 2008, it announced that it would begin 12-inch wafer fabrication activities by joining Elpida to build a US$2 billion 12-inch memory fab in Suzhou with help from the local government (Sung and Chan 2008; Reuters 2008a).

HeJian's production capacity has also grown steadily. In December 2003, it produced 14,000 8-inch wafers per month. In March 2004, it fabricated 16,000 8-inch wafers per month. By September 2005, it was churning out 35,000 wafers per month.[192] The number of wafers reached 41,000 per month by late April 2009 (UMC 2009). Between 2004 and 2006, its sales increased (Figure 5.5), and its market share grew accordingly. In 2004, it had 10 per cent of the Chinese market share, and 1 per cent of the global market share (Lapedus 2005; IC Insights 2006). In 2006, its worldwide market share reached 2 per cent. More importantly, its increasing market share has generated profits, thereby outperforming SMIC and Grace financially.

Meanwhile, the company has also cooperated with domestic chip players through MPW services and local talent training programs. "To attract potential local customers unfamiliar with our processes, we offer the MPW service to allow them to conduct either process verification or new product verification. HeJian pays for the masks used; after all, masks are very costly," said the company's president.[193] Key participants in the given MPW program have included not only indigenous IC design houses, but also the EDA Center under CAS. As the MPW program expands the company's local customer base, it may help to upgrade China's indigenous IC design capabilities.

In addition, the company has offered its inexperienced local engineers much-needed training programs because of the lack of experienced process engineers in China and the high cost of employing experienced engineers from abroad. Ultimately, these programs may help to cultivate local engineers, thereby enlarging

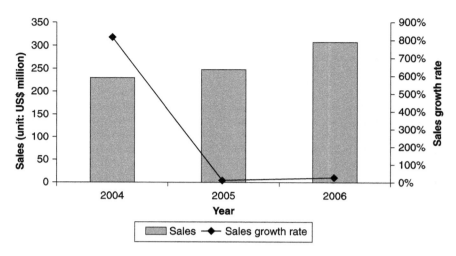

Figure 5.5 HeJian Technology's sales and growth rate in sales, 2004–06.

Source: IC Insights 2006.

the indigenous talent pool. As the company's president predicted in 2005, "We will see the emergence of a group of experienced local engineers in two to three years, and they will be viewed as treasures in China's semiconductor manufacturing industry."

In sum, the emergence of HeJian has been one of the most controversial cases in the migration. At the height of the UMC's fight with the Taiwanese authorities over the assistance it provided to HeJian, some of HeJian's Taiwanese engineers were requested by their families in Taiwan to give up their careers in China to return home.[194] Despite the controversy, HeJian has grown steadily in China, advancing in its technological development, production capability, market share, and most importantly, financial performance. Its rise cannot be separated from UMC's contributions, particularly in terms of talent and technologies, although the concrete ways in which both companies have forged their "friendly" ties are not entirely clear to outsiders. The HeJian–UMC case nonetheless has demonstrated the inability of the Taiwanese state to control the migration. More importantly, the case shows that HeJian, like SMIC, has deepened its local ties by offering MPW services to its prospective customers and by training its locally hired engineers. Both measures contribute to the development of China's strategic IC sector.

TSMC (Shanghai) Company Ltd

The fourth case involves TSMC (Shanghai) Company Ltd, a wholly owned subsidiary of TSMC. It was incorporated in August 2003 with prior approval from Taipei. Its 8-inch fab is located on the outskirts of Shanghai. In 2006, it became China's fifth largest foundry company in terms of revenue (Xinhua-PRNewswire 2007).

TSMC was the first Taiwanese chip-maker approved by Taipei to invest in an 8-inch wafer fab in China. Its subsidiary in Shanghai has absorbed the transfers of talent, technology, and investment from its parent company. Before its China-bound investment was approved by Taipei, TSMC had dispatched a small team to Shanghai to help with the initiation of its China project.[195] Its fab construction in China had preceded Taipei's approval.[196] In June 2005, the subsidiary had more than 900 employees, including 800 mainland Chinese and about 100 Taiwanese.[197] By September 2005, its headcount reached 1,000, one-tenth of whom were Taiwanese.[198]

In terms of technology, TSMC Shanghai began to mass-produce 8-inch wafers using 0.35 and 0.25 micron processes in 2004. In late 2006, Taipei lifted the ban on the transfer of 0.18 micron process to China. Thereafter, TSMC started to transfer its 0.18 micron process to its subsidiary in Shanghai, with approval from Taipei (TSMC 2006b). Its direct investment in the subsidiary totaled US$276 million by the end of 2004. As of 28 February 2007, TSMC Shanghai had a capital stock of US$371 million (TSMC 2005a: 140; 2007a: 62).

As of 2008, TSMC Shanghai still trailed behind its main competitors in China in terms of process technologies, production capacity and market share. Although TSMC remains the world's leading foundry player, Taipei's restrictions have prevented TSMC from equipping its fab in China with its latest technology. As of early 2008, its Shanghai subsidiary still used process technologies (i.e. 0.35–0.18 micron) which lagged behind those available at some China-based foundries including SMIC, HeJian and Grace.

In terms of production capacity, TSMC Shanghai manufactured 5,000 8-inch wafers in March 2005 and 6,000–7,000 in June 2005.[199] In October 2005, it produced 12,000 wafers.[200] In early 2007, TSMC hoped that its China fab would triple its output for the coming year after it began to use its 0.18 micron process (EMSNow 2007). The performance of the subsidiary trailed behind this objective: it manufactured only 31,000 8-inch wafers in December 2007, and 42,500 in December 2008; its whole-year capacity reached 375,000 in 2007 and 543,000 in 2008.[201]

TSMC Shanghai's use of relatively low technologies compared to those available at China's leading foundry companies has undermined its market performance. According to the president of a Taiwanese IDM subsidiary in Shanghai in 2005, "TSMC Shanghai has followed the rules and has not used the 0.18 micron process in its fab here. Its market share has thus been taken away by others."[202]

The financial performance of TSMC Shanghai has been mixed. While its net loss narrowed by 73.65 per cent in 2006 compared to 2005, its net sales decreased by 45 per cent during the same period. Its loss-ridden performance is in sharp contrast to its parent company's outstanding financial record (TSMC 2006a: 52; 2007a: 65, 68).

Meanwhile, TSMC Shanghai cooperated with local chip players in China through three major programs. First, it has started a cooperation project with Tsinghua University to cultivate local talent. Second, it has offered a regular

MPW service to potential local customers in order to enlarge its local customer base. Third, it has trained its newly hired local engineers. As of September 2005, about 150 of its Chinese engineers had been dispatched to Taiwan for related training programs.[203]

In short, the China subsidiary of the world's strongest foundry player has received resources from its parent company in terms of talent, technology and investment. Although it uses technologies to the level permitted by Taipei and, therefore, has underperformed in the market so far, its presence in China is a long-term asset to the Chinese IC sector. TSMC's leading performance in the global foundry sector, its niche in global state-of-the-art process technology and managerial expertise, and its outstanding corporate governance record may help upgrade China's chip capability through MPW services, local talent training schemes and foundry services available at its Shanghai subsidiary.

I conclude this section with the following observations. First, among the four foundries I have examined, SMIC and Grace pioneered the migration in the sub-sector. Since then, the migration has continued steadily, as revealed in these cases and beyond. Second, some of the players involved in SMIC, HeJian and Grace have led the geographical shift by circumventing or violating Taiwanese regulations, leaving the state in tatters. These instances in the sub-sector and beyond are summarized in Table 5.5.

Third, these firms have benefited from varying degrees of input from Taiwan in terms of talent, technology, and investment. So far, their performance has been impressive in China. In 2006, all were among China's top eight foundry firms. As they quickly emerge as China's major foundry companies, their ability to contribute to Chinese foundry sales revenue and to supply chips to China's domestic market increases accordingly. In 2004 alone, the sales revenue of SMIC, HeJian and Grace accounted for 58 per cent of the total. If second-tier player CSMC is included, these four Taiwan-related foundries generated 61 per cent of Chinese foundry sales revenue in the same year (Lapedus 2005). More importantly, their rise has also indicated a gradual shift in IC fabrication capability from Taiwan, which houses the world's strongest foundries, to China. Fourth, if the migration of Taiwan's IC fabrication sub-sector to China continues, it will assist further in upgrading China's IC fabrication capabilities through the offer of MPW services to local IC design houses and the cultivation of local engineering talents.

Conclusion

The following represents my key findings on the migration. First, Taiwanese chip production activities, which largely concentrate on the commercial market, began to shift to China as early as the late 1980s and have gathered momentum since 2000 due to a confluence of enabling, push and pull factors. Second, the scope of the migration has been extensive across the supply chain, contributing to the new

Table 5.5 Six ways of circumventing Taiwanese state control by Taiwanese non-state actors spearheading chip sector migration to China

Patterns	Comments	Cases
In the name of private investment	Investments in China's chip sector by Taiwanese persons without prior approval from Taipei	• Chairman of an IC design house vs. an IC design firm in Shanghai • CEO of a PC company vs. a packaging and testing start-up in Shanghai • Shareholders of an IC design house vs. a start-up in Zhuhai
In the name of "technical support"	China-based Taiwanese subsidiaries engage in design tasks beyond the business scope permitted by Taipei	• A design house in Shanghai • A design house in Suzhou • A design house in Beijing
Secretive ownership change	Take-over of China-based firms without Taipei's approval to run operations banned by Taipei	Take-over of a Shanghai-based firm by a Taiwanese packaging and testing company in 2004, which applied for Taipei's approval in 2006
"I am a US citizen"	Emphasis on US instead of Taiwanese citizenship by dual US-Taiwanese nationals who are leaders in China's chip sector	• CEO of a foundry applied to give up his Taiwanese citizenship while keeping his US one • Chairman and president of a foundry put their nationality down as "US" in company prospectus
High-end production without Taipei's approval	Taiwanese subsidiaries engage in production activities forbidden by Taipei	• A packaging and testing firm in Shanghai • A packaging and testing firm in Suzhou
Pursuit of global state-of-the-art by firms whose inception circumvented Taipei's regulations	Taiwan-related fabs in China whose incorporation was deemed illegal by Taipei pursue technologies beyond the ceilings set by Taipei	• A foundry in Suzhou • A foundry in Shanghai • Both have used process technologies for feature sizes smaller than 0.18 micron and have invested in 12-inch wafer fabrication activities in China

Sources: Interviews and secondary data; Chu 2008.

outsourcing trend in the global industry. It has also challenged the conventional claim that the migration has been confined to IC manufacturing. Moreover, the direction of migration has been complicated. Varied degrees of Taiwanese technology, capital, and talent have fostered the development of the commercial and defense sub-fields of the Chinese chip industrial base.

Third, the geographical shifts in investment, talent, and technology have unfolded in the following manner. Technology transfers have followed different patterns in the three sub-sectors. In the IC fabrication, and packaging and testing sub-sectors, most of the technology transfers have involved mature technologies. However, there are important exceptions. Some of the Taiwan-related foundry start-ups in China have pursued the global state-of-the-art through technology transfers from global chip giants not confined to those headquartered in Taiwan and from semiconductor equipment firms. This is in spite of their reliance on Taiwanese technologies at their initial stages of development. Other companies have transferred core technologies from the firms' headquarters in Taiwan to their subsidiaries in China. In the IC design sub-sector, most of the Taiwanese-owned firms have outsourced low-end tasks to their subsidiaries in China while keeping the high-end and sensitive parts of design activities close to their firms' competencies at home. A substantial amount of IC design technological know-how has been transferred to China at the infra-firm level. In other cases, China-based start-ups led by senior Taiwanese engineers have also benefited from the shifts of human capital across the Strait. The PRC semi-official data indicate that Taiwanese capital represented the largest portion of the total foreign investment in China's IC industry in 2005. The Taiwanese investment in the sector is believed to be higher than has been recorded in official or semi-official data because the capital involved has often entered China through tortuous channels in order to circumvent scrutiny by Taipei. The flow of talent has been the most important dimension of the migration, as Taiwanese senior executives have become leaders in upgrading China's IC capabilities, particularly in the foundry business. As of November 2005, at least 1,200 Taiwanese engineers and managers were working in various semiconductor operations in China whose senior executives I interviewed. The resultant knowledge transfers have helped China to cope with a paucity of experienced local engineers, a condition which has long impeded the development of the sector.

The fourth finding involves the impact of the migration on the capacity of the Taiwanese state. Taiwanese semiconductor actors have led the migration in ways that circumvent or violate regulations at home, undermining the capacity of the Taiwanese state to curb the migration.[204] They have also eroded the power and autonomy of the Taiwanese state vis-à-vis that of the non-state actors involved in the sphere of semiconductors. Such a development demonstrates the first major impact of the globalization of IC production on security as far as the power and autonomy of the state is concerned, a theme I shall re-visit in Chapter 7.

The final finding concerns the impact of the migration on the development of the Chinese chip industry. Taiwan has helped China upgrade its indigenous semiconductor capability through the infusion of talent, technology, and capital from the Silicon Island. The emergence of SMIC and Actions as China's leading foundry and IC design player respectively have benefited from Taiwan, indicating that the migration has resulted in gradual shifts of IC design and fabrication capabilities to China. Moreover, Taiwanese-run or -owned firms operating in China have sown the seeds for the development of China's indigenous chip capability. Taiwan-related foundries offer their local customers MPW services, which may help to enhance China's indigenous chip design capabilities. Taiwanese-run or -owned hubs have helped to train China's inexperienced engineers, and have created pools of highly skilled local labor that are critical to China's long-term technical development. Taiwan-related foundries are pillars of the overall Chinese chip sector, cultivating local IC design houses with their advanced foundry services and feeding packaging and testing firms with orders.

The Taiwan factor has been univocally recognized by industry interviewees I have spoken with in the USA and Asia as an asset to China in its efforts to catch up in the global chip race. As Klaus Wiemer, former president of TSMC and former CEO of Chartered, observed in 2005:

> I don't think there would be much of a semiconductor industry in China today had it not being for Taiwan. So they really should be grateful . . . Taiwan essentially went across the Strait and started to participate in the Chinese semiconductor industry. They took money over there. They took people and management skills."[205]

According to Wang Qinsheng, Chair of HED and Chair of CIDC:

> Through various forms of "internationalization," caliber and capital from Taiwan have entered mainland China and played important roles. Grace, SMIC, TSMC and HeJian, for example, cannot shake off their links to Taiwan . . . Chen Shui-bian is unable to control the trend. Taiwan has already exerted its impact here.[206]

Some analysts have highlighted linguistic and cultural proximity as the reason why input from Taiwan has played a distinctive role in helping China catch up in the global IC race in comparison with their counterparts from Europe, Japan, and the USA. In a sector characterized by "learning by doing," such a factor has accelerated the speed and efficiency of knowledge transfers from senior Taiwanese engineers and managers to their junior Chinese counterparts. Speaking from his experience in charge of foreign technology transfers in China's Ministry of Electronics Industry, an industry player said, "The Taiwanese have contributed to the indigenous chip industry in China . . . As far as efficiency is concerned, we speak the same language. It is faster for them to communicate with us than the French or the British."[207]

As Taiwanese chip actors have helped China move a step further towards developing a fully integrated semiconductor capability, what are the security implications of these commercial activities? Chapter 6 attempts to answer this question.

Notes

1 Interview, 12 January 2005, Dallas, TX, USA.
2 Interview, 30 August 2005, Beijing, China.
3 Also see Mathews and Cho 2000; Tung 2001; Saxenian 2002; Wade 2004: 103–8; Saxenian 2006; Breznitz 2007: 12–40, 97–145; Shih *et al.* 2007; Brown and Linden 2009: 177–82.

4 Also see interview with F.C. Tseng, vice-chairman of TSMC, 29 June 2005, Hsinchu, Taiwan; interview with Klaus C. Wiemer, former President of TSMC and former CEO of Chartered Semiconductor, 12 January 2005, Dallas, TX, USA.

5 Also see interview with the CEO of UMC, 11 July 2005, Hsinchu, Taiwan.

6 Suppliers include government-related institutes and local commercial firms. See interview with the former president of a Taiwanese IC design house that has sold a wireless communications commercial chip to the Taiwanese military, 5 July 2005, Taipei, Taiwan; interview with the president of a Taiwanese IC design house involved in a project tailored to improve the guidance system of an indigenously developed missile, 19 August 2005, Hsinchu, Taiwan; interview with a senior industry player who was involved in the fabrication of military chips for Taiwan's missiles, 29 June 2005, Hsinchu, Taiwan; interview with Abe C. Lin, Director General of Integrated Assessment Office at Ministry of National Defense, 27 June 2005, Taipei, Taiwan; interviews with former presidents of CSIST, 14 July 2005 and 9 August 2005, Taipei, Taiwan; interview with a member of Taiwan's defense industry who formerly worked at CSIST, 27 October 2005, Taipei, Taiwan.

7 They include TSMC, UMC, Winbond, SPIL, etc. For the role of Taiwan in providing IC manufacturing services to US military subcontractors, also see Ciufo 2004; Fabula and Padovani 2004; O'Neil 2004; Defense Science Board Task Force 2005: 24; Office of the Under-Secretary of Defense Acquisition 2005: 12, 14; interview with the CEO of UMC, 11 July 2005, Hsinchu, Taiwan; interview with a senior director at TSMC, 25 August 2005, Beijing, China.

8 The existence of other regulatory regimes involving the three major state actors further bespeaks the strategic nature of the industry. See General Accounting Office 2002; Bureau of Industry and Security 2003; The Wassenaar Arrangement on Export Controls for Conventional Arms and Dual-Use Goods and Technologies 2006; 2008; Bureau of Industry and Security 2007a; 2007b; Chu 2008: 55–6; interview with Matthew S. Borman, Deputy Assistant Secretary of Commerce, 1 February 2005, Washington, DC, USA; interview with a working-level official at the Department of Commerce, 1 February 2005, Washington, DC, USA; interview with officials at the Ministry of Economic Affairs (MOEA), 8 July 2005, Taipei, Taiwan.

9 Due to space constraints, case studies in the packaging and testing sub-sector are not included here.

10 Fuller's study (2008), however, does focus on both IC fabrication and design sub-sectors.

11 Also see interview with the president of a Taiwanese IC design house, 19 August 2005, Hsinchu, Taiwan; interview with a senior Taiwanese industry player, 24 March 2005, Shanghai, China.

12 Also see interview with a senior executive at a Taiwanese subsidiary, 27 September 2005, Shanghai, China.

13 Interview with the president of SEMI China, 24 March 2005, Shanghai, China. In 2004, a visiting delegation of Taiwanese IC design houses disclosed this information to the interviewee. Fuller's study (2008) concludes that seven of Taiwan's top ten firms have design teams in China.

14 Interview, 12 July 2005, Taipei, Taiwan. His firm runs numerous subsidiaries in China.

15 Interview, 4 August 2009, Taipei, Taiwan.

16 In December 2006, Taipei approved applications from Powerchip and ProMOS, for their planned 8-inch fab operations in China. See Chiang 2006; Digitimes 2007.

17 Interview, 8 December 2004, San Jose, CA, USA.

18 Interview, 21 September 2005, Suzhou, China.

19 They were from TSMC, UMC, SMIC, Huahong NEC, CSMC, and GSMC.

20 ASE acquired GAPT in 2004 without approval from Taipei. It then engaged in wire bond-based IC packaging and testing in China under the GAPT umbrella. In 2006, it applied for Taipei's approval of the deal. See interview with a manager at GAPT,

28 September 2005, Shanghai, China; interview with a manager at a Taiwanese-run foundry, 27 September 2005, Shanghai, China.

21 Interview, 24 August 2005, Beijing, China. Wang became the CEO of Shanghai's Huahong Group in September 2005. In 2009, he became the CEO of SMIC.

22 Interview, 14 September 2005, Ningbo, China. Also see Lin 2002; Electronic Engineering Times Asia 2004.

23 In late 2008, however, both firms began to seek government bailout assistance from the Taiwanese authorities to address mounting losses amid the ongoing global economic crisis. See Semiconductor Industry Association 2009: 6.

24 This contrasts with Powerchip's deferred actual investment in China, which arguably results from its mounting financial losses in recent years.

25 This is evidenced in the cases of ASE and SPIL. See interview with a manager at GAPT, 28 September 2005, Shanghai, China; interview with a manager at a Taiwanese-run foundry, 27 September 2005, Shanghai, China; interview with the president of SPIL's subsidiary, 22 September 2005, Suzhou, China; ASE 2007a.

26 Interview with Peter Cheng-yu Chen, the chairman of CSMC, 25 September 2005, Shanghai, China.

27 Interview, 27 September 2005, Shanghai, China.

28 Interview with the president of a fabless design house, 29 June 2005, Taipei, Taiwan. Field applications engineers give technical support to their customers to ensure that their products meet customer demand.

29 Interview with a vice-president of a Taiwanese IC design house, 15 July 2005, Hsinchu, Taiwan; interview with the president of another Taiwanese IC design house, 27 October 2005, Taipei, Taiwan.

30 Interview, 5 September 2005, Beijing, China.

31 Interview with the president of the firm, 21 September 2005, Suzhou, China.

32 Interview, 27 October 2005, Taipei, Taiwan. Also see interview with a senior executive of a Taiwanese IC design house, 12 July 2005, Taipei, Taiwan.

33 Interview with a general manager at a US venture capitalist subsidiary, 31 September 2005, Beijing, China.

34 Interview, 20 October 2005, Hsinchu, Taiwan. Also see interview with the president of a Taiwanese fabless design house, 29 June 2005, Taipei, Taiwan.

35 Interview with a vice-president of the firm, 15 July 2005, Hsinchu, Taiwan. Also see phone interview with the spokesperson of another firm, 25 October 2005, Hsinchu, Taiwan.

36 Interview with a senior Taiwanese industry player, 24 March 2005, Shanghai, China.

37 Interview with the president of a Taiwanese subsidiary, 27 September 2005, Shanghai, China; interview with a director at a US semiconductor subsidiary, 28 March 2005, Shanghai, China; interview with the former president of a Taiwanese IC design house, 4 July 2005, Taipei, Taiwan.

38 Interview, 5 September 2005, Beijing, China.

39 Interview, 25 September 2005, Shanghai, China. At the time of the interview, Zou was an academician of CAS and supervised students at the Shanghai Institute of Metallurgy of CAS, which is now reorganized as Shanghai Institute of Microsystem and Information Technology.

40 For the role of expatriate Chinese entrepreneurs in China's economic and technical development, see Weidenbaum and Hughes 1996; Leng 2002; Saxenian 2007.

41 For other estimates, see Howell *et al.* 2003; interview with David N.K. Wang, executive vice-president of Applied Materials, 24 August 2005, Beijing, China. No official data are available on the number of Taiwanese professionals who are working in the Chinese chip sector. My interview data allow us to estimate a lower limit to the number in question.

42 These companies had more than 17,000 employees, the majority of whom were hired locally.

43 Interview with a public relations manager at SMIC, 30 September 2005, Shanghai, China; phone interview with a manager at HeJian, 21 March 2005, Shanghai, China; interview with the president of TSMC Shanghai, 26 September 2005, Shanghai, China; interview with the chairman of GSMC, 27 September 2005, Shanghai, China; interview with the chairman of NSSI, 14 September 2005, Ningbo, China; interview with the chairman of CSMC, 25 September 2005, Shanghai, China.

44 Interview with the president of Jiangsu Changdian Electronics Technology Co. Ltd., 19 September 2005, Jiangyin, China; interview with the president of a Taiwanese packaging and testing subsidiary, 27 September 2005, Shanghai, China

45 Interview with the senior vice-president of a Taiwanese IC design house, 4 August 2009, Taipei, Taiwan; interview with the vice-president of a Taiwanese IC design house, 5 August 2009, Hsinchu, Taiwan; interview with the spokesman of a Taiwanese IC design house, 6 September 2009, London, UK.

46 Interview, 30 September 2005, Shanghai, China. Also see interview with the president of Jiangsu Changdian Electronics Technology Co. Ltd., 19 September 2005, Jiangyin, China.

47 Interview with the president of a Taiwanese subsidiary in China, 9 September 2005, Beijing, China; interview with the chairman of a Taiwanese IC design house, 21 September 2005, Suzhou, China; interview with a senior executive of a Taiwanese IC design house, 12 July 2005, Taipei, Taiwan; interview with the former president of a Taiwanese IC design house, 20 July 2005, Taipei, Taiwan; interview with the president of a Taiwanese IC design house, 19 August 2005, Hsinchu, Taiwan; interview with the president of a Taiwanese IC design house, 20 October 2005, Hsinchu, Taiwan.

48 Interview with a senior executive at a Taiwanese IC design house, 12 July 2005, Taipei, Taiwan.

49 Interview with the president of a Taiwanese IC design subsidiary, 9 September 2005, Beijing, China.

50 Interview, 21 September 2005, Suzhou, China.

51 Interview, 19 August 2005, Hsinchu, Taiwan.

52 Interview with a senior executive at a Taiwanese IC design house, 12 July 2005, Taipei, Taiwan; interview with a vice-president of a Taiwanese fabless design company, 15 July 2005, Hsinchu, Taiwan; interview with a vice-president of a Taiwanese IC design firm, 10 August 2005, Hsinchu, Taiwan; interview with the president of a Taiwanese foundry subsidiary, 26 September 2005, Shanghai, China; interview with Chintay Shih, former president of ITRI, 15 July 2005, Hsinchu, Taiwan.

53 This support mostly includes field applications engineering services. It also, at times, involves a comparable level of R&D assistance to customers who have moved their high-end R&D activities to China. Interview with a senior executive at a Taiwanese IC design house, 12 July 2005, Taipei, Taiwan.

54 Interview with a vice-president of a Taiwanese IC design house, 15 July 2005, Hsinchu, Taiwan. See also interview with a vice-president of another Taiwanese IC design house, 10 August 2005, Hsinchu, Taiwan.

55 Interview with the CEO of SMIC, 30 September 2005, Shanghai, China; interview with the vice-chairman of TSMC and the chairman of TSMC Shanghai, 29 June 2005, Hsinchu, Taiwan; interview with the president of TSMC Shanghai, 26 September 2005, Shanghai, China; interview with a senior director at TSMC, 8 December 2004, San Jose, CA, USA; interview with the CEO of UMC, 11 July 2005, Hsinchu, Taiwan; interview with the president of HeJian, 21 September 2005, Suzhou, China; phone interview with a manager at HeJian, 21 March 2005, Shanghai, China.

56 Interview, 11 July 2005, Hsinchu, Taiwan.

57 Interview, 21 September 2005, Suzhou, China.

58 Interview, 29 June 2005, Hsinchu, Taiwan.

59 Interview, 26 September 2005, Shanghai, China.
60 Interview, 28 March 2005, Shanghai, China.
61 Interview with a public relations manager at SMIC, 30 September 2005, Shanghai, China; interview with an official-turned industry player, 10 October 2005, Shenzhen, China; interview with the president of a Taiwanese IC design house, 20 October 2005, Hsinchu, Taiwan.
62 Interview with Nasa Tsai, the former president of Grace, 14 September 2005, Ningbo, China; interview with the chairman of Grace, 27 September 2005, Shanghai, China; Sung and Chan 2007.
63 Interview with the president of HeJian, 21 September 2005, Suzhou, China.
64 Interview with the chairman of TSMC Shanghai, 30 June 2005, Hsinchu, Taiwan.
65 Interview with the chairman of CSMC, 25 September 2005, Shanghai, China.
66 Interview with the CEO of SMIC, 30 September 2005, Shanghai, China; interview with the chairman of CSMC, 25 September 2005, Shanghai, China.
67 Interview with the president of the subsidiary, 27 September 2005, Shanghai, China.
68 Interview with a director at the company, 28 September 2005, Shanghai, China.
69 Interview with the president of a Taiwanese packaging and testing subsidiary, 22 September 2005, Suzhou, China.
70 Interview with the president of HeJian, 21 September 2005, Suzhou, China.
71 Phone interview, 21 March 2005, Shanghai, China.
72 Ibid.
73 Interview with the president of HeJian, 21 September 2005, Suzhou, China. Also see phone interview with a manager at HeJian, 21 March 2005, Shanghai, China; interview with the president of a Taiwanese packaging and testing firm, 22 September 2005, Suzhou, China.
74 Interview with the former president of a Taiwanese IC design house who led the company's expansion to China, 20 July 2005, Taipei, Taiwan; interview with the president of a Taiwanese fabless design firm, 27 October 2005, Taipei, Taiwan; phone interview with the spokesman of a leading Taiwanese IC design house, 25 October 2005, Hsinchu, Taiwan; interview with the head of a design house subsidiary, 9 September 2005, Beijing, China; interview with a vice-president of a Taiwanese IC design house, 15 July 2005, Hsinchu, Taiwan; interview with a vice-president of a Taiwanese fabless design firm, 10 August 2005, Hsinchu, Taiwan; interview with a vice-president of a Taiwanese IC design house, 19 August 2005, Hsinchu, Taiwan; interview with the president of TSMC Shanghai, 26 September 2005, Shanghai, China; interview with Chintay Shih, former president of ITRI, 15 July 2005, Hsinchu, Taiwan.
75 Interview with the president of a Taiwanese fabless design firm, 27 October 2005, Taipei, Taiwan; interview with the former president of a Taiwanese IC design house, 20 July 2005, Taipei, Taiwan.
76 Interview with the former CEO of a Taiwanese chip design house, 20 July 2005, Taipei, Taiwan.
77 As discussed in Chapter 3, software tasks are not viewed as part of a conventional chip design flow, although they have become increasingly important to today's IC design houses.
78 Interview with the former president of a Taiwanese IC design house who spearheaded the company's operations in China, 20 July 2005, Taipei, Taiwan; interview with the president of a Taiwanese IC design house, 27 October 2005, Taipei, Taiwan; interview with a Taiwanese American venture capitalist, 31 August 2005, Beijing, China.
79 Interview, 15 July 2005, Hsinchu, Taiwan.
80 Interview, 9 September 2005, Beijing, China.
81 Interview with the president of a Taiwanese fabless design firm, 27 October 2005, Taipei, Taiwan.

82 Interview with a vice-president of a Taiwanese IC design house, 15 July 2005, Hsinchu, Taiwan.
83 Interview with the president of HeJian, 21 September 2005, Suzhou, China. For the case of SMIC, see interview with the CEO of SMIC, 30 September 2005, Shanghai, China.
84 Interview, 29 September 2005, Shanghai, China.
85 Interview, 30 September 2005, Shanghai, China.
86 Interview, 21 September 2005, Suzhou, China.
87 Interview, 29 June 2005, Hsinchu, Taiwan. In particular, he highlighted a preferential tax treatment concerning a five-year, tax-free import of semiconductor manufacturing equipment.
88 Interview with a senior director at TSMC, 8 December 2004, San Jose, CA, USA.
89 Interview with a vice-president of a Taiwanese IC design house, 15 July 2005, Hsinchu, Taiwan.
90 Interview with a vice-president of a Taiwanese IC design house, 15 July 2005, Hsinchu, Taiwan.
91 Interview with the president of a Taiwanese fabless design house, 19 August 2005, Hsinchu, Taiwan; interview with a vice-president of a Taiwanese IC design firm, 19 August 2005, Hsinchu, Taiwan.
92 Interview, 15 July 2005, Hsinchu, Taiwan.
93 Speech by Chang at a conference in San Jose on 7 December 2004. SMIC's manufacturing costs were about 5 per cent lower than those of comparable fabs in Taiwan. Interview with Xue Zi, Deputy Secretary General of Shanghai Integrated Circuit Industry Association, 29 September 2006, Shanghai, China.
94 Interview, 8 December 2004, San Jose, CA, USA.
95 Interview, 27 September 2005, Shanghai, China. Also see interview with the president of a Taiwanese packaging and testing subsidiary, 22 September 2005, Suzhou, China.
96 For the orthodox view, see Flamm 1985; Brown and Linden 2005; Ernst 2005; Fabless Semiconductor Association and Industry Directions Inc. 2007: 3, 7–8; interview with the president of SIA, 8 December 2004, San Jose, CA, USA; interview with a vice-president of a Taiwanese IC design house, 19 August 2005, Hsinchu, Taiwan.
97 Interview, 14 September 2005, Ningbo, China.
98 Interview with the president of a Taiwanese IC design house, 20 October 2005, Hsinchu, Taiwan. Also see interview with the president of a Taiwanese packaging and testing firm in China, 22 September 2005, Suzhou, China; interview with David N.K. Wang, Executive Vice-President of Applied Materials, 24 August 2005, Beijing, China.
99 For an earlier version of the case studies, see Chu 2013.
100 VIA's microprocessors accounted for 1 per cent of the worldwide market. See interview with a senior executive at the company, 15 July 2005, Taipei, Taiwan; interview with the former president of a Taiwanese IC design house, 5 July 2005, Taipei, Taiwan.
101 The *IC Insights'* Strategic Reviews Database provided by the CEO of UMC.
102 Interview with the head of VIA Technologies (China) Ltd., 9 September 2005, Beijing, China.
103 Interview with the president of a Taiwanese IC design house, 12 August 2005, Hsinchu, Taiwan.
104 Interview with Qinsheng Wang, Chair of HED, and Chair of CIDC, 30 August 2005, Beijing, China.
105 They include Beijing, Hangzhou, Shanghai, and Shenzhen.
106 Company data on the Taiwan Stock Exchange website.
107 Interviews with staff at the subsidiary, 5 September 2005, Beijing, China. For a different estimate, see interview with a vice-president of a Taiwanese IC design house, 10 August 2005, Hsinchu, Taiwan.

108 Interviews with senior executives of the IC design house, 4 August 2009, Taipei, Taiwan.
109 Interview, 12 July 2005, Taipei, Taiwan.
110 These include subsidiaries in Beijing, Shanghai, and Hangzhou.
111 Interview with a senior executive of the company, 12 July 2005, Taipei, Taiwan. The subsidiary in Beijing had engaged in "some projects" concerning IC design, most of which were software related. See interview with the head of the subsidiary, 5 September 2005, Beijing, China.
112 Interview with a vice-president of a Taiwanese IC design house with experiences running the firm's operations in China, 19 August 2005, Hsinchu, Taiwan. VIA's operation center in Shanghai was a result of its acquisition of S3 Graphics, a specialist in graphics ICs.
113 Interview, 5 September 2005, Beijing, China.
114 Interview with the executive vice-president of Realtek, 19 August 2005, Hsinchu, Taiwan.
115 Interview with Chin Tan Huang, Executive Secretary of MOEA's Investment Commission, 18 August 2005, Taipei, Taiwan.
116 Interview with the president of a Taiwanese IC design house, 12 August 2005, Hsinchu, Taiwan.
117 Interview with the president of NSSI, 14 September 2005, Ningbo, China. Also see interview with the spokesman of another Taiwanese IC design house, 25 October 2005, Hsinchu, Taiwan.
118 Interview, 19 August 2005, Hsinchu, Taiwan.
119 Phone interview, 20 September 2005, Suzhou, China.
120 In January 2008, another former executive at Realtek joined Actions as the vice-president of marketing. See ASC 2008.
121 Interview with a co-founder and the former president of Ali who had run the firm's operations in China, 20 July 2005, Taipei, Taiwan.
122 The MOEA data show that between 2000 and 2004, Taipei approved the company's investment plans in five subsidiaries located in four Chinese cities (i.e. Suzhou, Zhuhai, Shanghai, and Beijing). Neither interview data nor MediaTek's 2005 annual report mentioned the actual operation of the subsidiary in Suzhou. The co-founder identified one operation center in Xi'an, which was not mentioned in the MOEA data. See Mediatek Inc. 2006: 175; Ministry of Economic Affairs 2007; interview with a co-founder of Ali, 20 July 2005, Taipei, Taiwan; interview with a senior executive of Ali, 27 October 2005, Taipei, Taiwan.
123 Interview, 20 July 2005, Taipei, Taiwan.
124 Telephone interview with a spokesman at MediaTek, 25 October 2005, Hsinchu, Taiwan.
125 Interview, 27 October 2005, Taipei, Taiwan.
126 Interview with the president of Faraday Technology, 20 October 2005, Hsinchu, Taiwan; interview with a senior executive of Ali, 27 October 2005, Taipei, Taiwan.
127 As discussed in Chapter 4, QPL identifies the types of military electronic parts and components produced by qualified onshore manufacturers for end-uses by the PLA.
128 A senior industry player regarded Kao as a partner of Silan. See interview, 14 September 2005, Ningbo, China.
129 Also see interview with a senior executive of a Taiwanese subsidiary, 27 September 2005, Shanghai, China.
130 In another instance not covered in the case studies, the chairman of a leading Taiwanese IC design house has invested, in his personal capacity, in Spreadtrum Communications Inc., which runs a subsidiary in Shanghai, Spreadtrum Communications (Shanghai) Co., Ltd. See interview with the CEO of Spreadtrum, 6 December 2005, Cambridge, England. In 2006, Spreadtrum was China's tenth largest IC design

house. With sales of 1.1 billion RMB in 2007, it became China's third largest IC design powerhouse. See Spreadtrum Communications Inc. 2008.

131 The emergence of Spreadtrum as one of China's top ten IC design houses in 2006 and beyond has benefited from Taiwanese investment.

132 Interview, 8 December 2004, San Jose, CA, USA.

133 Speech by Chang at a conference in San Jose on 7 December 2004.

134 Interviews with a public relations manager and the CEO office chief at SMIC, 30 September 2005, Shanghai, China.

135 For the linkage between Shanghai municipal government and Shanghai Industrial Holdings Ltd., see Howell *et al.* 2003: 93; phone interview with a manager at HeJian, 21 March 2005, Shanghai, China.

136 On Datang's success in developing China's first indigenous core network router in 2001 and its military utility, see Cheung 2009: 200–21.

137 Interview with the chairman of GSMC, 27 September 2005, Shanghai, China; interview with Qinsheng Wang, Chair of HED, and Chair of CIDC, 30 August 2005, Beijing, China; interview with the president of SIA, 8 December 2004, San Jose, CA, USA.

138 Also see interview with a senior director at TSMC, 18 December 2004, San Jose, CA, USA; interview with the CEO of a fabless design house, 7 September 7 2005, Beijing, China.

139 Also see interview with the vice-chairman of TSMC and the chairman of TSMC Shanghai, 29 June 2005, Hsinchu, Taiwan; interview with the president of a Taiwanese IC design house, 20 October 2005, Hsinchu, Taiwan.

140 Interview with a manager at SMIC, 30 September 2005, Shanghai, China.

141 Interview with the president of one of SMIC's major competitors, 26 September 2005, Shanghai, China.

142 Interview with an involved venture capitalist, 31 August 2005, Beijing, China.

143 Interview, 17 September 2005, Shanghai, China.

144 Interview, 9 September 2005, Beijing, China.

145 Also see interview with a senior trade official, 8 July 2005, Taipei, Taiwan.

146 Interview, 30 September 2005, Shanghai, China.

147 Interview, 26 September 2005, Shanghai, China.

148 Interview, 10 September 2005, Beijing, China.

149 Interview, 4 February 2005, Washington, DC, USA.

150 Interview with the vice-president of a Taiwanese IC design house that has used foundry services at TSMC and SMIC, 19 August 2005, Hsinchu, Taiwan.

151 Interview, 8 December 2004, San Jose, CA, USA.

152 Interview, 12 July 2005, Taipei, Taiwan.

153 Also see interview with the CEO of a design house, 7 September 2005, Beijing, China; interview with the chairman of GSMC, 27 September 2005, Shanghai, China.

154 They were the Global Strategic Investment Fund and Prudence Capital Co. Both cancelled their funding commitments to SMIC due to pressure from Taipei. They were fined NT$1 million each for their illegal involvement in SMIC.

155 Also see interview with the CEO of an IC design house, 7 September 2005, Beijing, China.

156 In addition, Taiwan was the largest market for the company in 2002 in terms of sales revenue, which contributed to 47.7 per cent of the company's revenue. See SMIC 2005b: 99; 2006b.

157 Interview, 6 December 2004, Palo Alto, CA, USA. Nishi was instrumental in facilitating TI–SMIC technological cooperation in his former capacity as a senior executive at TI.

158 Interview with the CEO of UMC, 11 July 2005, Hsinchu, Taiwan; interview with a vice-president of a Taiwanese IC design house which has used foundry services at TSMC and SMIC, 19 August 2005, Hsinchu, Taiwan; interview with C. Mark Melliar-

Smith, former President and CEO of International SEMATECH, 4 January 2005, Austin, TX, USA.

159 Interview, 4 January 2005, Austin, TX, USA.

160 Interview, 2 September 2005, Beijing, China.

161 Interview, 7 December 2004, San Jose, CA, USA.

162 For Washington's misgivings, see interviews with DTSA officials, 2 February 2005, Alexandria, VA, USA; interview with a US delegation, which comprised Pentagon officials and members of government-related defense microelectronics agencies, during its inspection of semiconductor facilities in China, 16 March 2005, Shanghai, China.

163 Interview with a public relations manager at SMIC, 30 September 2005, Shanghai, China; presentation of a SMIC manager at SEMI China 2005, 16 March 2005, Shanghai, China.

164 For a typical complaint about the US export control regulations, see interview with Walden C. Rhine, the CEO of Mentor Graphics Inc., 7 December 2004, San Jose, CA, USA.

165 Interview with a manager at SMIC, 30 September 2005, Shanghai, China.

166 Interview with the CEO of a leading foundry, 11 July 2005, Hsinchu, Taiwan; interview with a vice-president of a Taiwanese IC design house which uses foundry services at TSMC and SMIC, 19 August 2005, Hsinchu, Taiwan; phone interview with a deputy director of the electronic industry intelligence division of the IEK, ITRI, Taiwan, 2 November 2005, Hsinchu, Taiwan; interview with a senior director at TSMC, 27 August 2005, Beijing, China.

167 Interview with the CEO of a leading foundry, 11 July 2005, Hsinchu, Taiwan; interview with the president of Taiwan Mask Corp., 3 November 2005, Hsinchu, Taiwan.

168 Interview, 27 September 2005, Shanghai, China.

169 He was given a US$2 million, five-year contract as a director and adviser of Grace.

170 The number of employees was provided by the company's chairman.

171 Interview with the chairman of GSMC, 27 September 2005, Shanghai, China.

172 These include Chinese returnees.

173 Interview, 14 September 2005, Ningbo, China.

174 In August 2005, Grace appointed Shi-chung Sun as its new CTO. In January 2006, Kuan-yang Liao was recruited as its COO. Both were senior Taiwanese engineers. See GSMC 2005; 2006a.

175 Also see interview with the chairman of GSMC, 27 September 2005, Shanghai, China.

176 Interview, 27 September 2005, Shanghai, China.

177 Phone interview with a senior industry player, 21 March 2005, Shanghai, China; interview with a senior director at TSMC, 8 December 2004, San Jose, CA, USA; interview with a company insider, 27 September 2005, Shanghai, China.

178 Interview with a senior industry player, 20 March 2005, Shanghai, China; interview with the chairman of a Taiwanese IC design house, 21 September 2005, Suzhou, China.

179 Phone interview with a manager at HeJian, 21 March 2005, Shanghai, China.

180 Interview with the president of HeJian, 21 September 2005, Suzhou, China.

181 Interview with a manager at HeJian, 21 September 2005, Suzhou, China.

182 Interview with the president of HeJian, 21 September 2005, Suzhou, China.

183 Interviews, 21 September 2005, Suzhou, China.

184 Phone interview, 21 March 2005, Shanghai, China.

185 Interview, 4 December 2004, San Jose, CA, USA.

186 Over 80–90 per cent of the company's customers are headquartered in Taiwan. Phone interview with a sales manager at HeJian, 21 March 2005, Shanghai, China.

187 Interview, 21 September 2005, Suzhou, China.

188 Interview, 5 July 2005, Taipei, Taiwan.

189 Phone interview, 21 March 2005, Shanghai, China.

190 Interview, 24 June 2005, Taipei, Taiwan.
191 Interview with the president of HeJian, 21 September 2005, Suzhou, China.
192 Interview with the president of HeJian, 21 September 2005, Suzhou, China.
193 Interview, 21 September 2005, Suzhou, China.
194 Interviews with Taiwanese employees at HeJian, 21 September 2005, Suzhou, China.
195 Interview with pertinent TSMC staff, June 2003, Hsinchu, Taiwan.
196 Interview with a senior executive at one of TSMC's major rivals, 21 September 2005, Suzhou, China.
197 Interview with the chairman of TSMC Shanghai, 29 June 2005, Hsinchu, Taiwan.
198 Interview with the president of TSMC Shanghai, 26 September 2005, Shanghai, China.
199 Interview with the chairman of TSMC Shanghai, 29 June 2005, Hsinchu, Taiwan.
200 Interview with the president of TSMC Shanghai, 26 September 2005, Shanghai, China.
201 Email communication with a senior director at the company, 29 March 2009.
202 Interview, 24 March 2005, Shanghai, China.
203 Interview with the president of TSMC Shanghai, 26 September 2005, Shanghai, China.
204 For studies of the ineffectiveness of Taipei's restrictive policy towards Taiwanese investment in China, see Naughton 2004; Steinfeld 2005.
205 Interview, 12 January 2005, Dallas, TX, USA.
206 Interview, 30 August 2005, Beijing, China.
207 Interview, 10 October 2005, Shenzhen, China. For a similar view, see interview with the president of Jiangsu Changdian Electronics Technology Co. Ltd., 19 September 2005, Jiangyin, China; interview with an R&D vice-president of a Taiwanese IC design house, 19 August 2005, Hsinchu, Taiwan; interview with the president of a Taiwanese IDM subsidiary, 24 March 2005, Shanghai, China.

6 The security implications of the sectoral migration

> For any country that wants to develop military capability, having advanced consumer, commercial, and computing integrated circuits helps them . . . Having a strong commercial chip base certainly helps you build sophisticated electronics equipment, and you can't stop that.
>
> (Walden C. Rhine, CEO of Mentor Graphics Inc.[1])

> As China would upgrade its semiconductor capability to the level that its intelligence service could perform nefarious IC compromise acts against the USA and Taiwan, such a prospect is possible, but is extremely improbable. It is extremely hard to do. China is still far from being able to do it. CIA or FBI [Federal Bureau of Investigation] surely will play such a trick, but the intelligence organizations in China have no energy for such a task. But this does not mean that the Chinese won't do it in 20 years' time.
>
> (A senior IC designer experienced in military chip design in the USA[2])

This chapter analyzes the extent to which the migration has created or may create threats and vulnerabilities for the state actors in Washington–Beijing–Taipei relations. The chapter comprises three sections. The first section examines the economic security ramifications of migration, and it is argued that empirical data to date have not substantiated the hollowing-out thesis and that the long-term economic security impact of migration remains undetermined. The second section identifies the long-term technological and defense security repercussions of migration. These include: (1) the Chinese chip industrial base, the PLA modernization, and the balance of power; (2) technological risks resulting from the narrowing chip technology gap and China's semiconductor-targeted IW attacks; and (3) foreign dependency vulnerabilities. The third section concludes with a summary of the findings.

Economic security implications

It is argued below that empirical findings indicate that the migration has resulted in mixed repercussions in economic security terms. While the data so far have not completely substantiated the hollowing-out thesis, there are signs that Taiwan's economic security has been challenged by the migration and that the long-term economic security impact of migration remains undetermined.[3]

An examination of the hollowing-out thesis

In 2001 and 2002, during heated policy debates over Taipei's decision to permit domestic investment in 8-inch wafer foundries in China, two schools of thought emerged regarding Taiwan's economic security because of migration.[4] Skeptics feared that the migration would cultivate competitors in China, hollow out Taiwan's chip sector and cause additional job losses at home. They predicted that the clustering nature of the industry would push the supporting segments of the supply chain, following the relocation of the idle 8-inch fabs to China, to move across the Strait, thus eroding the Taiwanese IC industrial base. Such a hollowing-out argument envisioned an apocalyptic scenario for Taiwan, whose economy has long depended on the semiconductor industry as the single most important engine of growth (Market Intelligence Center 2002; Yang and Hung 2003; Cheng 2005).[5] "Some Taiwanese politicians claimed that as soon as we move our facilities there, Taiwan's economy will be in trouble," recalled F.C. Tseng, vice-chairman of TSMC and chairman of TSMC Shanghai.[6] Supporters of the open-up policy, however, contended that Taiwanese chip firms would prosper by moving their idle plants to China.

My research findings have not completely supported the hollowing-out argument. To date, Taiwan has avoided many of the perceived adverse consequences of migration by shifting to new competencies and maintaining existing competitiveness. The following empirical data show little sign that Taiwan's industry has become eroded because of the migration. Between 1999 and 2008, the industry's revenue has grown steadily with the exception of 2001 and 2008 when the global industry underwent a massive recession. While its revenue was NT$424 billion in 1999, it reached NT$1,347 billion in 2008 (Taiwan Semiconductor Industry Association 2004; 2006; 2007; 2008a; 2009). Its IC manufacturing sub-sector, which comprises foundry and memory businesses, has increased its share of worldwide capacity from 14.5 per cent in 2002 to 19.2 per cent in 2006 (Taiwan Semiconductor Industry Association 2007). As of July 2009, Taiwan's fab capacity by product type was second only to Japan, although it is expected that Taiwan would overtake that position soon (IC Insights 2009).

Moreover, employment in the Taiwanese IC industry remained high between 2003 and 2007 (Figure 6.1) (Institute for Information Industry 2008). But no data are available on the number of job losses caused directly by the migration. Even if the relocation of some Taiwanese IC production activities to China supposedly resulted in job losses at home in the short run, the employment data above seem to indicate that such a cost has not severely diminished Taiwan's economic security. So far, Taiwan's IC industry has maintained its sales growth lifting industry employment levels. The industry has continued to be a high-wage job generator in the Taiwanese economy. Thus, the migration has not resulted in a substantial decline in domestic semiconductor jobs, all things being equal.

Taiwanese chip-makers have also continued to invest heavily in state-of-the-art 12-inch wafer foundries and memory fabs at home, in order to build new competencies. As of June 2007, in Taiwan, 13 12-inch wafer plants were churning out

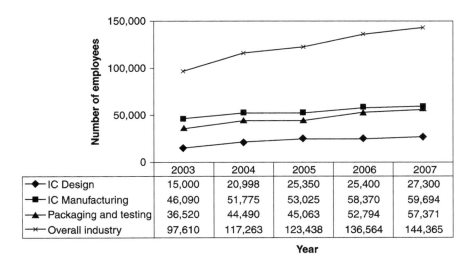

	2003	2004	2005	2006	2007
◆ IC Design	15,000	20,998	25,350	25,400	27,300
■ IC Manufacturing	46,090	51,775	53,025	58,370	59,694
▲ Packaging and testing	36,520	44,490	45,063	52,794	57,371
✕ Overall industry	97,610	117,263	123,438	136,564	144,365

Year

Figure 6.1 Employment in the Taiwanese semiconductor industry, 2003–07.

Source: Institute for Information Industry 2008.

chips, 7 were under construction, and 19 more were at the design stage. The island had the highest density of 12-inch wafer plants in the world (Electronic Engineering Times Asia 2007b). By the end of 2008, 19 12-inch wafer fabs were in operation, 6 were under construction, and 16 more were at the planning stage (Industrial Development Bureau 2009: 21). In 2010, Taiwan remained the world's largest spender on semiconductor equipment, representing nearly 30 per cent of the global market for equipment sales at US$11.19 billion, followed by South Korea (US$8.33 billion), North America (US$5.76 billion), and Japan (US$4.44 billion) (SEMI 2011).

Compared to their major China-based competitor, SMIC, and their distant follower, Singapore's Chartered, both TSMC and UMC have left their rivals behind by continuing to increase their revenues, to command large shares of the worldwide market, and to invest heavily in R&D (Figure 6.2 and Figure 6.3).[7]

To summarize, the above analysis does not support the argument that the sectoral globalization has severely hollowed out the Taiwanese industry, nor has it caused a serious decline in domestic semiconductor jobs. So far, Taiwan's economic security has not been seriously eroded by the migration.

However, the migration has, to some extent, challenged Taiwan's economic security in three major ways. First, migration has helped cultivate competitors in China through the transfers of technology, talent, and investment particularly in the foundry business and lower-end consumer IC design. In the foundry segment, where the migration has exerted the most powerful influence on the Chinese

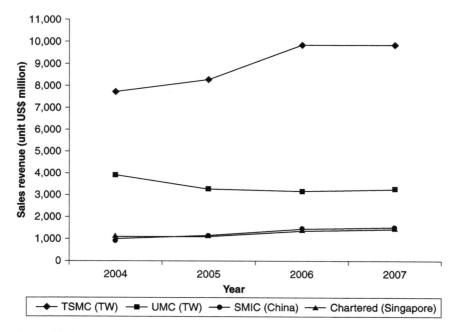

Figure 6.2 Sales revenue of the world's top four foundries, 2004–07.

Sources: IC Insights 2006; 2008; company reports.

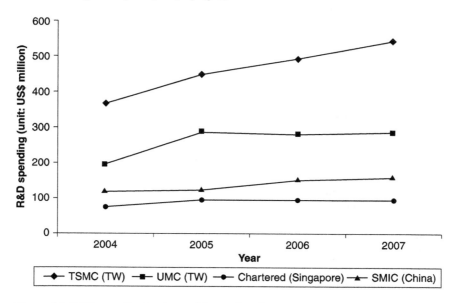

Figure 6.3 R&D spending by the world's top four foundries, 2004–07.

Sources: IC Insights 2008; company reports.

sector, as detailed in Chapter 5, there has been a steady decline in market share by Taiwanese firms in recent years (Figure 6.4) (Taiwan Semiconductor Industry Association 2004; 2007; 2008b; IC Insights 2006; TSMC 2007b; Lapedus 2008; Industrial Development Bureau 2009). The entry of China-based up-and-coming challengers has partly resulted in such a decline. These include SMIC, HeJian and Huahong NEC, all of which have benefited from the migration. In the IC design segment, ambitious Chinese latecomers, some of which are the offspring of the migration, are catching up from behind, taking over the market share of lower-end consumer ICs from their Taiwanese counterparts. The primary example includes Actions, which became China's No. 1 fabless design firm in 2004 and 2005 thanks to its link with Realtek, as detailed in Chapter 5. Taiwan-related Hangzhou Silan has also defeated its Taiwanese counterparts in the lower-end consumer IC design market.

Second, fieldwork data indicate that at least one of Taiwan's top 10 IC design firms began to hire more engineers in China than in Taiwan in 2008. As of August 2009, the company hired more than 2,000 engineers in its operation hubs in Beijing, Shanghai, Hangzhou, and Shenzhen.[8]

Third, the migration has covered the three major sub-sectors of the industry, resulting in the formulation of clusters of the Taiwanese chip supply chain in China, notably in Shanghai, Suzhou, Beijing, and Wuxi.[9] These clusters, if

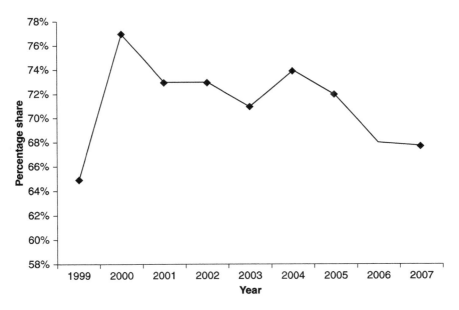

Figure 6.4 Worldwide market percentage share of Taiwan foundry business, 1999–2007.

Sources: For data from 1999 to 2003, see IEK/ITRI, ITIS Project (April 2004) in Taiwan Semiconductor Industry Association 2004. For 2004 and 2005 data, see (IC Insights 2006). For 2006 data, see Taiwan Semiconductor Industry Association 2007. For 2007 data, see Taiwan Semiconductor Industry Association 2008b.

continued to develop, may help China duplicate Taiwan's success some time in the future to the extent that China's gain may become Taiwan's loss, making fear of hollowing-out Taiwan's IC capability a less distant reality. "If the migration . . . involves IC design, manufacturing, and packaging and testing, it would be horrible [for Taiwan]," said Abe C. Lin, director general of Integrated Assessment Office at Taiwan's Ministry of National Defense.[10] William A. Reinsch, former US Under-Secretary of Commerce, described the migration of the important industry as a key issue for Taiwan's economic security and expressed his concerns as follows:

> If the manufacturing goes, then the R&D goes, too. Ultimately, if R&D goes, then everything is going out there across the Strait to the mainland, which means the value-added is there and the value is there . . . Unless they are putting that back into new R&D in Taiwan and develop new sectors and products there, the economy is not benefiting, job creation is not benefiting, and Taiwan will lose its role as a technology leader in the region.[11]

In spite of these challenges to Taiwan's economic security, some of the interviewees remain optimistic, arguing that migration has actually enhanced Taiwan's economic security. This is because it has enabled Taiwanese firms to exploit location-specific resources in China and pushed them to continue climbing up the technology ladder in order to ensure their competitiveness. According to a senior executive at a leading Taiwanese IC design house, access to engineers in China through company subsidiaries there have helped Taiwanese firms to advance their competitiveness, thereby contributing to Taiwan's economic power and security.[12] The president of a Taiwanese IC design house considered migration as positive to Taiwan's economic security for a different reason. In his view, since migration has helped cultivate competitors in China, Taiwanese fabless design firms have been sandwiched by two forces in the global race. On the one hand, ambitious Chinese latecomers, some of which are the offspring of the migration,[13] are catching up from behind, taking over the market share of lower-end consumer ICs from their Taiwanese counterparts. On the other hand, leading American firms continue to innovate in order to retain their leadership positions, posing challenges to their Taiwanese fast followers. The two joint forces have pushed the Taiwanese firms sandwiched in between to engage in high value-added production activities in order to retain Taiwan's position as the world's second largest IC design powerhouse.[14]

Uncertain long-term impact

The long-term impact of the migration in economic security terms has nonetheless been undetermined. Some of the interviewees stressed that so long as Taiwanese IC firms continue to gain new competencies, migration in the years ahead will not necessarily undermine Taiwan's economic security. As the CEO of UMC put it:

Outsourcing is not necessarily a bad thing. So long as Taiwan accelerates its engagement in value-added activities in the semiconductor industry pursuing high-end technologies, we do not need to be afraid of the prospect that mainland China will catch up in the IC sector.

He added that even though Taiwan's IC design sector generated about 27 per cent of worldwide revenue today thanks, in part, to Taiwanese returnees from the US, the US remained the world's most powerful IC design hub because of its continuous innovation.[15]

Others have expressed concerns that the continuous migration may help to build a technologically capable supply base in China, cultivating an increasing number of competitors that may dominate more and more segments of the industry in due course. As a senior Taiwanese trade official admitted, "We certainly do not want to see Taiwan help with the Chinese take-off to the extent that Chinese firms will become our competitors."[16] Such a trend may intensify, as global IC design firms, including those in Taiwan, continue to shift part of their production to China, fostering the development of a viable domestic industry in due course.[17]

Whether the migration will move in a direction that will ultimately undermine Taiwan's economic security depends on a confluence of factors, many of which go beyond the future depth and scope of the migration *per se*.[18] Additionally, the consensus among my interviewees is that migration has contributed to China's economic security because Taiwanese chip actors on the mainland have quenched the PRC's thirst for ICs and have contributed to that country's semiconductor sales revenue. In 2004, SMIC, HeJian, Grace, and CSMC manufactured semiconductors that accounted for over 61 per cent of China's foundry sales revenue. In the same year, the sales revenue of IC design firms with partial or 100 per cent Taiwanese capital in Shanghai, China's epicenter for semiconductor production activities, accounted for 6.1 per cent of the total in the country. However, the speed at which Chinese domestic chip capability will develop is unclear in spite of the expectation that indigenous firms will develop the capacity to produce a growing share of the high value-added chips that are now imported.

Arguably, what matters to Taiwan's economic security is not so much the continuous flow of investment, talent, and technology to China. It is, instead, whether Taiwan can increase value-added activities at home to move up the food chain in a competitive global chip race while continuing to exploit China as part of its globalized production base. Since the migration is an ongoing process and many factors may affect its future course of development, the long-term economic security ramifications are undetermined.

Technological and defense security implications

The long-term technological and defense security ramifications of the migration encompass the following three dimensions. They include: (1) concerns over the PRC chip industrial base, its contribution to PLA modernization and the ensuing shift in the balance of power; (2) technology-related misgivings over the narrowing chip technology gap or IW operations; and (3) fear of disruption or denial of supply due to foreign dependency.

The chip industrial base, the PLA, and balance of power

The first security implication of the migration concerns the extent to which it has contributed to the build-up of a viable chip industrial base in China, thereby helping to modernize the PLA and to shift the balance of power in Washington–Beijing–Taipei trilateral relations. We begin with an analysis of the Taiwanese contribution to the Chinese commercial as well as defense chip industrial base. What follows is a brief summary of the Chinese military IC acquisition strategy since 2001: it institutionalizes the process of spin-on to expand the original dual supply pool (i.e. the domestic defense chip sub-sector[19] and the international market) in order to advance the PLA's reform objectives. Against such a policy backdrop, the Taiwanese contribution to the PLA's modernization is examined, and the resultant impact on the balance of power among the three actors involved is explored.

Taiwan's contribution to the PRC's commercial chip industrial base

Chapter 5 documented Taiwan's contribution to the Chinese commercial chip industrial base because of the migration. To recapitulate such a contribution, Taiwanese investment, talent, and technology have permeated the commercial chip supply chain in China, a trend that has gathered momentum since 2000. This has contributed to the recent take-off of the Chinese commercial chip industry. The rise of several start-ups as forerunners in the IC design and manufacturing sub-sectors in China has benefited from the Taiwanese migration.

Moreover, some of these Taiwanese-owned or -run firms have partially quenched China's thirst for reducing dependency on imported ICs. Besides, Taiwanese chip operations in China have introduced several programs which will facilitate the creation of a fully-integrated commercial chip industrial base on the mainland in the end. These include local talent training programs, the offer of MPW services at foundries, and the function of foundries as pillars of the overall supply chain.

Taiwan's contribution to PRC defense chip industrial base: ten cases

Taiwan's contribution to the Chinese defense chip industrial base over the course of the migration, a neglected area of study, has also been noticeable, as evidenced by the following ten cases (Table 6.1). Each of them shows business links between Taiwanese chip actors and their Chinese counterparts formerly or currently associated with the Chinese defense production system. These ties involve the flows of hardware and expertise from the Taiwanese side to the Chinese side. The former refers to the supply of semiconductors for Chinese military systems. The latter comprises IC manufacturing service, MPW service, local talent training

Table 6.1 Links between Taiwanese chip actors and their Chinese counterparts associated with the PRC defense semiconductor industrial base

Case no.	Taiwanese chip players as security actors	Chinese chip players associated with the local defense chip industrial base	Comments
1	CSMC, a foundry	Huajing, a QPL supplier of ICs	1998–2000: CSMC improved the production capacity of Huajing Fab 1
2	HeJian, a foundry	CAS, with a record of supplying chips to the PLA	HeJian offers MPW services to CAS
3	CSMC, a foundry	CAS	2005: construction of CAS–CSMC 6-inch wafer fab began
4	GSMC, a foundry	SIM of CAS, with a record of supplying chips to the PLA	By 2005: SIM trained students at GSMC
5	TSMC, a foundry	SWID, a spin-out IC design house from the 24th Research Institute which supplies ICs to the PLA	2005: SWID used TSMC's 0.35 micron process to fabricate ICs for GPS receivers
6	An IDM	No. 771 Research Institute, a QML and QPL supplier of ICs	Mid-1990s: the institute sought to gain access to the Taiwanese firm's IC design service to advance the *Hangtian* project
7	TAMOS, a chip firm	Institute of Microelectronics, Tsinghua University, which has engaged in military IC R&D	1999–2003: TAMOS helped the university's fab improve clean room environment and process technology
8	A senior engineer at an IC design house subsidiary	A junior engineer at Fudan Microelectronics which has engaged in at least one PLA-commissioned CPU project	2002: the junior Chinese consulted the senior Taiwanese for the PLA project; the company won a national defense technology award
9	An IDM subsidiary	A spin-out company from a research institute which was part of the Chinese defense industrial base	2005: the Chinese company was keen to spur tech transfer from the Taiwanese firm
10	VIA, an IC design house	Shenzhen I-Lacs Technology Co., Ltd., China's first qualified civilian supplier of military PCs to the PLA	Since 2004: the Chinese firm has used VIA chips in its military PC products

Sources: Interviews, secondary data, and company websites.

program, fab management service, knowledge transfer at an individual-to-individual level, and joint ventures. They indicate Taiwan's direct or indirect contribution to China's tangential defense chip industrial base, thereby challenging the conventional view that the migration has had little to do with the PLA.[20]

A summary of these cases follows. The first case refers to the link between Huajing, a member of the Chinese defense chip industrial base, and CSMC, a semiconductor company established by Taiwanese executives with hybrid sources of capital. Huajing's association with the Chinese defense industry has been long-standing. In 1989, the State 742nd Factory was merged with the Wuxi branch of the 24th Research Institute to form Huajing (China Semiconductor Industry Association IC Design Branch 2005). As detailed in Chapter 4, Huajing began to build a 6-inch wafer fab in the early 1990s as part of the 908 Project but to no avail (Fang and Cao 2002; Howell *et al.* 2003: 24). Despite this, Huajing's military IC work has continued, primarily in its Central Research Institute (Luo *et al.* 1992).[21] In 1991, it won three military science and technology awards (China Machinery and Electronics Industry Yearbook Editing Committee 1991: IV–12-IV–13; China Machinery and Electronics Industry Yearbook Editing Committee 1992: IV–14-IV–15). It also appeared in the June 2002 version of the QPL, which identifies the types of military electronic parts and components produced by qualified onshore manufacturing lines for the PLA (Yu 2002).[22] In 1998, Huajing leased its Fab 1 to CSMC. The Taiwanese-run company offered a two-year "foundry management service" to the fab: it improved the fab's production capacity and customer base.[23] While CSMC succeeded in reviving Huajing's Fab 1, the fab under their helm had fabricated chips for commercial rather than military end-uses.[24] However, around the same time the project unfolded, Huajing's Central Research Institute churned out chips for the PLA. Although it is unclear whether Taiwanese engineers running the fab had indirectly contributed to the military IC work under Huajing's roof, such a possibility is conceivable.

The second to fourth cases concern the CAS as part of China's defense chip industrial base, on the one hand, and three respective Taiwan-related foundries in China (i.e. HeJian, CSMC, and GSMC), on the other. The CAS, a pioneer in China's semiconductor technology research, has supplied semiconductors to the PLA. "The CAS is known to be involved in the military," said an official in charge of export controls at the US Department of Commerce.[25] As detailed in Chapter 4, Soviet input in the 1950s enabled a captive plant attached to the Institute of Physics of CAS to fabricate transistors for the 109 Model III Computer, a large all-purpose digital computer.[26] These computers with CAS-fabricated semiconductors were inserted in China's atomic bombs between 1967 and the early 1980s (Chinese Academy of Sciences 2000: 36–7; Zhang and Zhang 2007: 29–30, 33). In the 1960s, SIM of the CAS also fabricated transistors for China's atomic bombs. In the 1980s and early 1990s, it developed military GaAs VHSIC components (Hua 1996). In recent years, CAS spin-offs have made inroads in various semiconductor arenas. For instance, BLX IC Design Co. Ltd., a CAS spin-off, introduced its 266MHz Godson–1A chip to the market in 2002 (Clendenin 2003;

Lemon 2004). Empirical evidence indicates that CAS has established business ties with China-based Taiwanese commercial chip players. Aided by HeJian educational programs, the EDA Center of CAS has taken part in HeJian's MPW services to upgrade the IC design capabilities of participants in the schemes.[27] Given the long-standing ties between the CAS and the defense establishment, these enhanced IC design capabilities are potentially exploitable by the Chinese military.

Moreover, the Institute of Microelectronics of the CAS signed a pact with CSMC in October 2005 to build a 6-inch wafer fab in Beijing. In June 2006, the fab started its qualification and production ramp up. The CAS potentially benefits from the joint venture because of CSMC's niche in 6-inch wafer fabrication. Given the research academy's intimate link with the defense sector, it is conceivable that the Chinese military or its spin-off companies will use the fab to make chips for end-uses in PLA systems, a step further towards the military's digitalization objective. Although the technology available at the fab lags behind the global commercial state-of-the-art, it may be sufficient to enable the fabrication of ICs that are better quality than those used in existing Chinese military systems.[28]

As detailed in Chapter 5, graduate students at the SIM of the CAS have received training at Grace. At the commercial fab, these inexperienced local engineers have gained access to advanced semiconductor equipment unavailable at their research institute.[29] They have also absorbed IC manufacturing expertise from their senior counterparts, including those from Taiwan. Originally trained in the Chinese defense chip industrial base, they may apply their improved expertise, gained at Grace, to Chinese defense microelectronics work.

The fifth case involves the business link between TSMC and SWID, a spin-out IC design house from the 24th Research Institute. As detailed in Chapter 4, the 24th Research Institute has specialized in the design and fabrication of analog ICs for Chinese military and commercial end-uses. From 1986 to 1990, it pioneered China's military foundry projects, running small-volume military ASIC experimental lines. In 1989, it fabricated military-grade analog semiconductors for end-use in Long March–2E. It also supplied semiconductors for the Chinese *Hangtian* project including Shenzhou V, China's first manned spaceship which debuted in 2003. As of 2002, the institute had housed two qualified manufacturing lines to provide the PLA with bipolar analog ICs and bipolar high-frequency wideband low-noise amplifiers. In 2003, it established China's National Laboratory of Analog Integrated Circuits. In recent years, it has spun out SWID (China Machinery and Electronics Industry Yearbook Editing Committee 1991: IV–12-IV–13; 1992: IV–14-IV–15; Wu 1991; Chen 1992; Guo and Liu 1992; He 2003; Lei 2003; Xu 2004; Zheng 2007).[30] Nevertheless, the institute has lagged behind its counterparts in leading semiconductor countries in the area of military analog ICs in terms of reliability and fabrication technology.[31]

The gap began to narrow when SWID started to use IC fabrication service at TSMC. As of 2005, it had used TSMC's 0.35-micron SiGe BiCMOS process technology to fabricate a radio frequency IC for GPS end-uses (Song 2003; Yang *et al.* 2005).[32] Given SWID's military connection, it is conceivable that some of these TSMC-fabricated ICs might have been inserted in Chinese military gadgets. The case has at least demonstrated that the Taiwanese foundry's fabrication service has contributed to the performance of a local Chinese IC design house intimately linked with the domestic defense industry.

In the sixth case, No. 771 Research Institute, which forms part of the Chinese defense industry, had lured a Taiwanese IDM to provide the institute with rad-hard IC design service in support of China's *Hangtian* project.

As detailed in Chapter 4, the research institute is China's flagship state-owned unit specializing in the research, development, and manufacture of airborne computers and semiconductors for missiles and satellites. It has supplied space-qualified and military-grade ICs and IC-based computers for end-uses in various spacecrafts and missiles. In 1965, it made China's first bipolar microcomputer using small-scale ICs. In 1980, it fabricated China's first CMOS, application-specific rad-hard microcomputer using large-scale ICs (No. 771 Research Institute n.d.a). According to a 2002 *neibu* publication, it had been a qualified QPL supplier using its 4-inch CMOS process technology to churn out analog and digital ICs for military and aerospace systems. As a QML supplier, it had, as of 2002, run two certified manufacturing lines to fabricate thin-film and thick-film hybrid military ICs (Hua 1996; Yu 2002). By 2005, its military-certified CMOS IC manufacturing line had produced an average of 5,000 to 10,000 wafers per month (No. 771 Research Institute n.d.a). China's second manned spacecraft, Shenzhou VI, carried 22 IC-based space-qualified computers made at the institute (Du 2005).

In the mid-1990s, however, the institute turned to Taiwan for rad-hard IC design services. According to a then senior executive at a Taiwanese IDM, the head of the institute contacted him during his business trip to China: a rad-hard IC design service was sought in support of the *Hangtian* project. The Taiwanese company declined the deal. One plausible explanation for the incident was that the level of the in-house semiconductor capability at the institute was such that its chief had to seek Taiwanese input to advance its military/aerospace undertakings.[33]

In the seventh case, a junior Chinese engineer at Shanghai Fudan Microelectronics, a local IC design house, endeavored to absorb technical expertise from his senior Taiwanese counterpart for a PLA-commissioned project. The Taiwanese, the head of a China-based subsidiary of a leading Taiwanese IC design house, had acquainted himself with the junior Chinese engineer in question through job interviews. Around 2001, the local firm, which is a spin-out firm from Fudan University, was engaged in a PLA-commissioned project to develop a 32-bit CPU similar to the Intel 80386 microprocessor. However, internal discord arose concerning the project: senior members of the company preferred a reverse engineering approach, whereas junior ones preferred innovative solutions. The junior engineer at the company contacted his previous Taiwanese job interviewer to seek his input in the project.[34] By late 2002, his company debuted Shenwei I, a 32-bit embedded CPU that represented the highest level of China's home-grown microprocessor technology at the time (Li 2002). The firm thus won a first-class national defense technology award (Shanghai Fudan Microelectronics n.d.).[35] It is difficult to assess the amount of contribution the Taiwanese engineer made to the PLA project through his informal contact with the Chinese engineer. However, the case illustrates the possibility that following the movement of senior Taiwanese semiconductor engineers to China, junior Chinese engineers engaged in military projects might attempt to absorb expertise from their experienced Taiwanese counterparts through informal arrangements.

In the eighth case, the Institute of Microelectronics of Tsinghua University, which has run military semiconductor and civil-military R&D projects, absorbed know-how from a Taiwanese firm to improve its IC manufacturing operations. The 863 Program, China's first significant effort to pursue coordinated civil-military R&D (Feigenbaum 1999; 2003: 141–3, 248–58; Gabriele 2002; Mulvenon 2005; Ministry of Science and Technology n.d.), financed the R&D projects at the institute.

In 1988, the head of the institute contacted a Taiwanese industry player to seek his help with the institute's IC manufacturing operations. In 1999, the Taiwanese set up a company, TAMOS, to run a five-year project to smooth the university's 5-inch wafer manufacturing operation. The Taiwanese team improved the clean room environment and the process technology used at the facility, which developed and manufactured a vast array of small-volume ICs.[36] Although the exact end-use of these ICs is unclear, it is conceivable that some of them might have been used in Chinese military systems, given the institute's involvement in military IC and civil-military R&D undertakings.

In the ninth case, a Chinese IC company which has been associated with the Chinese defense chip industrial base sought to absorb technology from a Taiwanese IDM. According to the president of the Taiwanese subsidiary in Shanghai, the Chinese firm had intended to spur technology transfers from his firm in early 2005, resulting in his visit to the Chinese company for a discussion. As he recalled, "As soon as I read the company brochure during the business meeting, I realized the unit has had ties with the military . . . It was formerly a military research institute."[37]

In the tenth and final case, Shenzhen I-Lacs Technology Co., Ltd., a systems house which has been a qualified supplier to the Chinese military, purchased ICs from Taiwan's VIA for its military PC products.

Established in 1992, Shenzhen I-Lacs has specialized in military and commercial PC products. Various Chinese military authorities have identified the company as a qualified military supplier because many of its products[38] have met pertinent military standards.[39] In 2005, GAD identified the firm as a qualified military supplier.[40] In 2006, the company received a certificate from a certification center under the Electronics and Information Base Department (*Dianzi Xinxi Jichu Bu*) of the GAD, which confirmed its status as a qualified military product supplier.[41] It has emerged as China's first civilian computer company to become a qualified supplier of military-grade PCs to the PLA.

The company has used Taiwanese semiconductor components, primarily from VIA, in some of its military products. In 2004, its president revealed that his company was VIA's customer (Gongkong.Com 2005). One of its products is equipped with a VIA Eden CPU (Shenzhen I-Lacs Technology Co. 2004),[42] whereas another has a VIA Eden CPU and a VIA chipset. Its website has identified both gadgets as military PCs. It has even displayed photos that show VIA's brand name marked on the surface of its CPUs. When asked if VIA had supplied chips to the Chinese firm in question, a senior executive at VIA said, "I am not sure . . . I don't think the military business is that important" for the company because it comprised a small percentage of the company's revenue.[43] Briefly, the case has demonstrated that a qualified supplier to the Chinese military has purchased semiconductors from a Taiwanese firm for its military PC.

In sum, these ten cases have indicated that members of the Chinese defense chip industrial base have endeavored to absorb resources from the Taiwanese semiconductor sector. Some of these Taiwanese resources have become accessible to the Chinese chip actors following the migration. The empirical evidence thus substantiates the argument that Taiwanese chip actors have directly or indirectly contributed to the development of China's defense chip industrial base over the course of the migration.

Policy context: spin-on and the enlargement of military chip supply

To what extent does such a contribution help the PLA modernize, thereby accelerating changes in the balance of power in the trilateral relations? To answer this question, it is important to understand China's military IC acquisition strategy since 2001: its recent inclusion of the domestic commercial chip sub-sector as another source of military IC supply, which has enlarged its conventional supply pool.

As discussed in Chapter 2, dual-use and spin-on have complicated the defense significance of a country's chip sector. Both defense and commercial sub-fields that comprise the national chip production base are of varying degrees of security importance, and a policy to ensure the military's access to these two sub-fields is key to national security.

As detailed in Chapter 4, China lacked such an IC acquisition strategy until 2001, when Beijing introduced a series of new policies to institutionalize the trend of spin-on, which was encouraged by a notion of *Yujun Yumin* (Locating Military Potential in Civilian Capabilities) embodied in a new dictum in the 10th Five-Year Plan to chart China's long-term economic and military modernization. These new initiatives aimed to integrate the defense industry into a broader civilian economy with a rich array of technological and industrial capabilities that the PLA can absorb to modernize its forces. Consequently, China's domestic supply of military chips has no longer been confined to its traditionally enclosed defense chip industrial base; it has expanded to include non-state-owned enterprises that operate in the growing commercial chip sub-sector.

If the *Yujun Yumin* policy package intent on institutionalizing the trend of spin-on succeeds in the arena of semiconductors, in spite of foreseeable challenges ahead,[44] two important outcomes may follow. First, hardware and knowledge from the technologically superior commercial chip industrial base

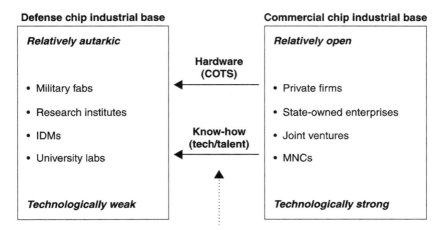

Defense chip industrial base

Commercial chip industrial base

Relatively autarkic

Relatively open

Hardware (COTS)

- Military fabs

- Research institutes

- IDMs

- University labs

- Private firms

- State-owned enterprises

Know-how (tech/talent)

- Joint ventures

- MNCs

Technologically weak

Technologically strong

Transfers pushed by the PRC reform to institutionalize spin-on since 2001

Figure 6.5 The PRC defense and civilian chip industrial base and the spin-on reform.
Sources: Interviews and secondary data.

may be systematically transferred to the technologically inferior defense sub-sector (Figure 6.5). Second, it may contribute to the PLA's modernization by minimizing the security threats that Beijing has faced because of its dependence on the domestic defense chip sub-sector and the international market for military IC supplies.

The first outcome is that a viable Chinese commercial chip sub-sector may supply high-performance and cost-effective COTS items to the PLA for end-use in defense systems operating in a relatively benign environment. This would help the PLA modernize as these military systems benefit from the COTS components with increasing performance, expanded functionality and a decreasing cost per function.

A viable commercial chip sub-sector may also transfer knowledge to the defense side by spurring talent and technology flows across the interface. Having acquired state-of-the-art expertise at the commercial forefront, local engineers may decide to engage in defense microelectronics work later in their career. Arguably, such a possibility is not a question of whether or not it will happen but a question of when.[45] Empirical data show that some of the local engineers have left their jobs at military research institutes to work for commercial firms in pursuit of financial gains. If supply meets demand, a reverse talent flow may occur as well.[46] Technology transfer possibilities are manifold. Members of the defense industry may form joint ventures with civilian firms to absorb expertise. They may outsource defense ASIC design and fabrication projects to civilian firms, as illustrated by Fudan Microelectronics' PLA-commissioned CPU project.[47] Mainstream chip firms may transfer their technologies to the defense industrial base to help advance military projects. The Intel–Sandia case and the Trusted Foundry Program in the US context, as detailed in Chapter 2, may be duplicated in China in due course.

Given the dual-use nature of IC manufacturing technology, moderately modified commercial process technologies may be used to make military chips. As explained in Chapter 2, it would take the world's leading foundry player a year to modify 10 per cent of its state-of-the-art process technology for civilian chip making in order to fabricate military semiconductors. The same may apply to leading China-based foundry firms, such as SMIC. So far, SMIC has denied any business link with the Chinese military. If the Taiwanese-run company decides, or is forced, to become part of the Chinese spin-on process to serve the PLA in due course, the firm would find it technologically feasible to modify slightly its current technology to make chips for the Chinese military. These possible resource transfers, if realized, would benefit the PLA.

Highlighting the contribution of a successful spin-on in the Chinese context to Chinese national security does not negate the fact that an improved domestic defense chip industrial base may also contribute to national security. The latter, improved over time through internal engineering, may do so by providing ICs unavailable in the mainstream commercial market, home, and abroad alike, to the PLA. The more successful such a combined approach is and the more viable a domestic chip supply base is, the more rewards the PLA is likely to reap.

The second outcome of a successful spin-on operation in China is that it helps Beijing minimize vulnerabilities and risks caused by reliance on the domestic defense chip sub-sector and the international market for military IC supplies.

As discussed in Chapter 4, the PLA's reliance on the domestic defense chip sub-sector for military IC supply has caused national security concerns because of technological, economic, political, and institutional constraints that the insular production system has faced in spite of the fact that the system has partially contributed to China's security by supplying small-volume ICs for indigenous military end-uses. Because of export controls,[48] the lack of sufficient investment and national industrial infrastructure, the chronic weakness of a centrally planned economy, and the lack of a well-functioned regulatory regime governing domestic defense production activities, the system has often returned low yields in production, churned out backward and/or unreliable semiconductors, and trailed far behind its Western counterparts in terms of defense chip capability. It has thus had a record of undermining China's security when domestically grown rather than imported ICs have been in demand for certain PLA systems.[49]

The Chinese military's endeavor to purchase ICs from the international market has also been problematic. Beijing's failure to understand Western military standardization systems led to its purchase of bad quality, counterfeit or improper components which cannot fit into Chinese military systems. In one account, several examination and testing centers under the Chinese aerospace and aviation sectors concluded that 50 per cent of imported electronic parts and components were disqualified for end-use in Chinese military aerospace and aviation systems (Guo and Gu 1996; Wang 1998; Zhang and Guo 2004). Worse yet, foreign dependency has deepened Beijing's fear of cuts in or denial of supply. Post-COCOM export controls have obstructed China's endeavors to gain access to these supplies[50] in spite of a tendency towards loosening controls.[51] China also lacks allies as its trustworthy suppliers. All of the above scenarios have threatened China's national security.

A viable spin-on operation in the Chinese context may reasonably mitigate the aforementioned security risks China has faced. Because high-tech civilian firms are technologically stronger than their counterparts in the domestic defense economy, to absorb the former into the PLA supply pool helps to alleviate potential risks the military has faced due to its reliance on the latter. It also helps China mitigate risks it has faced because of its reliance on the international market for military chip supplies. Hence, Beijing's success in spin-on will be an asset to the PLA's march towards digitalization.

In the context of China's most current military IC acquisition policy as discussed above, the Taiwanese migration contributes to the PLA modernization in two ways. First, the migration has nurtured the development of the Chinese commercial chip sub-sector, thereby helping to facilitate the process of spin-on. In particular, IC design capabilities of local firms, which have grown partly because of Taiwanese contribution, and inexperienced Chinese engineers trained in Taiwanese-run foundries and IC design houses may become potential assets for dedicated Chinese military chip operations through the process of spin-on.[52] Second, the migration has improved the Chinese defense chip industrial base. Taiwanese-owned or -run fabs in China have offered services to their customers to help them upgrade their IC design capabilities, which include IC design entities closely linked with the Chinese defense industry. TSMC's foundry has made chips for at least one spin-out firm of China's leading military analog IC research institute, although the exact end-uses of these chips remain unclear. At Grace, inexperienced Chinese engineers, originally trained at an institute that engages in military semiconductor work, have absorbed knowledge from their senior Taiwanese mentors; they have also gained access to advanced semiconductor equipment unavailable at their research base. CPUs and chipsets designed by Taiwan's VIA have appeared in military PCs sold to the PLA. In the future, it is conceivable that these PLA gadgets equipped with Taiwanese-designed or -fabricated chips may become part of the PLA war-fighting systems to be deployed during a Strait contingency. All the hard evidence above indicates that these Taiwanese IC players have, knowingly or unknowingly, augmented the Chinese defense chip capability, contributing to the PLA's modernization.

In both ways, Taiwanese expertise transfer is a primary asset in the Chinese military modernization over Taiwanese hardware supply. Given the semiconductor technology gap between Taiwan and China, the diffused Taiwanese knowledge, once in the hands of the Chinese military and defense sector, will contribute to the PLA modernization. According to James Mulvenon of DGI, "I am much more concerned about know-how in production and manufacturing technology transfer than I am about outputs . . . [although] I am sure the [Chinese] military searches carefully to buy these products, too."[53] The Taiwanese hardware for the

PLA is less important than knowledge transfer, partly because it constitutes a larger supply pool that may include government-owned defense entities, local commercial firms permitted to engage in military production activities, and the international market through diversion, espionage or legal channels in ways similar to those seen in the case of the USSR (Central Intelligence Agency 1983; 1985; Overend 1988; Cox 1999; Department of Defense 2007: 37–8).

Still, even this aspect of potential Taiwanese contribution to the PLA is not to be ignored because the level of technologies that Taiwanese chip-makers, such as TSMC and UMC, have used in their Chinese subsidiaries as of 2007 has been almost comparable to that of those they have used to make chips for the US military.[54] The PLA may regard technologies available at China-based Taiwanese operations as potential assets and decide to rely on these internal sources of chip supply if it does not consider these firms' Taiwanese background as a potential security threat.

However, it remains unclear whether Taiwan-related chip players in China would be eligible spin-on partners for the PLA, especially for sensitive Chinese military microelectronics projects. Some of my interviewees have argued that national security considerations would make wholly Taiwanese-owned firms the least likely spin-on partners for the design or fabrication of chips for sensitive or mission-critical Chinese military systems, in spite of their technological edge. The Chinese defense establishment would view Taiwan as a foreign adversary rather than "a part of China" in this instance, thereby preferring indigenous firms, where possible.[55] Such a rationale is conceivable, considering the fact that national security considerations have largely propelled Beijing to exclude China-based Taiwanese subsidiaries, in spite of their excellent chip-making performance, from making ICs for non-military national projects, such as the ID card scheme.[56]

China's stated regulation on the supply of dual-use goods to the military may offer another clue. As of 2000, the policy had stipulated that national security considerations had driven the authorities to confine eligible suppliers to purely state-owned agencies or civilian firms with at least 50 per cent Chinese government shares.[57] If the rule holds true today and is well implemented, it is logical to conclude that, at least in theory, only purely state-owned firms and companies with at least 50 per cent government shares[58] would be qualified suppliers to the PLA. However, several cases we have discussed earlier[59] seem to indicate that the rule may no longer hold true or is only partly implemented.

It remains unclear whether Taiwanese-owned or -run foundries in China have already churned out chips for end-uses in PLA systems. Senior executives at SMIC, Grace, HeJian, TSMC Shanghai, and CSMC have claimed that, to the best of their knowledge, their firms had not served any Chinese military customers.[60] A senior executive at TSMC said that TSMC Shanghai would not knowingly make chips for the PLA. This was because the company's awareness of the political antagonism between Taipei and Beijing, and the close link between the Taiwanese government and the firm, since the latter was a brainchild of the former.[61] Nevertheless, other industry interviewees admitted that it is possible that these Taiwanese foundries may have fabricated chips for end-use in Chinese military systems. They may have done so without prior knowledge of the exact end-use of the chips in question because it could be hard to tell the difference between a military IC and a commercial one. They may have knowingly engaged in these military businesses because of low demand from the commercial market. Alternatively, these businesses may have become possible because both parties involved agreed to keep the deals confidential.[62]

In sum, against the backdrop of Beijing's recent military IC acquisition policy and practice, Taiwan has accelerated China's pace in the acquisition of a fully-fledged integrated semiconductor capability, which, in turn, contributes to China's pursuit of a viable digitalized military force to advance its defense power and capability.

Broader implications: PLA catch-up and shifting balance of power

What implications does a solid Chinese chip industrial base, if realized, have for PLA modernization and the balance of power among Washington–Beijing–Taipei security ties? The answer is fivefold, given the premise that advanced ICs fuel information-dependent military systems and that IT helps to define power in international relations today, as discussed in Chapter 2.

Before I elaborate on the answer, it is noteworthy that most of my interviewees and secondary data have underscored the strategic importance of a viable domestic chip industry to PLA modernization, whereas a minority have not. The former have done so by emphasizing China's lack of allies as trustworthy suppliers of military ICs[63] and China's clear military agenda behind its support for the industry.[64] The latter have contended that China, without a viable domestic industry, can still manage to obtain military ICs from overseas because of loosening export controls, though they have admitted that the country is still constrained by these regulations.[65] Both schools of thought, however, have converged in viewing a viable domestic chip capability in China as a contributory rather than a deterministic factor in PLA modernization.[66]

The first three security implications echo what Chinese policy-makers have aspired to achieve in their support for developing a strong domestic chip industry, as elaborated in Chapter 4. First, such a domestic production base would mitigate the long-standing vulnerabilities Beijing has faced because of foreign dependency (Zhang and Guo 2004). It would also alleviate any negative impact on China's security due to export controls of militarily sensitive semiconductor items, equipment and materials to China, which aim at slowing down PLA modernization. According to Zou Shichang, chairperson of Grace, China's strong IC industry would be able to churn out security-related IC devices unavailable in the international market because of export controls.[67] This contributes to China's security particularly because of its lack of trustworthy allies as guaranteed suppliers of sensitive military ICs and technologies.

Second, China's competitive commercial semiconductor industry would increase its security by removing a chronic obstacle to the PLA's pursuit of digitalization, namely, the limited ability of its insular defense production system to supply high quality and highly reliable semiconductors to the PLA. This, in turn, would accelerate the formation of a modern digitalized Chinese military force. The president of a Taiwanese IC design house envisaged that while China had depended on state-run factories and research institutes to churn out low-end semiconductors for the PLA, its rapidly improved commercial chip sub-sector would produce high-quality ICs for both commercial and military end-use at home, accelerating the Chinese military take-off.[68] According to Liu Chenghai (2004: 2), deputy minister of GAD's Electronics and Information Base Department, whether China could manufacture high-quality and reliable electronic components (including semiconductors) could have a direct impact on the development and production of the PLA's electronic defense systems.

Walden C. Rhine, CEO of Mentor Graphics Inc., the world's third largest EDA firm in 2004, envisioned what he saw as the natural outcome of a strong commercial chip sub-sector in China: "For any country that wants to develop military capability, having advanced consumer, commercial and computing integrated circuits helps them . . . Having a strong commercial chip base certainly helps you build sophisticated electronics equipment, and you can't stop that."[69] Roger Cliff of Rand Corporation echoed this view, arguing that China's possession of a state-of-the-art civilian semiconductor capability is "a necessary step on the way to have the capability to produce advanced military systems."[70] As long as China builds a good industrial infrastructure at home, including a solid chip industry, its defense technologies would benefit through the process of spin-on, according to the head of a Taiwanese subsidiary in Beijing.[71] The president of HeJian argued that Beijing's "obvious military ambition" would drive the PLA to exploit the domestic commercial IC industry to modernize its forces, although officials had stressed in public only the economic driver behind Beijing's endeavor to develop a domestic IC industry. In his view, Beijing would utilize part of its chip industrial base, by pouring in state money, to produce ICs for the military as it continues to attract foreign investments to develop the industry.[72] As a Pentagon official argued that it would be "important to have a reliable and vibrant industry domestically" in the US because the country had relied on chips to field its weapons systems, the same observation would apply to China.[73]

Taken together, the two aforementioned security repercussions of a solid chip industrial base in China would lead to a third security outcome: an improvement in the PLA's capability to wage war using conventional and non-unconventional measures.

On the one hand, China's improved semiconductor capability will provide the PLA with cutting-edge and reliable semiconductors with strong computational capabilities to build its precision-guided weapons systems, and C4ISR force integration operations, an area in which the PLA has made impressive improvements in recent years (Mulvenon 2005). This helps the Chinese military pursue information superiority over its enemies and contributes to the performance of these military systems, thereby helping improve the PLA's traditional war-fighting capabilities.

On the other hand, China's upgraded IC industrial base will increase the PLA's unconventional war-fighting capabilities in defensive and offensive terms.

As discussed in Chapter 2, information is a double-edged sword in any modern high-tech war. To maximize the prospect of victory, a military force, aside from pursuing information superiority over its enemy, has to protect its information systems from enemy attacks and to destroy the enemy's information infrastructure. Semiconductor-dependent electronic systems are susceptible to IW operations such as chipping and EMP strikes. A country's strong chip industry helps the nation accumulate in-depth semiconductor-related knowledge and technologies, thereby contributing to its defensive and offensive IW capabilities. It also provides sufficient onshore supplies of ICs for any critical national infrastructure, thereby minimizing the prospect of chipping that may occur at offshore production sites as chip production continues to globalize. The above observation may apply to the Chinese case, particularly in view of the PLA's stated pursuit of a strong domestic IC industry in order to strengthen its IW capabilities. PLA strategists have aspired to launch asymmetrical warfare operations that are intent on destroying the technological edge of its adversaries (Mulvenon 1999: 175–6, 185; 2005: 250–1; Qiao and Wang 2002; Bergsten *et al.* 2006: 14–15).[74] These unconventional war-making tactics can take the form of chipping, anti-satellite attacks, or EMP strikes intent on simultaneously paralyzing networked electronic devices that underpin the US and Taiwanese economy, military, and society.[75] They have also planned to strengthen defensive IW capability by protecting its semiconductor-dependent critical national infrastructure from any enemy attacks (Fu and Li 2004). As China's IC industry improves over time, Beijing will become closer to achieving these objectives, thereby improving its unconventional war-fighting capabilities. We shall further explore this theme later.

The fourth security implication involves changes in the balance of power. As China improves its conventional and unconventional war-fighting capabilities, its coercive military capacity with respect to Taiwan may increase accordingly. This contributes to changes in China's relative military capability vis-à-vis that of Taiwan largely and that of the US to a lesser extent. Consequently, the balance of power in the trilateral ties may be tilting in Beijing's favor. As Segal (2006: 305–10) argues, the globalization of IT manufacturing and R&D could help the Chinese military improve its capabilities, which, in turn, may change the relative capability of the Chinese military vis-à-vis other powers, including Taiwan and the US. Since Taiwan's semiconductor capability has gradually transferred to China because of the migration, this globalization force has helped to upgrade China's semiconductor capability, especially in IC manufacturing and design. This may help the PLA remove a long-standing bottleneck to the development of its strategic weapons systems, thereby affecting the balance of power in question.[76] Matthew S. Borman, Deputy Assistant Secretary at the US Department of Commerce, summarized what he saw as a real security concern for the US and Taiwan arising from a solid Chinese chip industry as follows.

There is certainly a concern – at the Defense Department particularly – about the Chinese drive to significantly upgrade its own military and particularly its [capability] vis-à-vis that of Taiwan . . . how much they are able to project forces, what kind of missile system they have, what kind of naval capability they have, what kind of command and control they have.[77]

However, given the US military supremacy, few believed that the PLA will match its American counterpart anytime soon by developing weapons systems as sophisticated as those produced by the US, based on Beijing's acquisition of the advanced IC and IT technology.[78] According to C. Mark Melliar-Smith, former president and CEO of International SEMATECH:

> There is a big gap between having access to integrated circuits and being able to build weapons systems. In other words, systems engineering in weapons systems can be very significant . . . you can worry about the world's best integrated circuit automatically leading to the world's best defense system. I don't think the fact supports that statement.[79]

Others have been more concerned than Melliar-Smith. As an American defense industry player contended, China's improved chip capability, with assistance from Taiwan and other external players, may "help them improve their military capabilities rather significantly by having electronic systems comparatively competitive to the West."[80] And if China were to acquire an electronics capability comparable to that of the US, then, in theory,

> that would have enabled Chinese defense manufacturers to produce weapon systems that are comparable in capability to those produced by the US . . . if we were to enter into conflict with China, then the risk to US forces would increase significantly.[81]

In brief, China's upgraded chip capability will contribute to its comprehensive national power in defense, technological, and economic terms (Shambaugh 1999/2000; Shen *et al.* 2007). It may further improve, over time, China's relative capabilities vis-à-vis those of Taiwan largely, and those of the US to a lesser extent, thereby shifting the balance of power in its favor.

The fifth and final security outcome concerns international security. An improved Chinese semiconductor capability would enable the country to supply critical chips and other relevant defense systems to its friendly allies such as Iran and North Korea, thereby exacerbating potential proliferation problems. Lisa Bronson, Deputy Secretary of Defense for Technology Security Policy and Counterproliferation, offered a typical view following this line of argument in late 2003. According to Bronson, the Pentagon did not want MNCs and local firms on the soil of a potential strategic competitor, which refers to China, to acquire design capabilities for advanced microprocessors. After all, "China has been a source for missile, nuclear and chemical weapons proliferation to unstable countries as well as countries that support terrorism" (Sullivan 2003).

So far, I have analyzed the first major defense security outcome of a viable domestic chip industry in China, which has benefited from the Taiwanese migration.

Conversely, if China fails to build a solid chip industry in spite of the Taiwanese input and other contributing factors,[82] most of the aforementioned scenarios may not occur or may occur to a lesser degree. This could undermine Chinese national security, instead of translating into a downward spiral of compromised security for Taipei and Washington.

Summary

To sum up the first defense security repercussion of migration, empirical data show that the migration has contributed to the build-up of a viable chip industrial base in China, which comprises commercial and defense chip production sub-fields. Since 2001, China has expanded its conventional sources of military semiconductor supply by including the domestic commercial sub-sector through the process of spin-on. Against such a policy backdrop, the Taiwanese chip actors, by enabling the Chinese indigenous semiconductor capability to grow, have indirectly helped to increase the prospect of the PLA's reliance on the growing domestic chip production base in order to increase its conventional and unconventional war-fighting capabilities. The mainstream view holds that a viable chip industry in China contributes to rather than determines Chinese military capabilities. However, such a domestic capability may help mitigate security risks Beijing has faced because of export controls and foreign dependency. It may also facilitate the process of spin-on. This development alleviates vulnerabilities the PLA has faced because of its reliance on the backward domestic defense industry for chip supplies. In due course, all of the aforementioned developments will improve China's relative military capability, tilting the balance of power in the trilateral ties in its favor. In assessing the security repercussions of China's rising semiconductor industry, Pecht (2004: 618) argues, "A growing loss in US high technological capability and leadership could radically change the balance of world power." Although China's enhanced semiconductor capability will not enable the PLA to match the US military any time soon, it is a small step in that direction. China's strong IC industry may also enable Beijing to export critical semiconductor goods and technologies to countries like Iran and North Korea, exacerbating the challenge of proliferation to international security.

To sum up, Taiwan's semiconductor actors have helped China to build a viable semiconductor industry in recent years. If China succeeds in building a strong domestic chip capability, it will enhance the PLA's capabilities to develop advanced military electronics systems – including its ballistic missiles targeting the Silicon Island – which may be used against Taiwan. It will also help improve the PLA's relative military capabilities vis-à-vis those of its Taiwanese counterpart, thereby contributing to the continuous shift of the military balance in the Strait in Beijing's favor (Department of Defense 2009: VIII). Thus, these Taiwanese non-state players have become security actors: the outcome of their largely profit-driven globalization activities in a strategically important industry in China may potentially put their country of origin under an enhanced coercive threat from across the Strait in the end. This would challenge Taiwanese security and even US security if Washington were to be dragged into a wider conflict with China over Taiwan.

Technological security risks

The second defense ramification of the migration concerns two types of technological security risks. The first involves the extent to which the migration has helped narrow the chip technology gap between China and Taiwan and that between China and the US, first in the area of manufacturing, and then in the arena of IC design, which is critical to the development of future military systems. This, in turn, may potentially threaten Washington and Taipei, thereby changing the balance of power in the trilateral relationship. The second concerns the extent to which both Washington and Taipei will be susceptible to technological insecurity following the migration due to the untrustworthy supply of China-made semiconductors for their military systems, as these ICs may become victims of the PLA's chipping practices. Although US government agencies and officials, especially those at the Pentagon, have envisaged these two security risks based on unexamined Realist vulnerability assumptions, the analysis below demonstrates that even if these threats occur, they may do so in ways that are more complex than these claims have suggested.

The semiconductor technology gap and its security repercussions

This section examines the thesis of the shrinking chip technology gap. Scholars and policy-makers have often regarded the narrowing semiconductor gap between a country and its potential or perceived adversaries as a prelude to ensuing changes in the relative capabilities and the distribution of power among the countries involved (Hanson 1982: 186–7; Central Intelligence Agency 1983; Friedman and Martin 1988: 106). The section below examines such a thesis in the context under discussion.

To begin with, various studies and interviewees have delineated a rapidly diminishing gap between Chinese and US semiconductor manufacturing technology and between the Chinese and Taiwanese. According to a study by General Accounting Office (GAO) (2002: 9–10), Beijing was five generations of technology behind Washington's then-current commercial production capability in 1986, whereas its most advanced semiconductor manufacturing facilities (i.e. SMIC, an epitome of the Taiwanese migration) could produce ICs which were one generation or less behind then state-of-the-art in 2002 (Figure 6.6). In less than 10 years, China went from being five generations behind global state-of-the-art IC process technology to being current with the state of the art by 2004 (Figure 6.7) (Pecht 2004).

Yoshio Nishi of Stanford University estimated the shrinking chip gap between the US and Taiwan and that between Taiwan and China as follows. As of late 2004, Taiwan's TSMC and UMC were one year[83] behind Intel for microprocessor-type technology and one year behind TI for DSP-type and communications-type

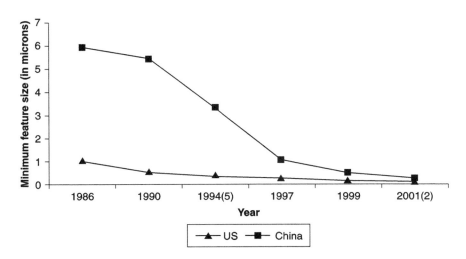

Figure 6.6 Semiconductor manufacturing technology gap between China and the US, 1986–2002.

Source: General Accounting Office 2002: 10.

Note: Complete data for the period between 1986 and 2002 were not available. The time scale was altered to show the years where data were available. Data for 2002 were based on estimate.

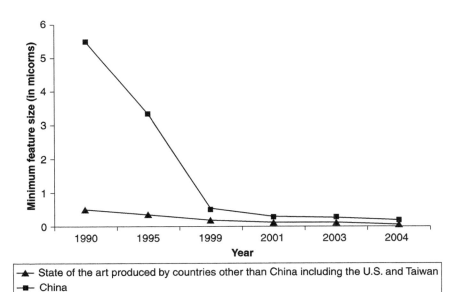

Figure 6.7 China's attempt to close the semiconductor manufacturing technology gap, 1990–2004.

Source: Pecht 2004.

technology. "SMIC in Shanghai is probably one year to, by most conservative estimation, probably two years behind TSMC," he said.[84]

Even in the arena of military semiconductor technology, where comparative data are relatively scarce, circumstantial evidence indicates that China has made incremental gains over the past decades. As of March 2005, GaAs process technology available at the 13th Research Institute, one of China's leading military microelectronics research institutes, was comparable to what was available in the US in the mid-1990s.[85]

Some have forecast that the trend of the aforementioned narrowing chip gap, which has primarily resulted from the global diffusion of semiconductor technology (Pecht 2004),[86] may challenge the security of the US and Taiwan in the long run.

In the worst-case scenario envisaged by the Pentagon, China may turn the tables to lead in IC design, which is critical to the design of future weapons systems, after having reached parity with leading semiconductor countries in IC manufacturing. This may erode US and Taiwanese leadership positions in IC design. According to the DSB study (Defense Science Board Task Force 2005: 12), "in light of the global dispersion underway in the semiconductor industry," US leadership in the design of programmable ICs may be challenged by ambitious latecomers in the global chip race. This would deprive the Pentagon of its long-standing "superior position to potential adversaries whose systems rely on US based suppliers and/or inferior parts procured abroad."

Some of the interviewees associated with the Pentagon have further envisaged the security risk both Washington and Taipei will face once China leads in IC design. These new Chinese-grown IC designs may become unavailable to weapons designers in both countries who rely on them for the R&D of next-generation weapons. This development would put the PLA in a superior position to its Taiwanese and American counterparts, thereby becoming an asset to China and a liability to the US and Taiwan in security terms. As a former Pentagon official put it:

> If you come up with the new chip in five years when they start doing the R&D for the future weaponry systems, these will be based on that kind of technology. And if the Chinese are developing that R&D, not the United States, our next-generation systems will be based on that technology that does not derive from here. And that is the potential problem.[87]

Others, however, have disagreed sharply about the extent to which the aforementioned closing gap between China and the US and Taiwan in IC manufacturing technology would engender significant security threats to Washington and Taipei. Nor have they seen eye-to-eye on whether China would soon be challenging the US and Taiwan in the semiconductor arena, particularly in IC design.

When asked if the shrinking gap between the US and China in IC manufacturing technology would potentially challenge the US superiority in defense terms, the president and CEO of SEMI said, "Well, that could happen." Generally, "there should be a concern in any region if they lose their base of high technology," particularly a loss in the R&D and manufacturing segments of the semiconductor industry.[88]

However, skeptics think differently. According to C. Mark Melliar-Smith, former president and CEO of International SEMATECH, it may sound "patriotic" to amplify the security challenges Washington may face because of the narrowing chip gap between the US and China, but he doubted the significance of that gap alone in security terms.[89]

He regarded it important to consider the inclination and constitutive nature of semiconductor technology to military performance when assessing the significance of the chip gap in security terms. The consideration of inclination should be more important than that of technology in evaluating the centrality of leadership in chip technology to military gains, although, in the case under discussion, China did arguably desire military dominance. "You [should] put the inclination first, and then the technology second ... Does the country want to have a dominant military position? If it does, then I would say technology is important," he argued. "Maybe the difference between Japan and China is that China really wants that dominant military position." Our discussion in Chapter 4 supports his observation regarding the Chinese aspiration to exploit its semiconductor technology to accomplish its military objectives. It follows that as soon as Beijing closes the technology gap, it would use its improved semiconductor technology to help advance its desire to outperform its adversaries militarily. Besides, Melliar-Smith did not view the semiconductor technology as a strategic differentiator of military performance; instead, he stressed the constitutive nature of chip technology to national military capabilities. Following this logic, China's superiority over Taiwan and even the US in IC technology, if realized, contributes to rather than determines its relative military capability vis-à-vis that of Taiwan and the US.

Several US defense systems designers have made similar points by referring to the Tour de France analogy, which we discussed in Chapter 2. A senior weapons designer disagreed with officials in Washington who had regarded the "shrinking chip gap" between China and the US as a real cause of alarm for the US, as evidenced by the GAO statement. He highlighted the relative difficulty in system engineering and system integration as opposed to the use of IC technology alone during the entire life cycle of developing a military system. "I am not sure if I think that one statement is so important," he concluded.[90]

Even a Pentagon official in charge of export controls doubted the importance of the "simple drawing" by the GAO, which, in his view, did not reflect the complex reality concerning China's chip technology. The fact that China's flagship chipmaker had been near the global state-of-the-art, as reflected in the drawing, did not mean that all chip-makers in China, including PLA suppliers, had reached the same level of technology.[91]

Numerous interviewees have doubted whether China would soon outperform Taiwan and the USA in IC design. They have done so by highlighting the various constraints Chinese latecomers have faced to upgrade their IC design capability,[92] and the innovation efforts by US and Taiwanese firms to retain their leadership.

According to Rupert J. Hammond-Chambers, the president of the US–Taiwan Business Council, the top-notch US IC design firms have rarely stopped innovating because they want to retain their leadership. As of early 2005, the US leadership in IC design has shown little sign of erosion, whereas China's chip technology had focused on IC manufacturing, not product development. Hence, he concluded that China would not surpass the US in IC design in the immediate future[93] making any purported security challenges Washington might face owing to Chinese supremacy in IC design premature.[94]

A comparison of the number of US semiconductor patents received by the US, China and Taiwan annually between 2003 and 2007 indicates the innovation gap among the three players.[95] China still lags far behind Taiwan and the US in patenting in chip manufacturing process technology and the design and analysis of circuit or semiconductor mask, although it has experienced explosive growth in patenting in process technology during the period studied (Figure 6.8 and Figure 6.9).

To sum up, although the Taiwanese migration has helped China upgrade its IC technology, as evidenced by the ability of the Taiwanese-run SMIC to be near global state-of-the-art process technology by 2004, the security implication of the closing chip gap between China, the US, and Taiwan has been controversial. Some of the Pentagon-related agencies and interviewees have asserted that China, following its success in almost reaching parity with the US and Taiwan in IC

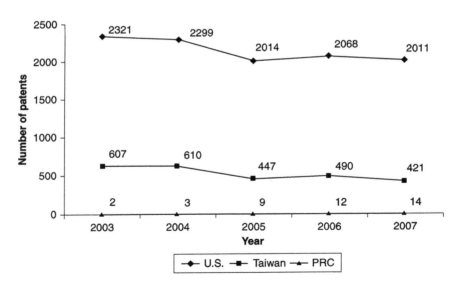

Figure 6.8 Patents in semiconductor device manufacturing process: the US, Taiwan, and China, 2003–07.

Source: US Patent and Trademark Office, Electronic Information Products Division, Patent Technology Monitoring Branch.

program or other unauthorized design inclusions, which are common in today's public software networks, in unclassified ICs used in US or Taiwanese military applications. China may accomplish chipping when designing or fabricating US- or Taiwanese-demanded critical non-COTS ICs on Chinese soil. These "contaminated" ICs thus become "time bombs" in US or Taiwanese defense gadgets. "Such backdoor features could be used by an adversary to disrupt military systems at critical times," warned the DSB report (Defense Science Board Task Force 2005: 23). Second, China may engage in subtle shifts in process parameters or layout line spacing, thereby shortening the lives of the ICs in question. Additionally, the security of classified information embedded in chip designs may become endangered following the shift from onshore to foreign IC manufacturers in China. The fourth risk scenario concerns COTS items. Although the use of COTS implies less risk to the extent that their destination in US or Taiwanese defense systems could be kept anonymous, the DSB study pointed out "even use of COTS components may not offer full protection from parts compromise." The fifth variation concerns the supply of counterfeit ICs to Taiwan or the US, inflicting damage on their military operations. In recent years, the issue of potential national security threat from counterfeit semiconductors in the US military supply chain has continuously attracted the attention from industry associations, US Congress and the press. Some of the studies indicate that many of these fake ICs found in US defense systems are traced back to China (Toohey 2011; Committee on Armed Services 2012).

The DSB report pointed out two realities that would further compound the aforementioned threats. One is technological in nature, and the other involves the hierarchical nature of defense semiconductor acquisition processes in the US and Taiwan. On the technological front, "neither extensive electrical testing nor reverse engineering is capable of reliably detecting compromised microelectronics components." Of the two classes of parts whose combined performance underpin most information process in defense system electronic hardware today, programmable chips, which operate on the data, have more intricate designs than memory ones, which store data. Compared to memory chips, programmable chips are more "difficult to validate (especially after manufacturing) and thus are more subject to undetected compromise" (ibid.: 4, 12).[96] On the acquisition front, the study (ibid.: 5, 26) charged that the Pentagon had not acquired components at the IC level and thus had little clue about the sources of chips used in US defense systems. Individual ICs "are most often specified by the designers of subsystems; even system primes have little knowledge of the sources of the components used in their system-level products." The same can be said of Taiwan's military IC acquisition process.[97]

To what extent are these security threats envisioned by the Pentagon viable in the eyes of heavyweight insiders in the semiconductor, defense, and government sectors? The consideration of vulnerability, motivation and capability factors enables us to assess the extent to which these vulnerabilities will occur in reality.[98]

The confluence of vulnerability and motivation factors has propelled some of the interviewees to argue that the aforementioned chipping threats are possible.

The vulnerability factor refers to the fact that as semiconductor firms continue to move their production activities to China, a window of opportunity is created for Beijing to launch semiconductor-targeted attacks against its foes. Amplifying such a factor at play, a pioneer in the Taiwanese semiconductor industry described chipping threats from China as "valid national security concerns" for Taiwan and the US.[99]

The motivation factor refers to China's aspiration to perform the given nefarious acts at the IC level against its adversaries, as discussed in Chapter 4, which further increases the prospect of the actual occurrence of chipping on PRC soil. To launch these attacks against the US and Taiwan constitutes a part of the PLA's stated aspiration to use IW tactics against its enemies, according to Joe Yu-wu Chen, former president of CSIST. "This is absolutely possible. It falls into the IW arena," he contended.[100]

By factoring in the capability element, however, some have doubted whether these vulnerability scenarios would occur in the immediate future, whereas others have viewed these threats as genuine long-term concerns for the US and Taiwan because China's chip capability may reach a certain level at some later time enabling its Information Warriors to successfully launch these attacks.

Emphasizing what he saw as China's trailing chip technology, Yu Zhongyu, president of CSIA, shrugged off the possibility that his country would insert any Trojan Horse programs in China-made ICs for the US and Taiwanese military applications. "Our technology still trails far behind. They [referring to the US] are overtly sensitive to China's development," he said. "Our current objective is to satisfy the demand of our domestic market."[101] A Taiwanese-American president of a foundry in China also rejected the Pentagon's portrayal of chipping scenarios with respect to China. "This is like a knave who thinks of others in terms of his own desires. China is yet to become formidable in technological terms," he said. "This reflects pure paranoia on the part of the hawkish group in the US."[102]

According to the CEO of a Beijing-based IC design house, China's semiconductor capability will not enable its intelligence service to launch chipping attacks against its foes any time soon. But the country would eventually upgrade its chip capability to a level that would enable its Information Warriors to launch these attacks, causing real security concerns for the US and Taiwan. As he put it:

It is possible, but it is extremely improbable. It is extremely hard to do. China is still far from being able to do it. CIA or FBI [Federal Bureau of Investigation] surely would play such a trick, but the intelligence organizations in China have no energy for such a task. But this does not mean that the Chinese won't do it in 20 years' time.[103]

If China used chipping to strike at the core of US and Taiwanese military capabilities, who are the most and least likely agents of threats and what is the most likely scope of such threats? The answers are three-fold considering, respectively, the nature of industry operations, the types of semiconductor operations, and the Taiwanese and American military IC acquisition policies and practices.

First, senior industry players, by highlighting the nature of pure-play foundry operations, have deemed it almost impossible that any China-based foundries would insert unauthorized designs in unclassified chips for US or Taiwanese military systems. Because foundries produce chips based on masks offered by their customers, they would be unable to insert any unauthorized design in chips without the original circuitry at hand. However, they have admitted that China-based IDMs or IC design houses, where IC design activities occur, may become possible agents for launching chipping attacks (Adee 2008).[104]

Second, China-based commercial semiconductor companies are the least likely agents to perform chipping against Taiwan and the US because their priority – under normal circumstances – is to make profits, not to become an extended arm of the Chinese state machinery for IW attacks against the country's foes. In contrast, government-run or -owned semiconductor operations in China are the most likely agencies to launch these attacks. They may include spin-out Chinese IC design houses of military research institutes that are engaged in outsourcing US or Taiwanese military projects.

According to a senior industry player with experience in military IC design, "Under normal circumstances, most of the China-based commercial firms today are beyond the realm of state control. Nor do they belong to any Chinese ministry. Hence, they will not intentionally make any undesirable move that would put their customers at risk."[105] Nevertheless, this does not rule out the possibility that Beijing still could use, through bribery or coercion, some of these commercial firms to make IC parts compromises. Nor does it eliminate the possibility that individual Chinese Information Warriors may infiltrate some of these commercial firms to work as engineers in order to engage in chipping activities.[106]

Finally, US and Taiwanese military IC acquisition policies and practices may ideally determine the scope of impact of these chipping attacks. Because American and Taiwanese policies have continued to ensure that onshore suppliers would supply confidential defense chips for end-uses in their mission critical military systems, some interviewees have argued that the production of these sensitive military ICs would be isolated from any chipping attacks that may occur in China.

Various homebound agencies in the US have designed and fabricated sensitive chips for US military end-uses, as discussed in Chapter 2. If Washington continues to ensure that trustworthy facilities at home are solely responsible for the design and production of these ICs for US military applications, these devices will not become susceptible to any chipping that may occur in China. Taipei has also confined the supply of a limited amount of confidential military ICs to onshore suppliers. In theory, these ICs are immune from any chipping associated with offshore IC production.

Although the reasons considered above may limit the impact of chipping attacks, neither the US nor Taiwan is completely isolated from these potential threats. This is because an increasing number of ICs for their military applications are not constrained by their policies towards onshore production, and these components depend on offshore sources of supply, which possibly include those in China.

As discussed in Chapter 3, certain types of ICs for US military applications have depended on offshore suppliers, such as chips supplied by the QML 38535 firms. If these offshore production hubs are not strictly controlled, chipping may occur, undermining US national security. Moreover, because of the increasing use of COTS in the US military systems and the deeper globalization of COTS production than that of mil-spec ICs, the US is susceptible to chipping targeting at offshore-made COTS ICs for end-uses in its military systems.

Taiwan has also formulated a policy to acquire ICs for its military systems from both onshore and offshore sources, and it has increased the number of COTS ICs in its military systems because of a reform in 1997 that emulated the Perry Initiative.[107] Hence, ICs originating from offshore suppliers for Taiwanese military applications, including potential ones based in China, are susceptible to chipping.

When asked to comment on chipping risks that the US and Taiwan might face, assuming more and more ICs would be fabricated in China amid the intensification of semiconductor production globalization, Wang Qinsheng, Chair of HED and Chair of CIDC, said disapprovingly: "It is American and Taiwanese firms which voluntarily decide that they would place their orders in China, right?"[108]

Her remark reflects an inconvenient truth concerning emerging hardware threats any country faces as it deepens its dependence on an increasingly globalized supply pool of ICs for its military applications, especially when the pool covers its (potential) adversaries. From China's standpoint, globalization entails a security risk for the PLA because of its belief that "foreign production of chips, no matter how well disguised, nonetheless can create the potential for nefarious purposes, which is to put elements into these chips that actually would allow Westerners to control them and to shoot them . . . down in a war."[109] A similar risk also burdens the US and Taiwan as analyzed above in the context of the migration across the Strait.

In a nutshell, Taiwan and the US may face the risk of hardware tampering because of the migration, although these vulnerabilities may occur only under certain conditions. China may launch chipping attacks on its soil against the US and Taiwan in pursuit of its asymmetrical warfare objective. But the vulnerability, motivation and capability factors at play, the nature of IC industry operations, the types of semiconductor operations which are most likely to launch these attacks, and Taiwanese and American military IC acquisition policies and practices – all have complicated the extent to which Beijing can and will accomplish their supposed objective.

Foreign dependency risks

The third long-term defense security repercussion of the migration concerns the resultant foreign dependency risks the US and Taiwan may face. These include the disruption of chip supply from China due to wars or massive natural disasters and the denial of chip supply by China due to a governmental "reverse-ITAR" policy (Defense Science Board Task Force 2005: 24–5).[110] It is argued that the migration over time helps to strengthen China's chip capacity, which, in turn, may deepen Washington's and Taipei's reliance on offshore supplies of ICs from China, subjecting both countries to increasing vulnerabilities because of foreign dependency under certain conditions.

Supply chain disruption

As the world's IC capacity continues to shift towards China, the country runs a growing number of foreign- or Chinese-owned semiconductor operations to serve the domestic and international market with apparent input from Taiwan and elsewhere. If such a trend continues, it is highly likely that both Taiwan and the US over time will increasingly depend on a supply of China-based chip goods and technologies for end-use in their civilian and even defense systems. Any war across the Strait or natural disaster in China might obstruct the supply to both countries. Beijing may further initiate a reverse-ITAR policy denying Washington and Taipei access to critical chip goods and technologies (ibid.: 24).

Nevertheless, semiconductor players and security experts have offered mixed assessments of these foreign dependency risks. Some have argued that an armed conflict across the Strait or serious natural disasters in China will indeed disrupt the supply of ICs from China to the US and Taiwan, threatening the security of both countries.

In the case of the US, man-made or natural calamities in the region will disrupt the global supply chain, undermining the American economy and national security, according to Hammond-Chambers of the US–Taiwan Business Council. Such a disruption of the chip supply in East Asia would have dire consequences for the US economy because of the issue of inventory control and the concentration of semiconductor foundry operations[111] and related high-tech production activities in Taiwan and China. As he put it:

> Inventories have been kept so low now with companies as a function of controlling cost. If there is a disruption in supply chain, most businesses run out of parts or start running out of parts within 24 to 48 hours . . . If you look at the disruption in the supply chain for Taiwan-China, where you have a penetration for supplies of up to 80 per cent in respective areas, like motherboards, you start to see the potential damage to US technology companies if there is a war in the Taiwan Strait.[112]

As numerous US high-tech companies have huge investments in Taiwan and China for their growth, any war in the region is expected to disrupt their market performance threatening the overall US economy. Such vulnerabilities from outsourcing for the US may deepen if East Asia, particularly the Taiwan–China pair, continues to increase chip manufacturing capacities, he concluded.

When asked if the US has any genuine security concerns over the rise of a Chinese chip industry, Prof. Yoshio Nishi of Stanford University said: "I think there is a certain concern, because . . . fewer and fewer chips are being manufactured in the

US continent." Although Intel, TI, and IBM still make a lot of ICs in the continental US, an increasing number of US companies "are now depending upon foreign manufacturing or offshore manufacturing." Hence, "if one day all borders of all countries are shut to each other for some reason . . . the US has to rely on what we have today." Such a scenario, which might be an outcome of war, may pose a short-term security challenge to the US, albeit not a long-term one. This is because in the long aftermath of war, when there is still no offshore supply of ICs to meet the US demand, the US can still build fabrication facilities on its soil in order to make cutting-edge chips in due course, thanks to its high-quality semiconductor technology.[113]

According to Kathleen Walsh, Senior Associate of the Henry L. Stimson Center, the US, and Taiwan will face similar foreign dependency risks only if China emerges as a monopoly supplier of certain ICs for American and Taiwanese defense end-use. In the context of Sino-US relations, she said:

> If China becomes a monopoly supplier for semiconductors, which looks like it may be possible and because our relationship with China is not settled, then it becomes *de facto* automatically a security concern . . . If more and more of our suppliers for the defense industry come from one place, whether it is China or anywhere else, it is going to be a potential supply concern.[114]

Randall G. Schriver, Deputy Assistant Secretary for East Asian Pacific Affairs of the US Department of State, was doubtful of the degree of security threats to the US and Taiwan that would arise from the disruption of IC supplies from China during a Strait contingency. Although these security concerns would be legitimate ones for Washington and Taipei, he was unsure of their significance relative to other security challenges that would be of more immediate concern to both countries during the war. For example, supply-cut security concern in the event of a conflict would be lower on the list of US security concerns than Washington's worry about its being drawn into a broader conflict with China.[115]

Klaus C. Wiemer, former president of TSMC and former CEO of Chartered, did not see the disruption of foreign chip supply in a Strait contingency as a real security issue for Washington or Taipei because any such conflict would be a short one, shorter than the average 6-week cycle time for chip production. He described the related foreign dependency argument as dim-witted: "The cycle time to make a chip is about six weeks. The next war is not going to last that long."[116] Roger Cliff of Rand Corporation partly echoed his view, contending that since the war would most likely be short, it would not be "a war of production."[117]

Denial of supply

My interviewees have disagreed sharply over the possible denial of chip supply to the US and Taiwan because of a reverse-ITAR policy imposed by Beijing. Some have viewed such a possibility as "the biggest threat" to both countries. According to an American industry observer, the fear for the US is that as IC manufacturing continues to shift to China, the US will be without a primary source of certain IC supplies at home. Even though Intel and other IDMs still base a majority of their facilities in the US, not all of them will produce all the types of ICs the US military needs. It is possible that some of the specific ICs the US military needs for its weapons systems will only be produced in Taiwan or China. "The fear is that they will be cut off from the supply that they are gonna need" because of China's strategy to put the US military at a disadvantaged position, he argued. "I think that is the biggest threat."[118] More importantly, the reverse-ITAR threat scenario is probable because Beijing largely views Washington and Taipei as its potential adversaries in security terms.

James A. Lewis, Director of Technology Policy at the Center for Strategic and International Studies (CSIS), however, was doubtful of the aforementioned threat scenario. He argued that as China becomes a major producer of ICs in the world, it would be hard to fathom why any Chinese chip companies would deny ICs to the US under government instructions. After all, Chinese firms have to become reliable partners if they want to fully integrate into the global market. Moreover, even if Beijing did push for a reverse-ITAR policy, such a restriction on international trade against the US and Taiwan would be impossible to enforce, given a vast array of IC suppliers in the global market aside from those based in China.[119]

As the migration has arguably strengthened China's chip capacity, it may deepen the reliance by the US and Taiwan on IC supplies from China, subjecting both countries to foreign dependency risks in due course. These risks may result from a disruption of IC supply from China due to man-made or natural disasters or a Beijing-led reverse-ITAR policy. However, it is unclear whether these threats will become reality and, if so, how much they will seriously compromise the security of Taipei and Washington.

Conclusion

In brief, the security impact of the migration has been complex, convoluted, and case sensitive. The economic security repercussions of the migration have been mixed, although the long-term ramification is undetermined. As far as defense security is concerned, the migration has contributed to the development of China's chip industrial base, which, in turn, may boost PLA modernization and accelerate changes in the balance of power in the trilateral relationship in Beijing's favor. Taiwanese chip players have thus become security actors: the outcome of their business activities may help increase China's coercive capability vis-à-vis that of Taiwan and even that of the US. Moreover, the migration has also helped China narrow its chip technology gap with Taiwan and the US. Although this development causes certain security concerns in Washington and Taipei, whether that gap will quickly translate into military gains for China is unclear. Besides, as Taiwanese semiconductor production shifts to China, it creates an opportunity for the PLA to launch chipping against the US and Taiwan. This development, if successful, may undermine the security of both countries, a triumph for China's asymmetrical warfare practices. However, chipping may occur only under certain conditions, such as China's possession of a viable semiconductor capability. Finally, the sectoral globalization has contributed to the development of China's industry, raising foreign dependency concerns for some in Taiwan and the US, although it is unclear whether these concerns will become reality. Hence, the globalization of production activities in a strategic industry that has unfolded in one of the most explosive flashpoints in world politics today has generated complex fears and vulnerabilities for key players involved. To what extent does the case study analyzed so far shed light on the impact of globalization on security? This theme will be discussed in the concluding chapter.

Notes

1 Interview, 7 December 2004, San Jose, CA, USA.
2 Interview, 7 September 2005, Beijing, China.
3 For an earlier version of the analysis of the economic security repercussions of the migration, see Chu 2012.
4 For a discussion of how Taipei's anxieties over developments in Cross-Strait economic ties have driven the Taiwanese state to liberalize its economic policy towards China within a strong institutional and regulatory framework, see Dent 2003.
5 Also interview with Maw-kuen Wu, Minister of National Science Council, 24 June 2005, Taipei, Taiwan.
6 Interview, 29 June 2005, Hsinchu, Taiwan
7 In the latter half of 2009, AMD spin-off Global Foundries acquired Chartered. The four big players in the global pure-play foundry sector will thus include TSMC, UMC, SMIC, and Global Foundries.
8 Interviews with senior executives of the firm, 4 August 2009, Taipei, Taiwan.
9 According to my research, Shanghai is now home to Taiwan-related SMIC, Grace, GAPT, ASE, GTM Electronics, Ali, VIA, and Sunplus, among others. Suzhou houses Taiwan-related HeJian, SPIL, Anpec, and Realtek, etc. Wuxi is home to Taiwan-linked CSMC and Sigurd, among others. Taiwanese semiconductor operations in Beijing include SMIC, VIA, CSMC, Ali, and MediaTek, etc.
10 Interview, 27 June 2005, Taipei, Taiwan. Also interview with a former president of CSIST, 14 July 2005, Taipei, Taiwan.
11 Interview, 28 January 2005, Washington, DC, USA.
12 Phone interview, 25 October 2005, Hsinchu, Taiwan.
13 Actions is a primary example.
14 Interview, 20 October 2005, Hsinchu, Taiwan.
15 Interview, 11 July 2005, Hsinchu, Taiwan. Also interview with a managing director of a US IDM subsidiary, 28 March 2005, Shanghai, China.
16 Interview, 8 July 2005, Taipei, Taiwan.
17 Interview with the president of a Taiwanese packaging and testing subsidiary, 27 September 2005, Shanghai, China; interview with the president of a Taiwanese IC design house, 20 October 2005, Hsinchu, Taiwan.
18 These include the soundness of China's national innovation system, government policies, and the extent to which global giants retain their leadership positions in the industry.
19 As detailed in Chapter 4, China's domestic sources of defense microelectronics supply have included state-owned research institutes and academies, factories, and university labs since the mid-1960s. Despite various institutional restructuring and downsizing, part of the structure of the industrial base in question has continued to the present day. These research institutes are now under a diverse array of defense industry conglomerates.

20 For an example of the conventional view, see interview with the president and CEO of SEMI, 10 December 2004, San Jose, USA.

21 Also interview with the former chief of the institute, 19 September 2005, Shanghai, China.

22 This version identified qualified manufacturing lines from April 1992 to 31 May 2002.

23 Interview with the chairman of CSMC, 25 September 2005, Shanghai, China; interview with the president of CSMC, 31 August 2005, Beijing, China; interview with Yu Xie Kang, former vice-president of Huajing, 19 September 2005, Jiangyin, China.

24 Interview with the chairman of CSMC, 25 September 2005, Shanghai, China; interview with the president of CSMC, 31 August 2005, Beijing, China.

25 Interview, 1 February 2005, Washington, DC, USA.

26 Soviet assistance fostered the initial development of the Chinese defense semiconductor industrial base. For the role of Moscow in Chinese defense industries, see Zheng and Li 1989: 118; Editing Committee on China Electronics Industry in Fifty Years 1999; Shambaugh 2004: 226.

27 Since 2003, CAS has had similar arrangements with SMIC, Grace, and Chartered.

28 For a discussion of the backward semiconductors in Chinese military systems, see interview with Qinsheng Wang, Chair of HED, and Chair of CIDC, 30 August 2005, Beijing, China. For insiders' accounts of the technology gap between China's defense chip-makers and their counterparts in leading semiconductor countries, see Wu 1990; Xu 1991; interviews with veteran engineers of the 13th Research Institute, 17 March 2005, Shanghai, China.

29 Members of the Chinese defense industrial complex have appeared on the entity list updated regularly by the US government, which identifies companies that have been involved in or had close ties with weapons of mass destruction. Almost anything leaving the USA that otherwise would not need a license, does need a license if it is going to a listed entity. This partly explains why SIM does not have advanced semiconductor equipment, whereas Grace does. Interview with an official from the US Department of Commerce, 1 February 2005, Washington, DC, USA.

30 For the *neibu* materials consulted, see Hua 1996; Yu 2002.

31 In 1990 and 1991, it won China's military-use science and technology annual award for its work on CMOS analog ICs, rad-hard ICs, CAD development for rad-hard ICs. See China Machinery and Electronics Industry Yearbook Editing Committee 1991: IV–12-IV–13; 1992: IV–14-IV–15.

32 The two authors of the 2005 publication cited here are insiders. Yang is an engineer at SWID, whereas Wang works at the institute.

33 Interviews, 8 December 2004, San Jose, CA, USA, and 27 August 2005, Beijing, China.

34 Interview, 19 August 2005, Hsinchu, Taiwan.

35 For the university's effort to participate in defense research and production activities, see Department of Science and Technology 2006. The firm uses SMIC's foundry services. See Song 2003.

36 Interview with the Taiwanese industry player in question, 14 September 2005, Ningbo, China.

37 Interview, 24 March 2005, Shanghai, China.

38 They include military-grade industrial PCs, low-radiation PCs, and rad-hard PCs.

39 In 2004, its military-grade industrial PC product ACS–2426P-GJB passed environmental testing of electronic equipment respectively performed by the Chinese Air Forces and China Aerospace Corporation. In 2005, its PC product ACS–2908ATX passed the PLA's information security test. See Shenzhen I-Lacs Technology Co. 2006.

40 Since the reorganization of GAD and COSTIND in early 1998, GAD has supervised the establishment of a national military standardization system. See Bureau of Defense Industrial Base 1996: 18; Jencks 1999; Mulvenon and Yang 2002: 276–8, 304–8; Shambaugh 2004: 143–6.

41 For a review of the firm's efforts to become a military supplier, see Shenzhen I-Lacs Technology Co. n.d. The department develops weapons and technologies in electronic, counterelectronic and information warfare for the Fourth Department of the GAD and other services in the PLA. See Shambaugh 2004: 144–6.
42 The gadget also uses Ethernet chips from Realtek, another Taiwanese firm.
43 Interview, 12 July 2005, Taipei, Taiwan.
44 For the argument that the policy should ideally spur spin-on, see Drewry and Edgar 2005. However, the policy is faced with challenges, including the problem of confidentiality, and the compatibility of commercial and military standards. Another challenge is that indigenous civilian firms largely lack innovative capability and depend too much on foreign technologies. Once they are engaged in Chinese defense production activities, they might deepen the dependence of the defense sector on foreign technologies. See Zhong 2005; Song and Niu 2006; Zhang and Li 2006.
45 Interview with a senior semiconductor expert at IBM's Thomas J. Watson Research Center, 14 December 2004, San Francisco, CA, USA.
46 Interview with the vice-president of a Taiwanese IC design house who has recruited local engineers in China, 19 August 2005, Hsinchu, Taiwan; interview with the president of a Taiwanese-run foundry, 21 September 2005, Suzhou, China.
47 Jiangsu Changdian Advanced Packaging Technology Co., Ltd, China's indigenous leading player in the packaging and testing sector, had attempted to engage in defense production activities, but to no avail. See Wu 2004.
48 Interviews with senior engineers of the 13th Research Institute, 17 March 2005, Shanghai, China; interview with a vice-president of a Taiwanese IC design house who has established contacts with China's military IC research institutes, 19 August 2005, Hsinchu, Taiwan.
49 For alternative views, see Hua 1996; No. 214 Research Institute of China North Industries Group Corporation 2007.
50 China is often the targeted destination for unlawful exports of US semiconductor goods and technologies that have military applications. Hence, Washington controls these exports to China for national security reasons. However, the dismantling of COCOM enabled China to purchase high-end military components from the West more easily than ever. See Guo and Gu 1996.
51 In 2005, Washington decontrolled the exports of general-purpose microprocessors because it viewed these ICs as mass commodity items. A license would only be required to export these chips to terrorist countries or for military end-use or end-users in countries posing national security concerns.
52 Interview with Rupert J. Hammond-Chambers, the president of the US–Taiwan Business Council, 25 January 2005, Washington, DC, USA; interview with James A. Lewis, Director of Technology Policy at CSIS, 14 January 2005, Washington, DC, USA; interview with a senior semiconductor expert at IBM's Thomas J. Watson Research Center, 14 December 2004, San Francisco, CA, USA.
53 Interview, 21 January 2005, Alexandria, VA, USA. Also interview with Roger Cliff of Rand Corporation, 31 January 2005, Arlington, VA, USA.
54 While TSMC runs certified manufacturing lines using 0.35, 0.25, and 0.18-micron process technology to fabricate chips for QML–38535 firms for US military end-uses, it has transferred the same level of technologies to its subsidiary in Shanghai.
55 Interview with a senior Taiwanese IC designer, 19 August 2005, Hsinchu, Taiwan; interview with a vice-president of a Taiwanese IC design house who has established contacts with China's military IC research institutes, 19 August 2005, Hsinchu, Taiwan; interview with the chairman of CSMC, 25 September 2005, Shanghai, China; interview with the chairman of NSSI, 14 September 2005, Ningbo, China.
56 Interview with Qinsheng Wang, Chair of HED, and Chair of CIDC, 30 August 2005, Beijing, China.

57 Such a policy had been documented in China's official electronics industry yearbooks and *neibu* materials. See Development Research Center of State Council 2000.

58 These would include firms that have Taiwanese input primarily in the form of top managerial and engineering talent.

59 These include the case involving VIA and Shenzhen I-Lacs Technology Co., Ltd.

60 Interview with the CEO of SMIC, 7 December 2004, San Jose, CA, USA; interview with a former president of Grace, 14 September 2005, Ningbo, China; interview with a manager at Grace, 27 September 2005, Shanghai, China; interview with the president of HeJian, 21 September 2005, Suzhou, China; interview with the chairman of TSMC Shanghai, 29 June 2005, Hsinchu, Taiwan; interview with the president of TSMC Shanghai, 26 September 2005, Shanghai; interview with the chairman of CSMC, 25 September 2005, Shanghai, China.

61 Interview, 30 June 2005, Hsinchu, Taiwan.

62 Phone interview with a sales manager at HeJian, 21 March 2005, Shanghai, China; interview with a vice-president of a Taiwanese IC design house, 7 December 2004, San Jose, CA, USA.

63 Interview with Thomas Howell, an attorney at Dewey Ballantine LLP, 31 January 2005, Washington, DC, USA; interview with a DTSA official, 2 February 2005, Alexandria, VA, USA; interview with a senior member of the Chinese defense chip industry, 17 September 2005, Shanghai, China.

64 They include Chinese policy-makers, PLA strategists and chip industry players. For emphasis on the motivation factor in the Chinese context discussed, see interview with the president of HeJian, 21 September 2005, Suzhou, China; interview with C. Mark Melliar-Smith, former president and CEO of International SEMATECH, 4 January 2005, Austin, TX, USA; interview with Stephen D. Bryen, president of Finmeccanica North America, Inc., 31 January 2005, Washington, DC, USA.

65 Interview with the president of SIA, 8 December 2004, San Jose, CA, USA.

66 Interview with Roger Cliff of Rand Corporation, 31 January 2005, Arlington, VA, USA; interview with the president of a Taiwanese IC design house with experiences in military IC design in the US and Taiwan, 12 August 2005, Hsinchu, Taiwan; interview with the president of SIA, 8 December 2004, San Jose, CA, USA; interview with William A. Reinsch, former Undersecretary of Commerce, 28 January 2005, Washington, DC, USA; interview with a senior manager at AMD, 6 January 2005, Austin, TX, USA; interview with C. Mark Melliar-Smith, former president and CEO of International SEMATECH, 4 January 2005, Austin, TX, USA.

67 Interview, 27 September 2005, Shanghai, China. For an official view on the impact of export controls on China's IC industry, see interview with an official, 23 September 2005, Shanghai, China.

68 Interview, 20 October 2005, Hsinchu, Taiwan. Also interview with a senior IC designer, 10 August 2005, Hsinchu, Taiwan; interview with the president of a Taiwanese DRAM maker, 4 July 2005, Taipei, Taiwan; interview with a senior IC designer, 19 August 2005, Hsinchu, Taiwan.

69 Interview, 7 December 2004, San Jose, CA, USA.

70 Interview, 3 January 2005, Washington, DC, USA.

71 Interview, 9 September 2005, Beijing, China. Also interview with the former president of a Taiwanese IC design house, 5 July 2005, Taipei, Taiwan; interview with a former president of CSIST, 14 July 2005, Taipei, Taiwan.

72 Interview, 21 September 2005, Suzhou, China.

73 Interview, 2 February 2005, Alexandria, VA, USA.

74 On how China can capitalize the global diffusion of technologies in order to develop cyberwar capacities, see Kirshner 2006b: 17.

75 Interview with the president of a leading defense company, 31 January 2005, Washington, DC, USA; interviews with defense experts, 5 August 2005, Taipei, Taiwan. On China's EMP strikes against the USA, see Helprin 2007.

76 Interview with a member of Taiwanese defense industry, 27 October 2005, Taipei, Taiwan.
77 Interview, 1 February 2005, Washington, DC, USA.
78 For example, interview with Klaus Wiemer, former president of TSMC and former CEO of Chartered Semiconductor, 12 January 2005, Dallas, TX, USA.
79 Interview, 4 January 2005, Austin, TX, USA.
80 Interview with Stephen D. Bryen, president of Finmeccanica North America, Inc., 31 January 2005, Washington, DC, USA.
81 Interview with Roger Cliff of Rand Corporation, 31 January 2005, Arlington, VA, USA.
82 These include government policies, foreign investment, R&D environment, human resources, etc.
83 This means half a generation.
84 Interview, 6 December 2004, Palo Alto, CA, USA.
85 Interview with senior engineers at the institute, 17 March 2005, Shanghai, China. For a history of the institute, see Pillsbury 2005: 47; No. 13 Research Institute n.d.
86 Also interview with Roger Cliff of Rand Corporation, 31 January 2005, Arlington, VA, USA; interview with Denis Fred Simon, professor at State University of New York, 11 February 2005, New York, USA; interview with an American official involved in a study of the Chinese semiconductor industry, 28 January 2005, Washington, DC, USA.
87 Interview, 18 January 2005, Washington, DC, USA.
88 Interview, 10 December 2004, San Jose, CA, USA.
89 Interview, 4 January 2005, Austin, TX, USA.
90 Interview with an American engineer with more than 40 years of experience in military avionics system design, 22 February 2005, St. Louis, MI, USA. He referred the "one statement" to the GAO contention that the shrinking chip gap between the USA and China would challenge US superiority in defense terms.
91 Interview with a DTSA official, 2 February 2005, Alexandria, VA, USA.
92 These include China's poor record in IP protection and its insufficient investment in semiconductor R&D.
93 Interview, 25 January 2005, Washington, DC, USA. Also interview with a senior vice-president of R&D at TSMC, 30 June 2005, Hsinchu, Taiwan; interview with the former president of a Taiwanese IC design house, 5 July 2005, Taipei, Taiwan.
94 Interview, 25 January 2005, Washington, DC, USA. Also interview with Denis Fred Simon, professor at State University of New York, 11 February 2005, New York, USA.
95 This is to assume that international patent issuing is one of the most reliable proxies for industrial innovation.
96 Also see Adee 2008; Derene and Pappalardo 2008; interview with the CEO of an IC design house who had designed military ICs in the US, 7 September 2005, Beijing, China.
97 Interview with Abe C. Lin, director general of Integrated Assessment Office at Ministry of National Defense, 27 June 2005, Taipei, Taiwan.
98 The IDA study identifies these factors, although its assessment of these factors is not identical to that of my interviewees.
99 Interview with a former president of ITRI, 15 July 2005, Hsinchu, Taiwan. Also interview with Maw-kuen Wu, Minister of National Science Council, 24 June 2005, Taipei, Taiwan.
100 Interview, 9 August 2005, Taipei, Taiwan. Also interview with a vice-president of a Taiwanese chip design house, 10 August 2005, Hsinchu, Taiwan; interview with Hua-ming Shuai, retired Taiwanese army lieutenant general, 9 August 2005, Taipei, Taiwan.
101 Interview, 2 September 2005, Beijing, China.

102 Interview, 31 September 2005, Beijing, China.
103 Interview, 7 September 2005, Beijing, China. Also interview with the former president of a Taiwanese IC design house, 5 July 2005, Taipei, Taiwan; interview with a vice-president of a Taiwanese IC design house who used to head the firm's subsidiary in China, 19 August 2005, Hsinchu, Taiwan; interview with a senior executive at a leading Taiwanese IC design house, 12 July 2005, Taipei, Taiwan.
104 Also interview with a vice-president of a Taiwanese IC design house, 10 August 2005, Hsinchu, Taiwan; interview with the president of a Taiwanese IC design house, 19 August 2005, Hsinchu, Taiwan.
105 Interview, 20 July 2005, Taipei, Taiwan. However, the line between the commercial IT firms and the Chinese military has blurred. See Mulvenon 2005; interview with the president of a Taiwanese IC design house, 29 June 2005, Taipei, Taiwan.
106 In a recent attack, a Chinese factory making PIN entry devices (chip and pin machines) for UK shops was subverted. The machine contained miniature mobile phones that sent customers' card and PIN details to Karachi by short message service.
107 Interview, Abe C. Lin, director general of Integrated Assessment Office at Ministry of National Defense, 27 June 2005, Taipei, Taiwan.
108 Interview, 30 August 2005, Beijing, China.
109 Interview with James Mulvenon of DGI, 21 January 2005, Alexandria, VA, USA.
110 ITAR stands for International Traffic in Arms Regulations, which are a set of regulations that control the import and export of defense-related articles and services on the US Munitions List.
111 In 2007, Greater China represents 68 per cent of worldwide committed foundry fab capacity. See PricewaterhouseCoopers 2007: 50.
112 Interview, 25 January 2005, Washington, DC, USA.
113 Interview, 6 December 2004, Palo Alto, CA, USA.
114 Interview, 14 January 2005, Washington, DC, USA.
115 Interview, 26 January 2005, Washington, DC, USA.
116 Interview, 12 January 2005, Dallas, TX, USA.
117 Interview, 3 January 2005, Washington, DC, USA.
118 Interview, 25 January 2005, Washington, DC, USA.
119 Interview, 14 January 2005, Washington, DC, USA.

7   Conclusion

This book examines the migration of the Taiwanese semiconductor industry across the Taiwan Strait, one of the flash points in world politics today, and the security implications of this migration for US–China–Taiwan relations in order to understand the impact of globalization on security. This concluding chapter summarizes the major research findings of the study and points out its limitations and contributions.

Summary of research findings

The key findings of the study are as follows. Based on the premise that both economic and defense weapons are instruments of power, it was argued in Chapter 2 that a country's semiconductor industry is central to its national power and security as it contributes to national economy, high-tech development and military modernization. The industry has strong linkages to economic power and security for four reasons. To begin with, the semiconductor is an enabling technology, spurring breakthroughs in computer and information-based high technologies, which are often associated with economic growth. Moreover, the industry spurs innovation and growth in other industries. Besides, the sector contributes to economic growth and productivity, economic competitiveness and job creation. Finally, the industry is a cornerstone of economic security contributing to economic competitiveness and economic independence of the country in question.

Moreover, the profound and complex linkage between the industry and national defense power and security revolves around four inter-related arguments. To begin with, semiconductors, including mil-spec and COTS ICs, are building blocks of modern information-dependent military systems. The defense importance of advanced semiconductors is expected to continue assuming the move towards RMA progresses unabated. Furthermore, even though the making of national defense power goes beyond semiconductor technologies, a strong indigenous chip industry remains significant in defense terms – especially for countries with difficulties in ensuring offshore military chip supply because of the lack of allies, such as China. In addition, a strong indigenous chip capability contributes to national defense security by increasing the given country's defensive and offensive IW capabilities. Finally, inter-related technological changes in the chip sector (i.e. dual-use and spin-on) have complicated the linkage between the industry and defense power and security. Both defense and commercial sub-fields of the national chip production base are of varying degrees of security importance, and a policy to ensure the military's access to both sub-fields contributes to defense modernization.

As was contended in Chapter 3, the globalization of production in the semiconductor industry for both commercial and military end-uses is complex. Fabrication, packaging and testing segments of the industry for mainstream commercial end-uses have become deeply globalized, in contrast to IC design subsector. Of a lesser degree, the production of certain types of military graded chips has also become increasingly internationalized, as evidenced in the analysis of the US case. Others, however, have remained homebound, such as the accredited trusted IC supplies of chips for end-uses in mission critical US military systems.

Concerning the linkage between semiconductors and national security in the context of the PRC, it was argued in Chapter 4 that economic, technological, and defense security considerations have driven Beijing to establish a strong domestic chip industrial base to advance its economic and defense interests. On the one hand, the traditionally autarkic Chinese defense chip industrial base has only managed to partially contribute to China's national security by supplying small-volume semiconductors for indigenous military end-users because of various hurdles and constraints. Besides, Beijing's long-standing efforts to establish a comprehensive military standard systems in order to sufficiently manage its defense industrial base has merely partially advanced China's security because of pitfalls inherent in the implementation of pertinent initiatives. However, recent reforms to actively permit civilian firms, including those in the area of semiconductors, to take part in defense production activities have opened the door for spin-on in the PRC context; know-how, talent, and products from the relatively vibrant commercial chip industrial base have, in some cases, already benefited the Chinese military end-users.

On the other hand, the development of Chinese commercial chip industrial base, after decades of underdevelopment, has taken off since 2000, as measured by market size, production capacity, and the growth of IC design, fabrication, packaging, and testing sub-sectors. In particular, IC design has become the fastest growing segment since 2001, achieving notable qualitative improvements such as breakthroughs in developing homegrown CPUs and migration to the upper end of the technological ladder. Crucially, foreign firms have been instrumental in transforming the semiconductor landscape in the PRC context as they increasingly shift part of their production operations to China. However, the local industrial base is faced with numerous challenges such as the relatively weak technological capability and the ability to compete with the global giants in the fiercely competitive worldwide market.

As was argued in Chapter 5, the migration of the Taiwanese semiconductor industry to China has contributed to the increasing globalization of semiconductor production activities for commercial end-uses to a large extent and for defense end-uses to a lesser extent. The migration is also part of a recently emerged outsourcing trend whereby MNCs increasingly shift their production operations to China. The migration began as early as the late 1980s, although it has gathered momentum since 2000. Measured by the cross-border flows of Taiwanese technology, investment, and talent, the migration has been extensive in scope, covering IC design, fabrication, packaging, and testing. This finding challenges some of the previous claims about the given economic phenomena. The direction of the migration has been far more complex than was formerly understood. Subsidiaries of existing Taiwanese chip firms or newly founded Taiwanese enterprises in China are the major beneficiaries of the migration. However, the migration, to a lesser extent, has also nurtured other types of semiconductor operations in China, which form the backbone of the commercial and/or defense sub-fields of the indigenous industrial base. Given the technology gap between Taiwan and China, different patterns of technology transfer which have occurred across the Strait become assets to China's nascent IC industry. While Taiwanese investment is another important part of the migration, human capital flow is the most critical dimension of the geographical shift, as senior Taiwanese professionals bring in much-needed expertise and business connections to China and contribute to local workforce training. Most notably, all the major foundry firms in China are led by Taiwanese.

A confluence of enabling, pull and push factors has driven the migration. The enabling factor has centered on the economics of the semiconductor industry, such as the growing demand for IC design hardware and software engineers because of increases in design complexity. Primary pull factors have involved a desire to gain access to the Chinese market and an aspiration to exploit China's location-specific resources, such as engineering talent. Secondary pull factors have included cost reductions, policy incentives in China, and political factors. The major push factor has resulted from concerns over a long-term shortage of Taiwan's engineering manpower.

The migration has given rise to several leading start-ups in China's commercial IC industry, such as SMIC in the foundry sub-sector and Actions in the IC design segment. Cases analyzed indicate that the migration has resulted in a gradual shift of IC design and fabrication capabilities to China primarily through the infra-firm level and secondarily through the inter-firm level. Moreover, Taiwanese firms operating in China have sown the seeds for the long-term development of China's chip capability because of the following three functions they have performed. Taiwanese-owned or -run foundries have offered their local customers the MPW services which help to enhance indigenous chip design capabilities. Besides, Taiwanese-run or -owned semiconductor projects have trained local engineers, creating pools of highly skilled indigenous labor, which contributes to China's long-term technical development. Taiwan-related foundries have become pillars of the Chinese chip sector, cultivating local IC design houses with their advanced foundry services and feeding downstream packaging and testing firms with orders. Arguably, Taiwanese "new colonialism" has come into being in China's strategically important semiconductor industry. The lure of profit has largely driven related Taiwanese firms and individuals to join the silicon gold rush to China.

As a by-product of globalization, these non-state Taiwanese actors have helped to advance China's ambition to develop a fully integrated semiconductor capability.

Such a globalization force, as was contended in Chapters 5 and 6, has influenced security relations among the trilateral state actors involved in a complex fashion. First, in spite of Taipei's restrictive policy, some firms and individuals from the Silicon Island have violated Taiwanese regulations while spearheading the migration. They have eroded the capacity of the Taiwanese state to curb the geographical shift. The diminished power of the Taiwanese state vis-à-vis that of non-state actors who have orchestrated the migration has been the most obvious security repercussion of globalization in the case study. This finding has rendered the "state-in-retreat" school of thought viable. This has also echoed Raymond Vernon's (1998: 50) observation concerning the increasing difficulties any states have faced in their attempt to control cross-border movement of intangibles. As he puts it, "Controlling the movement of intangibles such as technology is like bagging rainbows."

Second, the ramification of the migration in economic security terms is mixed. The migration has not severely hollowed out Taiwan's chip industry nor diminished Taiwan's economic security, as evidenced by the local industry's continuous growth in revenues, employment levels, and technology developments. However, Taiwanese firms' share of the global foundry market has gradually declined since the migration gathered momentum as China-based competitors, most of which are the products of the migration, enter the race. Taiwan's IC design houses have also been faced with growing competition from their Chinese counterparts in the low-end consumer IC market. So far, the migration has cemented China's economic security because Taiwanese-run or -owned chip-makers and IC design houses in China have helped to quench the country's thirst for ICs and have contributed to its semiconductor sales revenue. In 2004 alone, SMIC, HeJian, Grace and CSMC churned out semiconductors that accounted for over 61 per cent of China's foundry sales revenue. In the same year, the sales revenue of IC design firms with partial or 100 per cent Taiwanese capital in Shanghai accounted for 6.1 per cent of the total in the country. Even so, the long-term ramification of the migration in economic security terms is uncertain.

The third security ramification of the migration concerns the extent to which the globalization forces at play have shaped the build-up of the Chinese chip industrial base, thereby contributing to the PLA's modernization, and changing the balance of power among states involved. I have shown that the migration has contributed to the chip industrial base in China, which comprises commercial and defense sub-fields. The Taiwanese expertise transfer can potentially benefit the PLA because the level of technologies some of the Taiwanese chip-makers, most notably TSMC, have used in their operations in China, as of 2007, has been almost comparable to that of those they have used to make chips for the US military. The PLA may arguably rely on these internal sources of foundry services if it does not consider these firms' Taiwanese background as a potential security threat. More importantly, the analysis of ten empirical cases in Chapter 6 has substantiated

Taiwan's contribution to the Chinese defense chip industrial base, an area that has never been systematically studied in any existing academic literature. Since the 2000s, China has expanded its conventional sources of military semiconductor supply by including the domestic commercial sub-sector through the process of spin-on. Against such a policy backdrop, Taiwanese chip actors, by enabling the Chinese indigenous semiconductor capability to grow, indirectly help to increase the prospect of the PLA's reliance on the growing indigenous chip production base to increase its conventional and unconventional war-fighting capabilities.

Admittedly, the mainstream view among my interviewees is that a viable chip industry in China contributes to rather than determines Chinese military capabilities, and that such an indigenous Chinese semiconductor power is a necessary but not a sufficient condition for China's defense clout. In the long run, however, such a domestic capability, partly nurtured by the Taiwanese firms and individuals spearheading the migration, may help mitigate the security risks Beijing has faced because of export controls and foreign dependency. It may also enable, through the process of spin-on, technologically superior indigenous civilian firms to supply IC goods and technologies to the PLA. This development helps the PLA to break the US-led embargo on military semiconductor imports and to alleviate vulnerabilities it has faced because of its reliance on the backward domestic defense industry for the supply of semiconductors.

In due course, all of the aforementioned developments may further improve China's relative military capability vis-à-vis that of Taiwan to a large extent and that of the US to a lesser extent, thereby tilting the balance of power in the trilateral linkage in its favor. Although China's enhanced semiconductor capability will not enable the PLA to be a match for the US military anytime soon, it is a small step in that direction. This line of reasoning thus echoes Robert Gilpin's observation, which I quoted at the beginning of the book, concerning contemporary states' growing anxiety about the shifting territorial distribution of high-tech production activities and its impact on the balance of power among the states involved. The shift of semiconductor production activities across the Strait may pose long-term security challenges to Taipei and Washington by altering the balance of power among the Washington–Beijing–Taipei triumvirate. The migration, by fostering the rise of a strong industry in China, can also enable Beijing to export critical semiconductors to countries such as Iran and North Korea in due course, thereby exacerbating the challenge of proliferation to international security.

Conversely, if China fails to build a solid chip industrial base in spite of the Taiwanese contribution and other contributory factors, most of the aforementioned security repercussions may not occur or may occur to a lesser degree, to the extent that they will undermine Chinese national security instead of translating into a downward spiral of compromised security for Taiwan and the US.

The fourth security implication of the migration involves the extent to which the globalization forces at play have affected the technological security of the states involved by narrowing the technology gap between China, on the one hand, and Taiwan and the US, on the other. I have argued that the migration has helped China upgrade its IC technology, as evidenced by the ability of the Taiwanese-run

SMIC to acquire near-global state-of-the-art process technology by 2004. However, the security implication of the perceived closing chip gap between China, Taiwan, and the US has been controversial. Some in Washington and Taipei, by emphasizing SMIC's achievements, have expressed genuine security concerns about the perceived narrowing chip gap in question. Others have argued that it remains unclear whether the perceived diminishing chip gap will quickly translate into military gains for China thereby undermining the security of the US and Taiwan. My analysis of the number of US semiconductor patents received by the US, China, and Taiwan annually between 2003 and 2007 indicates that although China has experienced an explosive growth in patenting in process technology during the period analyzed, it still lags far behind Taiwan and the US in patenting in the design and analysis of circuit or semiconductor masks.

The fifth security ramification of the migration concerns aspects of technological security pertaining to the practice of IW tactics of chipping. I have argued that, as the Taiwanese chip production shifts to China, it creates an opportunity for China to attack chip production operations on its soil by performing nefarious acts of chipping, as detailed in Chapter 6. US Secretary of Homeland Security Michael Chertoff raised the general issue of chipping in 2008: "Increasingly when you buy computers they have components that originate . . . all around the world . . . We need to look at . . . how we assure that people are not embedding in very small components . . . that can be triggered remotely" (Derene and Pappalardo 2008). If Beijing's chipping attacks are successful to the extent that compromised chips are used in US and Taiwanese military systems, they will sabotage the systems in question and undermine the security of both countries, thereby signaling a triumph for China's asymmetrical warfare practices. However, although chipping has emerged as part of the Chinese military's strategic thinking, the extent to which the PLA will be able to launch these attacks depends on how quickly China can possess a viable indigenous semiconductor capability. Aside from the vulnerability, motivation and capability factors at play, as analyzed above, the following three elements complicate the extent to which Beijing can and will accomplish its supposed IW objective. They include the consideration of the nature of IC industry operations, the types of semiconductor operations which are most likely to launch chipping attacks, and Taiwanese and American military IC acquisition policies and practices.

The sixth security implication of the migration concerns foreign dependency. As the migration has strengthened China's chip capacity, it may deepen US and Taiwanese reliance on IC supplies from China, thereby subjecting both countries to ensuing foreign dependency risks. These risks may result from the disruption of IC supply from China due to man-made or natural disasters in the region or a Beijing-led reverse-ITAR policy. These threats engender genuine long-term insecurity for Taipei and Washington. After all, 31 per cent of the world's leading edge 12-inch wafer fabrication capacity in 2007 was in Taiwan and China combined, whereas only 14 per cent was in America (Semiconductor Industry Association 2009: 17). Nevertheless, it is unclear whether these threats will become reality and, if so, how much they will translate into a downward spiral of compromised security for Taipei and Washington.

At a theoretical level, how does this case study shed light on the impact of contemporary globalization on security? Based on the empirical findings of the study as summarized above, I argue below that the impact of globalization on security concerns the following four aspects, which I introduced in the analytical framework of the study in Chapter 1. They include the autonomy and capacity of the state, the agency and scope of security, the balance of power, and the nature of conflict.

First, as shown in the case study, contemporary globalization has affected security by eroding the capacity and autonomy of the state vis-à-vis that of the non-state actors who lead cross-border production activities. This has echoed the findings in the existing literature on globalization–security interconnections, as reviewed in Chapter 1.

Second, the case study has shown that the traditionalist definition of security as far as the agency and scope of security is concerned is no longer viable. The agency of security has extended beyond state-centrism amid forces of globalization. Non-state players have become security actors and the outcome of their globalization activities may potentially alter inter-state security relations by creating insecurity for some countries involved. Moreover, the scope of security has extended beyond military-centrism. Under some conditions, economic insecurity (e.g. loss of economic competitiveness) and technological insecurity (e.g. chipping) because of production globalization may also undermine national security.

Third, production globalization forces, through the international diffusion of technologies, may change the distributive military and technological capabilities of states involved, thereby altering the balance of power among the states in question. This has echoed the findings in the existing literature on globalization–security interconnections, as reviewed in Chapter 1. Fourth, and finally, MNCs, in the process of spearheading production globalization, expand the geographical locations of manufacturing activities to go beyond the firms' home soil and may engage in offshore operations in a foreign country which is deemed an adversary by the firms' home country. As such, these offshore production hubs may create opportunities for the foreign government to launch new forms of conflict (e.g. chipping) against the firms' home country.

Limitations and contributions of the study

In hindsight, the limitations and contributions of the study to the relevant field of study are summarized as follows. The limitations are three-fold. The first set of limitations has resulted from topic sensitivity. Rule breakers do not normally disclose their wrongdoings to outsiders if they view such a revelation as detrimental to their interests, though exceptions did exist among many of my interviewees. But such an inhibiting factor potentially affects the result of the interviews. Besides, the non-disclosure agreements between semiconductor companies and their military customers would possibly limit the amount of information I could obtain from firms I interviewed regarding business orders they had received from the military. It is conceivable that these interviewees would presumably disclose to me only limited, vague or general information on their military business, as evidenced by my interviews with senior executives at a global foundry giant. Besides, the sensitivity of the topic has constrained my efforts to collect certain types of data for the study resulting in the unavoidable space of uncertainty concerning the security implications of the globalization of the industry. For instance, my lack of comprehensive access to information on the extremely sensitive topic of military semiconductor supply to the PLA has constrained my ability to examine the extent to which Taiwanese firms have become part of the Chinese military chip supply network.

The second set of limitations concerns several types of potentially problematic secondary data. First, information available on the websites of semiconductor companies should be treated with care because it may not be completely accurate, according to numerous industry interviewees. Moreover, what has been left unsaid in company press releases can be potentially more important than what has been included in these statements. Second, government statistics in China and in Taiwan should also be handled with care. In the case of China, senior officials and market analysts admitted that official statistics on the fast-moving semiconductor industry were largely incomplete. A senior analyst at CCID, China's flagship market analyst with close links to the central government, revealed that the CCID market reports could be far from comprehensive. As he put it, "We only counted those that were made available to us."[1] Taiwanese official statistics on cross-Strait economic flow have been marred by the widely accepted view that government figures cannot fully reflect market reality since many Taiwanese operations are believed to have extended into China insidiously in order to avoid government scrutiny. As a result, these official data should be treated with caution.

Third, my work is a probability probe due to the lack of a completely representative sample. Although critics of my work may raise the question of possible selection bias in my project, I did cover interviews across a reasonably wide spectrum because of my efforts to maximize cases. What I achieved enabled me to conduct a detailed within-case analysis with as many observations as possible. Nevertheless, these interviews with semiconductor firms were not a representative sample of the overall industry under investigation. Aware of the fact that I am not engaged in formal hypothesis testing, I am at best conducting a plausibility probe.

Despite these limitations, this research undertaken has made five main contributions to the relevant field of study. First, I have contributed to the field in methodological terms because I developed innovative methodological strategies for this sensitive project, especially in the areas of gaining access to prospective interviewees and effectively conducting interviews. These strategies, as detailed in Chapter 1, are potentially useful to those who are about to undertake sensitive research.

Besides, my access to heavyweight interviewees has enabled me to contribute to the chosen field of study. My interviewees were diverse in their professions and were well suited to offer me insights on a wide range of issues because a substantial number of them have occupied important positions in their respective institutions. Their insights have thus deepened my understanding of the given issues critical to my research questions. For instance, my interviews with senior Chinese chip industry players who began their work in the field as early as the 1950s have shed light on the link between national security and the industry in the Chinese context. Such knowledge has been essential to my analysis of the security implications of the migration. As a whole, the reference to these original interviews has permeated the book, a feature which distinguishes my work from that of others which has been penned largely based on secondary data.

My third contribution has resulted from access to useful materials from as wide a range of channels as possible, especially those not normally available to outsiders. These materials include Chinese-language materials on the history of Chinese defense microelectronics production activities and the Chinese perception of IW. They also include classified Chinese internal circulars that shed light on the strength and weakness of secretive Chinese defense chip production activities. Without doubt, my ability as a native Chinese speaker has helped me to efficiently collect and digest these sources, which are important pillars of the book. As David Shambaugh (1999: 15) admits, regarding access to raw materials on Chinese military affairs, "having the time to carefully read, digest, and use them in research is no small challenge for non-natives." In sum, all of these materials provided me with an abundant source of data, the analysis of which is often lacking in any existing literature.

Fourth, I have completed a challenging project on a sensitive and yet understudied topic, and the research undertaken has enabled me to present a nuanced, interdisciplinary analysis of the geopolitical outcome of the migration, which certainly negates the claim by some of the industry interviewees that nothing can be said about potential security problems arising from the migration. Fifth, and finally, the sector-based approach which I used to examine globalization–security interconnections in the book has enabled me to shed light on the complex impact of globalization on security, based on the empirical findings of the case study. The book thus makes a theoretical contribution to the relevant field of international relations.

Note

1 Interview, 9 September 2005, Beijing, China.

Bibliography

Abrams, Michael (2003) "Dawn of the E-Bomb," *IEEE Spectrum*, 40(11): 24–30.

Adamson, Fiona B. (2006) "Crossing Borders: International Migration and National Security," *International Security*, 31(1): 165–99.

Addison, Craig (2001) *Silicon Shield: Taiwan's Protection against Chinese Attack*, Irving, TX: Fusion.

Adee, Sally (2008) "The Hunt for the Kill Switch," *IEEE Spectrum*, 45(5): 34–9.

Aizcorbe, Ana *et al.* (2007) "The Role of Semiconductor Inputs in IT Hardware Price Decline: Computers Vs. Communications," in Berndt, E. R. and Hulten, C. R. (eds.) *Hard-to-Measure Goods and Services: Essays in Honor of Zvi Griliches*, Chicago, IL: University of Chicago Press, pp. 351–81.

Alberts, David S. *et al.* (1999) *Network Centric Warfare: Developing and Leveraging Information Superiority*, Washington, DC: DOD C4ISR Cooperative Research Program.

Allison, Graham (2000) "The Impact of Globalization on National and International Security," in Nye, J. S. and Donahue, J. D. (eds.) *Governance in a Globalizing World*, Washington, DC: Brookings Institution Press, pp. 72–85.

Almasi, G. *et al.* (2002) "Blue Gene/L, a System-on-a-Chip," *Proceedings of 2002 IEEE International Conference on Cluster Computing*, pp. 349–50.

Amato, Ivan (1998) "Micromachines: Fomenting a Revolution, in Miniature," *Science*, 282(5388): 402–5.

Anderson, J. Ross (2008a) "Re: Monique," email (19 December 2008).

—— (2008b) *Security Engineering: A Guide to Building Dependable Distributed Systems*, Indianapolis, IN: Wiley Publishing Inc.

Andrews, W. (2003) "Computers and Electronics Component of DoD Budget Continues to Rise," *COTS Journal*, 5: 54–8.

Anselmo, Joseph (1997) "U.S. Seen More Vulnerable to Electromagnetic Attack," *Aviation Week & Space Technology*, 147(4): 67.

ARM (2004) "ARM Opens Beijing Office to Support Strong Regional Growth," available at: http://www.arm.com/about/newsroom/6357.php (accessed 1 February 2005).

ASC (2007) *Annual Report 2006*, 31 December, available at: http://www.wikinvest.com/stock/Actions_Semiconductor_(ACTS)/Filing/20-F/2007/F1996831 (accessed 17 June 2008).

—— (2008) "Actions Appoints Vincent Lin as Vice President of Marketing," 29 January, available at: http://www.actions.com.cn/en/viewNews.aspx?type=1&id=200 (accessed 15 March 2008).

ASE (2007a) *2006 Annual Report*, available at: http://www.corpasia.net/taiwan/2311/annual/2006/EN/ASX_2006_Annual_Report_E.pdf (accessed 1 March 2008).

—— (2007b) "ASE Group and NXP Semiconductors Embark on a New Assembly and Test Joint Venture in Suzhou, China," 28 September, available at: http://www.nxp.com/news/content/file_1367.html (accessed 15 October 2007).

—— (2009) *2008 Annual Report*, available at: http://www.corpasia.net/canpanel/taiwan/listco/2311/annual/2008/EN/2008%20English%20Annual%20Report.pdf (accessed 19 December 2009).

Asian Private Equity Review (2005) "When the Chips Are Down," 22 June, Hong Kong: Centre for Asia Private Equity Research, available at: http://www.altassets.com/casefor/countries/2005/nz6972.php (accessed 5 March 2006).

Baldwin, David A. (1997) "The Concept of Security," *Review of International Studies*, 23: 5–26.

Balfour, Frederik *et al.* (2003) "A Bush in Hand Is Worth . . . A Lot," *Business Week*, 3862, 15 December: 56.

Bao, Keming (1989) "Zou Ziji De Lu (Break a Path of Our Own)," *Huigu yu Zhangwang: Xin Zhongguo de Guofang Keji Gongye (Retrospect and Prospect: New China's Defense Science and Technology Industry)*, 260–1.

Barbe, D. F. (1980) "VHSIC Technology Barriers," *Proceedings of the 1980 International Electron Devices Meeting*, 20–3.

Barnaby, Frank (1984) "Microelectronics and War," in Tirman, J. (ed.) *The Militarization of High Technology*, Cambridge, MA: Ballinger, pp. 45–61.

Barrett, Barrington M. (2005) "Information Warfare: China's Response to U.S. Technological Advantages," *International Journal of Intelligence and CounterIntelligence*, 18(4): 682–706.

Berger, Suzanne and Lester, Richard K. (2005) "Globalization and the Future of the Taiwan Miracle," in Berger, S., and Richard K. Lester (eds.) *Global Taiwan: Building Competitive Strengths in a New International Economy*, New York: M.E. Sharpe, pp. 3–32.

Bergsten, C. Fred *et al.* (2006) *China: The Balance Sheet: What the World Needs to Know Now about the Emerging Superpower*, New York: Public Affairs.

Berlin, Leslie (2005) *The Man Behind the Microchip: Robert Noyce and the Invention of Silicon Valley*, New York: Oxford University Press.

Betz, David J. (2006) "The More You Know, the Less You Understand: The Problem with Information Warfare," *Journal of Strategic Studies*, 29(3): 505–33.

Bishop, Matt and Goldman, Emily O. (2003) "The Strategy and Tactics of Information Warfare," *Contemporary Security Policy*, 24(1): 113–39.

Bolt, Paul J. and Brenner, Carl N. (2004) "Information Warfare across the Taiwan Strait," *Journal of Contemporary China*, 13(38): 129–50.

Booth, Ken (2005) "Critical Explorations," in Booth, K. (ed.) *Critical Security Studies and World Politics*, London: Lynne Rienner, pp. 1–18.

Borkan, H. (1982) "VHSIC Technology Approaches," *1982 IEEE International Solid-State Circuits Conference, Digest of Technical Papers*, pp. 226–7.

Borrus, Michael and Zysman, John (1992) "Industrial Competitiveness and American National Security," in Sandholtz, W., *et al.* (eds.) *The Highest Stakes: The Economic Foundations of the Next Security System*, New York: Oxford University Press, pp. 7–52.

Borsuk, Gerald M. and Coffey, Timothy (2003) "Moore's Law: A Department of Defense Perspective," *Defense Horizons*, 30: 1–8.

Brady, Henry E. and Collier, David (2004) *Rethinking Social Inquiry: Diverse Tools, Shared Standards*, Lanham: Rowman & Littlefield.

Brannen, Julia (1988) "Research Note: The Study of Sensitive Subjects," *Sociological Review*, 36(3): 552–63.

Branscomb, Lewis M. *et al.* (1992) *Beyond Spinoff: Military and Commercial Technologies in a Changing World*, Boston, MA: Harvard Business School Press.

Braun, Ernest and Macdonald, Stuart (1982) *Revolution in Miniature: The History and Impact of Semiconductor Electronics*, Cambridge: Cambridge University Press.

Breznitz, Dan (2007) *Innovation and the State: Political Choice and Strategies for Growth in Israel, Taiwan, and Ireland*, New Haven, CT: Yale University Press.

Broad, William J. (1991) "War in the Gulf: High Tech; War Hero Status Possible for the Computer Chip," *New York Times*, Section: Foreign Desk, 21 January.

Brooker, Matthew (2006) "China's Grace Seeks Debt Repayment Delay on Loss," *Bloomberg*, 17 March.

Brooks, Stephen G. (2005) *Producing Security: Multinational Corporations, Globalization, and the Changing Calculus of Conflict*, Princeton, NJ: Princeton University Press.

—— (2007) "Reflections on *Producing Security*," *Security Studies*, 16(4): 637–78.

Brown, Clair and Campbell, Ben (2001) "Technical Change, Wages, and Employment in Semiconductor Manufacturing," *Industrial and Labor Relations Review*, 54(2A): 450–65.

Brown, Clair and Linden, Greg (2005) "Offshoring in the Semiconductor Industry: A Historical Perspective," *Brookings Trade Forum*, 2005: 279–322.

—— (2009) *Chips and Change: How Crisis Reshapes the Semiconductor Industry*, Cambridge, MA: The MIT Press.

Browning, Larry D. and Shetler, Judy C. (2000) *Sematech: Saving the U.S. Semiconductor Industry*, College Station, TX: Texas A&M University Press.

Bryen, Stephen D. (2008) "Re: Monique's Thesis Chapter Draft," email (15 August 2008).

Bucy, J. Fred (1980–1) "Technology Transfer and East–West Trade: A Reappraisal," *International Security*, 5(3): 132–51.

Bureau of Defense Industrial Base, Ministry of Electronics Industry (1996) "Junyong Dianzi Yuanqijian Shengchanxian Guanche Guojunbiao Gongzuo De Tihui [Reflections on the Implementation of National Military Standards for the Manufacturing Lines of Electronic Parts and Components for Military End Uses]," *Junyong biaozhunhua [Military Standardization]*, 1: 16–19.

Bureau of Industry and Security (2003) "New Regulation Streamlines Export Controls," 14 January, Washington, DC: Department of Commerce, available at: http://www.bis.doc. gov/news/2003/DOCMicroProcessors01_14.htm (accessed 1 January 2005).

—— (2007a) "New BIS Program Changes Export Rules on Targeted Products for Select Companies in China," 18 October, Washington, DC: Department of Commerce, available at: http://www.bis.doc.gov/news/2007/china10182007.htm (accessed 30 October 2007).

—— (2007b) "Revisions and Clarifications of Export and Reexport Controls for the People's Republic of China (PRC); New Authorization Validated End-User; Revision of Import Certificate and PRC End-User Statement Requirements," *Federal Register*, 72(117): 33646–62.

Burnham, Peter *et al.* (2004) *Research Methods in Politics*, New York: Palgrave Macmillan.

Bush, Richard C. (2005) *Untying the Knot: Making Peace in the Taiwan Strait*, Washington, DC: Brookings Institution Press.

Business Week (2001) "The Stars of Asia: Entrepreneurs: Wen-Chi Chen, Chief Executive, Via Technologies," 3739, 2 July.

Buzan, Barry (1991) *People, States and Fear: An Agenda for International Security Studies in the Post-Cold War Era*, New York: Harvester Wheatsheaf.

Buzan, Barry and Herring, Eric (1988) *The Arms Dynamic in World Politics*, London: Lynne Rienner.

Buzan, Barry *et al.* (1998) *Security: A New Framework for Analysis*, London: Lynne Rienner.

Cable, Vincent (1995) "What Is International Economic Security?," *International Affairs*, 71(2): 305–24.

Cantwell, John (1995) "The Globalisation of Technology: What Remains of the Product Cycle Model?," *Cambridge Journal of Economics*, 19(1): 155–74.

Cao, Yuhe and Gao, Xiangzhu (1989) "Chengxiao Zhuozhu De Gaokekao Qizhuan Gongzuo [Remarkable Success of "Seven Specializations" High Reliability Program]," in Nie, L. and Huai, G. (eds.) *Huigu Yu Zhangwang: Xin Zhongguo De Guofang Keji Gongye [Retrospect and Prospect: New China's Defense Science and Technology Industry]*, Beijing: Defense Industry Press, pp. 445–7.

Carayannis, Elias G. and Gover, James (2002) "The SEMATECH-Sandia National Laboratories Partnership: A Case Study," *Technovation*, 22: 585–91.

Carlson, G. (2005) "Trusted Foundry: The Path to Advanced SiGe Technology," paper presented at Compound Semiconductor Integrated Circuit Symposium, Palm Springs, CA, 30 October–2 November.

Carlson, Peter (2003) "The Relatively Charmed Life of Neil Bush," *Washington Post*, D01, 28 December.

Carlston, Peter (2007) "Riding the Next Wave of Embedded Multicore Processors," *COTS Journal*, May, available at: http://www.cotsjournalonline.com/articles/view/100659 (accessed 2 October 2007).

Carr, E. H. (2001) *The Twenty Years' Crisis, 1919–1939: An Introduction to the Study of International Relations*, London: Palgrave.

Caverley, Jonathan D. (2007) "United States Hegemony and the New Economics of Defense," *Security Studies*, 16(4): 598–614.

CCID Consulting (2007) "Sales Revenue and Growth Rate of China's IC Market, 2003–2007," December, available at: http://www.ccidconsulting.com/upload/12548.jpg (accessed 10 January 2008).

Central Intelligence Agency (1983) "Soviet Acquisition of Western Technology," in Bertsch, G. K. and McIntyre, J. R. (eds.) *National Security and Technology Transfer: The Strategic Dimensions of East–West Trade*, Boulder, CO: Westview Press, pp. 92–112.

—— (1985) *Soviet Acquisition of Western Technology: An Update*, Washington, DC: US Government Printing Office.

Cha, Victor D. (2000) "Globalization and the Study of International Security," *Journal of Peace Research*, 37(3): 391–403.

Chang, Morris C. M. (1998) *The Autobiography of Morris C.M. Chang, Volume I 1931–1964*, Taipei: Commonwealth.

Chang, Pao-Long and Tsai, Chien-Tzu (2002) "Finding the Niche Position – Competition Strategy of Taiwan's IC Design Industry," *Technovation*, 22: 101–11.

Chase, Michael S. *et al.* (2004) *Shanghaied? The Economic and Political Implications of the Flow of Information Technology and Investment across the Taiwan Strait*, Santa Monica, CA: RAND.

Chen, Ethan and Chan, Rodney (2007) "Taiwan Packaging and Testing Industry," *DigiTimes*, 12 September.

Chen, Jingjing (2004) "Zhengfu Yaowei Minyong Gongye Canyu Guofang Jianshe Chuangzao Huanjing [The Government Should Create the Right Environment to Facilitate the Participation of Civilian Firms in Defense Productions]," *Zhongguo Junzhuanmin [Defense Industry Conversion in China]*, 3: 5–6.

Chen, Liang-Jung (2007) "Actions Semiconductor, a Taiwan–China Hybrid IC Design House Emerges as the World No. 1 in Its Niche Market," *Commonwealth*, 374, 22 June.

Chen, Qilin (1992) "Junyong Jichengdianlu Yanjiusuo ASIC Foundry De Gouxian [The Concept of Military ASIC Foundry]," *Weidianzixue [Microelectronics]*, 22(3): 1–4.

Chen, Shuhong (2005) *Jingji Quanqiuhua Qushi Xia De Guojia Jingji Anquan Yanjiu [The Study of National Economic Security Amid the Trend of Economic Globalization]*, Changsha: Hunan Renmin Chubanshe.

Chen, Zhongzhou (1992) "Cong QPL, QML Kan Bashi Niandai Meiguo Junyong Weidianlu De Zhiliang Yu Kekaoxing [A Study of Quality and Reliability of US Military-Use Microcircuits in the 1980s through QPL and QML]," *Dianzi Chanpin Kekaoxing Yu Huanjing Shiyan [Electronic Product Reliability and Environmental Testing]*, 4: 34–41.

Cheng, Jian *et al.* (1997) "Tansuo Wojun Zhiliang Jianshe De Xinmoshi [Exploring New Ways to Build up Our Military]," *Jiefangjun Bao [PLA Daily]*, 18 February.

Cheng, T. J. (2005) "China–Taiwan Economic Linkage: Between Insulation and Superconductivity," in Tucker, N. B. (ed.) *Dangerous Strait: The U.S.–Taiwan–China Crisis*, New York: Columbia University Press, pp. 93–130.

Cheung, Tai Ming (2007a) "The Remaking of the Chinese Defense Industry and the Rise of the Dual-Use Economy," Testimony before the United States–China Security Review Commission, Washington, DC: United States Senate.

—— (2007b) Leaping Tiger, Hybrid Dragon: The Search for Innovation and Civil–Military Integration in the Chinese Defence Economy, PhD Dissertation, King's College London, University of London.

—— (2009) *Fortifying China: The Struggle to Build a Modern Defense Economy*, Ithaca, NY: Cornell University Press.

Chiang, Rei-Chih (2006) "China-Bound Investments by Powerchip, ProMOS and ASE Approved," *China Times*, 19 December.

Chien, Chung (2000) "High-Tech War Preparation of the PLA: Taking Taiwan without Bloodshed," *Taiwan Defense Affairs*, October, 141–63.

China Integrated Circuit Yearbook Editing Committee (ed.) (2005) *China Integrated Circuit Yearbook, 2004–2005*, Hong Kong: Asian Information.

China Machinery and Electronics Industry Yearbook Editing Committee (ed.) (1989) *Zhongguo Jixie Dianzi Gongye Nianjian 1989 [China Machinery and Electronics Industry Yearbook, 1989]*, Beijing: Electronics Industry Publishing House.

—— (ed.) (1991) *Zhongguo Jixie Dianzi Gongye Nianjian 1991 [China Machinery and Electronics Industry Yearbook, 1991]*, Beijing: Electronics Industry Publishing House.

—— (ed.) (1992) *Zhongguo Jixie Dianzi Gongye Nianjian 1992 [China Machinery and Electronics Industry Yearbook, 1992]*, Beijing: Electronics Industry Publishing House.

China Semiconductor Industry Association (2002) *An Investigation Report of China's Semiconductor Industry*, Beijing: China Semiconductor Industry Association.

China Semiconductor Industry Association and China Center of Information Industry Development (2005) *A Report on Development Status of Semiconductor Industry in*

China (2005 Edition), Beijing: China Semiconductor Industry Association and China Center of Information Industry Development.

China Semiconductor Industry Association IC Design Branch (2005) *Zhongguo Jicheng Dianlu Sheji Chanye Shinian Licheng [A Ten-Year Progress Report on China's IC Design Industry]*, Beijing: China Semiconductor Industry Association IC Design Branch.

Chinese Academy of Sciences (2000) *2000 Gaokeji Fazhan Baogao [2000 High Technology Development Report]*, Beijing: Kexue Chubanshe.

Chu, Ming-Chin Monique (2008) "Controlling the Uncontrollable: The Migration of the Taiwanese Semiconductor Industry to China and Its Security Ramifications," *China Perspectives*, 1: 54–68.

—— (2012) "Globalization and Economic Security: The Case of Taiwan's Semiconductor Industry," in Blundell, D. (ed.) *Taiwan since Martial Law: Society, Culture, Politics, Economy*. Berkeley and Taipei: University of California, Berkeley and National Taiwan University Press, pp. 549–96.

—— (2013) "Semiconductor Interconnectivity across the Taiwan Strait: A Case Study Approach," in Chow, P. (ed.) *Economic Integration across the Taiwan Strait: Global Perspectives*, Cheltenham: Edward Elgar, pp. 197–238.

Chung, Chin (1997) "Division of Labor across the Taiwan Strait: Macro Overview and Analysis of the Electronics Industry," in Naughton, B. (ed.) *The China Circle: Economics and Technology in the PRC, Taiwan, and Hong Kong*, Washington, DC: Brookings Institution Press, pp. 164–207.

Ciufo, Chris A. (2004) "It Doesn't Get More Rugged Than Rad: Space and Earth-Based Systems Keep Electronics Demand Growing," *COTS Journal*, August, available at: http://www.cotsjournalonline.com/articles/view/100166 (accessed 2 November 2004).

Clarke, Peter (2006) "Fake NEC Company Found, Says Report," *Electronic Engineering Times*, 4 May.

Clendenin, Mike (2003) "China's BLX Making Headway with Godson CPU," *Electronic Engineering Times*, 24 November.

—— (2005) "Chinese Banks Agree to Loan SMIC $600 Million," *Electronic Engineering Times*, 26 May.

—— (2006) "Deflated Expectations in China's IC Biz," *Electronic Engineering Times*, 28 August.

—— (2007) "Updated: Intel Confirms $2.5 Billion Fab in China," *Electronic Engineering Times*, 26 March.

Cliff, Roger (2001) *The Military Potential of China's Commercial Technology*, Santa Monica, CA: RAND.

Cobb, Adam (1999) "Electronic Gallipoli," *Australian Journal of International Affairs*, 532: 133–49.

Cohen, Brian S. (2007) "Integrated Circuits Supply Chain Issues in a Global Commercial Market–Defense Security and Access Concerns, Statement of Dr. Brian S. Cohen, Assistant Director of Information Technology and Systems Division, Institute for Defense Analyses," 14 March, Washington, DC: United States House of Representatives, available at: http://armedservices.house.gov/pdfs/TUTCtech031407/Cohen_Testimony031407.pdf (accessed 15 June 2007).

Cohen, Fred (2000) "Chipping," *Network Security*, 2000, 9: 16–17.

Cohen, Sarah (1997) "Defense Dept. to Intel: Stay – the Defense Supply Center Asks Intel to Retain Its Qualified Manufacturers List Product Line," *Electronic News*, 3 February.

Commission on Science, Technology and National Defense Industry (ed.) (2004) *Deng Xiaoping Guofang Keji Gongye Jianshe Sixiang Yanjiu (Studies of Deng Xiaoping's Thoughts on the Construction of National Defense Technology Industry)*, Beijing: Beijing Hangkonghantian Daxue Chubanshe.

Commission of Science Technology and Industry for National Defense (2005) "Wuqi Zhuangbei Keyan Shengchan Xuke Shishi Banfa [Measures for the Implementation of Weaponry Research and Production Permit]," 16 September, Beijing: Commission of Science Technology and Industry for National Defense, People's Republic of China, available at: http://www.costind.gov.cn/n435777/n435943/n435949/n1593098/119138. html (accessed 5 May 2006).

—— (2007) "Guanyu Yinfa Feigongyouzhi Jingji Canyu Guofang Keji Gongye Jianshe Zhinan De Tongzhi [Circular on the Guide for Non-State-Owned Economic Entities to Participate in the Construction of the Defense Industry]," 30 July, Beijing: Commission of Science Technology and Industry for National Defense, People's Republic of China, available at: http://www.costind.gov.cn/n435777/n435779/n435922/n1243973/112347. html (accessed 2 February 2008).

Committee on Armed Services, US Senate (2012) *Inquiry into Counterfeit Electronic Parts in the Department of Defense Supply Chain*, Washington, DC: US Government Printing Office.

Cowan, Robin and Foray, Dominique (1995) "Quandaries in the Economics of Dual Technologies and Spillovers from Military to Civilian Research and Development," *Research Policy*, 24(6): 851–68.

Cox, Christopher (1999) *Report of the Select Committee on U.S. National Security and Military/Commercial Concerns with the People's Republic of China*, Washington, DC: US Government Printing Office.

Cox, Robert (1996) "Production and Security," in Cox, R. (ed.) *Approaches to World Order*, Cambridge: Cambridge University Press, pp. 276–93.

Crawford, Beverly (1993) *Economic Vulnerability in International Relations: The Case of East–West Trade, Investment, and Finance*, New York: Columbia University Press.

Critchlow, Robert (2000) "Whom the Gods Would Destroy," *Naval War College Review*, 53(3): 21–39.

Daly, James (1996) "Texas Instruments TMS320C80 Digital Signal Processor Development Team," *Forbes*, 157(4): S62.

Dean, Jason (2002) "Taiwan Government Investigates Investments in China Chip Firms," *Asian Wall Street Journal*, 25 July.

—— (2006) "Taiwan Indicts Two Ex-Officials at Big Chip Firm: Move Highlights Tensions as Many Executives Seek to Do Business in China," *The Wall Street Journal*, 10 January.

Defense & Aerospace Electronics (1991) "NATO C3 Must Use Commercial Equipment. (North Atlantic Treaty Organization Should Rely on Commercial Sources for Military Command, Control and Communications Components)," *Defense & Aerospace Electronics*, 1(4): 7–9.

Defense Science Board (1987a) *Defense Science Board 1986 Summer Study: Use of Commercial Components in Military Equipment*, Washington, DC: Office of the Under Secretary of Defense for Acquisition.

—— (1987b) *Report of Defense Science Board Task Force on Defense Semiconductor Dependency*, Washington, DC: US Department of Defense.

—— (1989) *Report of the Defense Science Board on Use of Commercial Components in Military Equipment*, Washington, DC: Office of the Under-Secretary of Defense for Acquisition.

—— (1999) *Final Report of the Defense Science Board Task Force on Globalization and Security*, Washington, DC: Office of the Under Secretary of Defense for Acquisition and Technology.

Defense Science Board Task Force (2005) *High Performance Microchip Supply*, Washington, DC: Office of the Under-Secretary of Defense for Acquisition, Technology, and Logistics.

Defense Supply Center – Columbus (2000) "Japanese Space Engineers Like DSCC's Quality Program," 22 June, available at: http://www.dscc.dla.mil/News/Releases/2000/00–06–023.html (accessed 12 November 2007).

—— (2007) *Qualified Manufacturers List of Products Qualified under Performance Specification MIL-PRF–38535 Integrated Circuits (Microcircuits) Manufacturing General Requirements For*, 2 February, available at: http://www.dscc.dla.mil/downloads/qplqml/38535/archive/38535rev020.pdf (accessed 1 January 2008).

—— (2008) "DSCC Supplemental Information Sheet for Electronic QML–19500," 31 July, available at: http://www.landandmaritime.dla.mil/downloads/qplqml/19500/archive/QPDSIS_19500_080731.pdf (accessed 1 January 2009).

De Fontenay, Catherine and Carmel, Erran (2004) "Israel's Silicon Wadi: The Forces Behind Cluster Formation," in Bresnahan, T. and Gambardella, A. (eds.) *Building High-Tech Clusters: Silicon Valley and Beyond*, Cambridge: Cambridge University Press, pp. 40–77.

Dellin, T. A. *et al.* (1998) "New Trends in the Commercial IC Industry and the Impact on Defense Electronics," Sandia National Laboratories, available at: www.osti.gov/servlets/purl/634063-ffRYkO/webviewable/ (accessed 13 February 2007).

Denning, Dorothy E. (1999) *Information Warfare and Security*, Reading, MA: Addison-Wesley Longman.

Dent, Christopher M. (2003) "Taiwan's Foreign Economic Policy: The 'Liberalization Plus' Approach of an Evolving Developmental State," *Modern Asian Studies*, 37(2): 461–83.

Department of Defense (2000) "Joint Vision 2020: America's Military: Preparing for Tomorrow," available at: http://www.fs.fed.us/fire/doctrine/genesis_and_evolution/source_materials/joint_vision_2020.pdf (accessed 1 March 2007).

—— (2002) "Department of Defense Directive 8500.1, Information Assurance," certified current as of 23 April 2007, available at: http://www.dtic.mil/whs/directives/corres/pdf/850001p.pdf (accessed 13 November 2007).

—— (2004) *Annual Report to Congress: Military Power of the People's Republic of China 2004*, Washington, DC: Department of Defense, available at: http://www.defenselink.mil/pubs/d20040528PRC.pdf (accessed 29 December 2004).

—— (2005) *Annual Report to Congress: Military Power of the People's Republic of China 2005*, Washington, DC: Department of Defense, available at: http://www.defense.gov/news/Jul2005/d20050719china.pdf (accessed 29 December 2005).

—— (2006) *Annual Report to Congress: Military Power of the People's Republic of China 2006*, Washington, DC: Department of Defense, available at: http://www.defense.gov/pubs/pdfs/China%20Report%202006.pdf (accessed 29 December 2006).

—— (2007) *Annual Report to Congress: Military Power of the People's Republic of China 2007*, Washington, DC: Department of Defense, available at: http://www.defense.gov/pubs/pdfs/070523-China-Military-Power-final.pdf (accessed 30 November 2007).

—— (2008) *Annual Report to Congress: Military and Security Developments Involving the People's Republic of China 2008*, Washington, DC: Department of Defense, available at: http://www.mcsstw.org/www/download/China_Military_Power_Report_2008.pdf (accessed 1 December 2008).

—— (2009) *Annual Report to Congress: Military Power of the People's Republic of China 2009*, Washington, DC: Department of Defense, available at: http://www.defense.gov/pubs/pdfs/China_Military_Power_Report_2009.pdf (accessed 30 December 2009).

—— (2010) *Annual Report to Congress: Military and Security Developments Involving the People's Republic of China 2010*, Washington, DC: Department of Defense, available at: http://www.defense.gov/pubs/pdfs/2010_CMPR_Final.pdf (accessed 31 December 2010).

Department of Science and Technology, Fudan University (2006) "Keji Jianbao [Science and Technology Briefing]," 20 October, available at: http://dst.fudan.edu.cn/UploadFile s/%e7%a7%91%e6%8a%80%e7%ae%80%e6%8a%a52006%e7%ac%ac%e4%ba%8c %e6%9c%9f.doc dst.fudan.edu.cn/newsview.aspx?id%15046 (accessed 14 February 2008).

Derene, Glenn and Pappalardo, Joe (2008) "The Manchurian Chip," *Popular Mechanics*, 185(4): 52–6.

Development Research Center of State Council (2000) *Goujian Xinshiqi Junshi Dianzi Gongye Zhengfu Guanli Moshi De Jiben Silu [Fundamental Thoughts on Constructing a Government Management System of the Military Electronics Industry in the New Era]*, Beijing: Development Research Center of State Council.

Devine, Fiona (2002) "Qualitative Methods," in Marsh, D. and Stoker, G. (eds.) *Theory and Methods in Political Science*, New York: Macmillan, pp. 197–215.

Dicken, Peter (2003) *Global Shift: Reshaping the Global Economic Map in the 21st Century*, London: Sage.

—— (2007) *Global Shift: Mapping the Changing Contours of the World Economy*, London: Sage.

Dickson, Keith (1983) "The Influence of Ministry of Defence Funding on Semiconductor Research and Development in the United Kingdom," *Research Policy*, 12(2): 113–20.

Digitimes (2007) "ProMOS Confirms China Fab Location; Pilot Run by 2008," 22 January.

Doering, Bob R. (2004) "Re: A Humble Request from a Cambridge PhD Student," email (10 December 2004).

Dombrowski, Peter J. and Gholz, Eugene (2006) *Buying Transformation: Technological Innovation and the Defense Industry*, New York: Columbia University Press.

Dornheim, Michael A. (1998) "Bombs Still Beat Bytes," *Aviation Week and Space Technology*, 148(3): 60.

Drewry, Seth and Edgar, William (2005) "China Gambles with Private Sector," *Jane's Defence Industry*, 1 November.

Du, Hongfu and Xie, Shaoming (1997) "Junyong Dianzi Yuanqijian Shoujian Jianyan Tantao [A Discussion on Inspection of the First Article of Military Electronic Parts and Components]," *Junyong biaozhunhua [Military Standardization]*, 1: 9–13.

Du, Junling (2005) "Shenliu Ershier Tai Jisuanji You Shanxi Zhizao Danji Zaojia Zuigao Guo Baiwan [Twenty-Two Computers in Shenzhou Six Made in Shanxi with the Single Most Expensive Unit Costing More Than One Million Dollars]," *Huashangbao [Chinese Business View]*, 17 October.

Dunn, Peter N. (1997) "China's Major Foundry: Project '909'," *Solid State Technology*, 40(5), 1 May.

Dunning, John (1988) *Explaining International Production*, London: Unwin Hyman.

Economic Security Forum (2002) *Zhongguo Guojia Jingji Anquan Taishi Guancha Yu Yanjiu Baogao [Survey Report on China's National Economic Security 2001–2002]*, Beijing: Jingji Kexue Chubanshe.

Editing Committee on China Electronics Industry in Fifty Years (1999) *Zhongguo Dianzi Gongye Yushinian [China Electronics Industry in Fifty Years]*, Beijing: Publishing House of Electronics Industry.

Editing Committee on Studies of Jiang Zemin's Thoughts on Defense Technology Industry Construction (2005) *Jian Zemin Guofang Keji Gongye Jianshe Sixiang Yanjiu [Studies of Jiang Zemin's Thoughts on Defense Technology Industry Construction]*, Beijing: Publishing House of Electronics Industry.

Edmonds, M. *et al*. (1990) "Defense Interdependence: UK and US Dependency on Foreign Technology in Defense Reserach and Development," *Science & Public Policy*, 17(3): 157–69.

Einhorn, Bruce (2002) "Taiwan's Silicon Invasion," *Business Week*, 3811, 9 December, 54–5.

Electronic Engineering Times (2006) "2006 Electronic Design Automation Branding Study: Chip Design," 31 July, available at: http://i.cmpnet.com/eet/news/06/07/surveyonline.pdf (accessed 15 March 2007).

Electronic Engineering Times Asia (2004) "NSSI Mass Produces Chips for Taiwan-Based Firms," 22 November.

—— (2005) "IC Design House Survey 2005: China," 10 May.

—— (2006) "IC Design House Survey 2006: China," 23 May.

—— (2007a) "IC Design House Survey 2007: China," 3 May.

—— (2007b) "MOEA: 12-Inch Fabs Drive Taiwan's Chip Industry," 26 June.

—— (2007c) "China Fabs Accounted for 12% of Global Market in 2006," 31 July.

—— (2008) "Datang Holdings to Invest in SMIC," 12 November.

Electronic Engineering Times Taiwan (2007) "China Puts the Development of IC Design Sector as a Priority with Consumer ICs Playing Key Roles," 4 September.

EMSNow (2007) "TSMC Shanghai Plans to Triple Output," 28 February.

Engardio, Pete (2005) "Where the Valley's Chips Are Born," *Business Week*, 3933, 16 May, 23.

Engardio, Pete and Einhorn, Bruce (2005) "No, You Can't Buy That Chip Gear," *Business Week*, 3934, 23 May: 54–5.

Ernst, Dieter (2005) "Complexity and Internationalisation of Innovation – Why Is Chip Design Moving to Asia?," *International Journal of Innovation Management*, 9(1): 47–73.

Fabless Semiconductor Association and Industry Directions Inc. (2007) *Building a Better Supply Chain: Successful Collaboration in the Fabless Semiconductor Industry*, Dallas, TX: Fabless Semiconductor Association and Industry Directions Inc., available at: http://www.gsaglobal.org/publications/scstudy/FSA_IDISupplyChain-Study.pdf (accessed 1 June 2008).

Fabula, Joe and Padovani, Rick (2004) "From Space to Base: Leading the Era of Transformation," paper presented at Microelectronics Reliability & Qualification Workshop, Manhattan Beach, CA, 7–8 December.

Fang, Hanting and Cao, Fang (2002) "Jichengdianlu Chanye Fazhan Zhong De Huahong Moshi Yu Xiangguan Jianyi [The Huahong Model in the Development of the IC Industry and Related Suggestions]," *Diaoyanbaogao [Survey Report]*, 11, 1–10 (*neibu* material).

Fang, Yu-Wei (2005) "Renminbi 30 Yuyuan IC Chanye Yanfa Yu Kaifa Zhuanxiang Jijin Chengli [3-Billion Special Fund for IC Industry Research and Development Established]," *DigiTimes*, 21 October.

Feigenbaum, Evan A. (1999) "Who's Behind China's High Technology 'Revolution'? How Bomb Makers Remade Beijing's Priorities, Policies, and Institutions," *International Security*, 24(1): 95–126.

—— (2003) *China's Techno-Warriors: National Security and Strategic Competition from the Nuclear to the Information Age*, Palo Alto, CA: Stanford University Press.

Finley, James I. (1998) "Industry Perspective on Battlespace Digitization," *Proceedings of SPIE Conference on Digitization of the Battlespace III*, pp. 14–19.

Flamm, Kenneth (1985) "Internationalization in the Semiconductor Industry," in Grunwald, J. and Flamm, K. (eds.) *The Global Factory: Foreign Assembly in International Trade*, Washington, DC: The Brookings Institution, pp. 38–136.

—— (1996) *Mismanaged Trade? Strategic Policy and the Semiconductor Industry*, Washington, DC: The Brookings Institution.

Flamm, Kenneth and Reiss, Peter C. (1993) "Semiconductor Dependency and Strategic Trade Policy," *Brookings Papers on Economic Activity. Microeconomics*, 1: 249–333.

Fong, Glenn R. (1986) "The Potential for Industrial Policy: Lessons from the Very High Speed Integrated Circuit Program," *Journal of Policy Analysis and Management*, 5(2): 264–91.

—— (2000) "Breaking New Ground or Breaking the Rules: Strategic Reorientation in U.S. Industrial Policy," *International Security*, 25(2): 152–86.

Forney, Matthew (2001) "Taipei's Tech-Talent Exodus," *Time Asia*, 157(20), 21 May.

Freedman, Lawrence (1999) "The Changing Forms of Military Conflict," *Survival*, 40(4): 39–56.

Freescale (2007) "Freescale Chairman and CEO Michel Mayer Inaugurates 300,000-Square-Foot Campus in Noida; Company Invested $50 Million Last Year; Plans to Grow India Headcount over Next Three Years," 21 March, available at: http://media. freescale.com/phoenix.zhtml?c=196520&p=irol-newsArticle&&ID=976101 &highlight= (accessed 19 December 2007).

Friedberg, Aaron L. (1991) "The Changing Relationship between Economics and National Security," *Political Science Quarterly*, 106(2): 265–76.

Friedman, Wendy and Martin, J. J. (1988) "Changing Military Technology and Its Impact on Asian-Pacific Security," in Scalapino, R. A., *et al.* (eds.) *Asian Security Issues: Regional and Global*, Berkeley, CA: University of California Press, pp. 79–107.

Frost & Sullivan (1981) *The Military Semiconductor Market in the U.S.*, New York: Frost & Sullivan.

Fu, Luyan and Li, Yong (2004) " 'Xinpianzhan' Qiaoran Zoulai ['Chip War' Quietly Creeps in]," *Jiefangjun Bao [PLA Daily]*, 26 May.

Fu, Wenbiao (ed.) (2005) *Research Report on Shanghai's IC Industry in 2005 [2005 Nian Shanghai Jichengdianlu Chanye Fazhan Yanjiubaogao]*, Shanghai: Shanghai Education Publishing House.

Fujian Provincial Development and Reform Committee (2007) "*Fujian Fushun Microelectronics Selected as Recipient of the First National Sponsorship of IC Enterprises*," 5 September, available at: http://www.fzic.gov.cn/!fzic/News_View. asp?NewsID=1316 (accessed 14 December 2007).

Fulghum, David (2005) "Antenna Angst," *Aviation Week & Space Technology*, 163(16): 52.

Fuller, Douglas B. (2008) "The Cross-Strait Economic Relationship's Impact on Development in Taiwan and China," *Asian Survey*, 48(2): 239–64.

Gabriele, Alberto (2002) "S&T Policies and Technical Progress in China's Industry," *Review of International Political Economy*, 9(2): 333–73.

Gain, Bruce (2003) "Military Mode – Will Homeland Security and the War with Iraq Lead to an Increase in Spending on Military Electronics?," *Electronics Buyers News*, 4 July.

Gansler, Jacques S. (1987) "Needed: A U.S. Defense Industrial Strategy," *International Security*, 12(2): 45–62.

—— (1988) "Integrating Civilian and Military Industry," *Issue in Science and Technology*, 5: 68–73.

Garver, John W. (1997) *Face Off: China, the United States, and Taiwan's Democratization*, Seattle, WA: University of Washington Press.

Gao, Guanxin and Guan, Shengxia (2004) "Shilun Sinxihua Zhanzheng Dui Tuidong Woguo Wuqi Zhuangbei Dongyuan De Qishi [Information War and Its Lessons for the Mobilization of Weapons and Equipment in Our Nation]," *Keji jinbu yu duice [Science & Technology Progress and Policy]*, 21(2): 11–13.

Gelpi, Christopher and Grieco, Joseph M. (2003) "Economic Interdependence, the Democratic State, and the Liberal Peace," in Mansfield, E. D. and Pollins, B. (eds.) *Economic Interdependence and International Conflict: New Perspectives on an Enduring Debate*, Ann Arbor, MI: University of Michigan Press, pp. 44–59.

General Accounting Office (2002) *Export Controls: Rapid Advances in China's Semiconductor Industry Underscore Need for Fundamental U.S. Policy Review*, Washington, DC: General Accounting Office, available at: http://www.gao.gov/new. items/d02620.pdf (accessed 1 November 2003).

George, Alexander L. and Bennett, Andrew (2004) *Case Studies and Theory Development in the Social Sciences*, Cambridge, MA: The MIT Press.

Geppert, L. (2005) "Silicon Gold Rush: Taiwan's Chip Makers Can't Resist the Lure of China, the World's Largest Consumer of Semiconductors," *IEEE Spectrum*, 42(6): 62–6.

Gerring, John (2004) "What Is a Case Study and What Is It Good For?," *American Political Science Review*, 98(2): 341–54.

Gholz, Eugene (2007) "Globalization, Systems Integration, and the Future of Great Power War," *Security Studies*, 16(4): 615–36.

Gilpin, Robert (2001) *Global Political Economy: Understanding the International Economic Order*, Princeton, NJ: Princeton University Press.

Giridhar, Chitra (2006) "India's Niche: Semiconductor Design Services," *Electronic Business*, 24 October.

Glasstone, Samuel (ed.) (1964) *The Effects of Nuclear Weapons*, Washington, DC: United States Atomic Energy Commission.

Global Semiconductor Alliance (n.d.) "Industry Data," available at: http://www.gsaglobal. org/resources/industrydata/facts.asp (accessed 3 November 2011).

Goering, Richard (2006a) "Design Outsourcing: Navigating a Maze of Decisions," *Electronic Engineering Times*, 9 January.

Gongkong.Com (2005) "Pathway to Military Industry: Interview with Zhang Shengrong, President of Shenzhen I-Lacs Technology Co., Ltd.," 25 April, available at: http://www. gongkong.com/news/detail.asp?id=6644 (accessed 4 May 2005).

Green, Eric Marshall (1996) *Economic Security and High Technology Competition in an Age of Transition: The Case of the Semiconductor Industry*, Westport, CO: Praeger.

Grindley, Peter *et al.* (1994) "SEMATECH and Collaborative Research: Lessons in the Design of High-Technology Consortia," *Journal of Policy Analysis and Management*, 13(4): 723–58.

GSMC (2004) "Cheung Kong and Hutchison Whampoa Invest US$90 Million in Grace Semiconductor," 23 November, available at: http://www.gracesemi.de/enhtml/press_news_detail.jsp?newsid=100000039&id=25 (accessed 2 March 2007).

—— (2005) "Grace Appoints Dr. Shi Chung Sun as CTO and Executive Vice President of Technology Development," 23 August, available at: http://www.gracesemi.com/enhtml/press_news_detail.jsp?newsid=100000045&id=25 (accessed 6 April 2006).

—— (2006a) "Grace Appoints Dr. Kuan Yang Liao as Executive Vice President & Chief Operations Officer," 19 January, available at: http://www.gracesemi.com/enhtml/press_news_detail.jsp?newsid=100000069&id=29 (accessed 23 March 2007).

—— (2006b) "Cypress Ships First Devices Made by China Foundry Partner, Grace Semiconductor, Several Months Ahead of Plan; Cypress Will Begin Transferring 0.13-Micron Process Technology to Grace in Q3," 20 July, available at: http://www.gracesemi.de/enhtml/press_news_detail.jsp?newsid=100000081&id=25 (accessed 30 March 2007).

—— (2006c) "Grace Semiconductor Extends Partnership in Advanced Technologies and Products with Oki," 11 September, available at: http://www.gracesemi.com/enhtml/press_news_detail.jsp?newsid=100000084&id=29 (accessed 30 March 2007).

—— (2007) "Dr. Ulrich Schumacher, New President and CEO of Grace Semiconductor," 20 September, available at: http://www.gracesemi.de/enhtml/press_news_detail.jsp?newsid=100000125&id=25 (accessed 19 October 2007).

—— (2008) "Grace Semiconductor Extends Partnership in Advanced Technologies and Products with Sanyo Semiconductor," 5 February, available at: http://www.gracesemi.com/enhtml/press_news_detail.jsp?newsid=100000135&id=29 (accessed 28 April 2008).

Guizzo, Erico (2005) "IBM Reclaims Supercomputer Lead," *IEEE Spectrum*, 42(2): 15–17.

Guo, Jiajun and Gu, Lihua (1996) "Jinkou Junyong Yuanqijian De Zhiliang Huafen Biaozhun Yu Jiankong Yaodian [Quality Differentiation Standards and Supervisory Controls of Imported Military Parts and Components]," *Junyong Biaozhunhua [Military Standardization]*, 6: 29–31.

Guo, Lin and Liu, Jianhua (1992) "Jianli Junyong Foundry De Tansuo [An Exploration for Establishment of the Military Foundry]," *Weidianzixue [Microelectronics]*, 22(3): 5–9.

Gurtov, Mel (1993) "Swords into Market Shares: China's Conversion of Military Industry to Civilian Production," *China Quarterly*, 134: 213–41.

Haavind, Robert (2003) "Let's Keep a Wary Eye on China," *Solid State Technology*, 46(9), 1 September.

Hansell, R. (1988) "GaAs as a Semiconductor," *IEEE Potentials*, 7(4): 9–12.

Hanson, Dirk (1982) *The New Alchemists: Silicon Valley and the Microelectronics Revolutions*, Boston, MA: Little, Brown and Company.

Hayward, Keith (2000) "The Globalisation and Defence Industries," *Survival*, 42(2): 115–32.

He, Xing (2003) "Long March Launch Services," paper presented at APSCC Special Conference, Hanoi, Vietnam, 8–10 October.

Held, David *et al.* (1999) *Global Transformations: Politics, Economics and Culture*, Palo Alto, CA: Stanford University Press.

Helpman, Elhanan and Trajtenberg, Manuel (1998) "Diffusion of General Purpose Technologies," in Helpman, E. (ed.) *General Purpose Technologies and Economic Growth*, Cambridge, MA: The MIT Press, pp. 85–119.

Helprin, Mark (2007) "The Nuclear Threat from China," *Washington Post*, B07, 4 March.

Henderson, Jeffrey (1989) *The Globalisation of High Technology Production: Society, Space and Semiconductors in the Restructuring of the Modern World*, London: Routledge.

Hilkes, Rob (2007) "Under the Hood: Uncovering Hidden Chip Cost," *Electronic Engineering Times*, 22 October.

Hille, Kathrin (2005) "UMC Names Hejian Investors," *Financial Times*, 18 February.

Hirst, Paul and Thompson, Grahame (1999) *Globalization in Question: The International Economy and the Possibilities of Governance*, Cambridge: Polity.

Hoffman, Stanley (2002) "Clash of Globalizations," *Foreign Affairs*, 81(4): 104–15.

Hopfner, Jonathan (2007) "Asia Chip Market to Hit \$203b in 2011," *Electronic Engineering Times*, 13 June.

Howell, Thomas R. *et al.* (2003) *China's Emerging Semiconductor Industry: The Impact of China's Preferential Value-Added Tax on Current Investment Trends*, San Jose, CA: Semiconductor Industry Association and Dewey Ballantine LLP.

Hsiang, Cheng-Chen (2007) "Richard Chang Defeated in His Attempt to Relinquish Taiwan Citizenship," *Liberty Times*, 16 February.

Hsu, Ren-Chuan and Wang, Shih-Chi (2002) "Resale of Tenancy in Hsinchu Science Park by ADT, Formerly the Second Semiconductor Company in the Park after the Pioneering UMC, Marked Its Withdrawal [in Taiwan] and Its Ambition to Open a New Chapter in Mainland China," *Commercial Times*, 12 September.

Hu, Qili (2001) *Zhongguo Xinxihua Tansuo Yu Shijian [Exploration and Implementation of Informatization in China]*, Beijing: Publishing House of Electronics Industry.

Hua, Keyun (ed.) (1996) *Zhongguo Junyun Dianzi Yuanqijian [Electronic Components and Devices for the Chinese Military End Use]*, Beijing: Publishing House of Electronics Industry (*neibu* material).

Hua, Keyun *et al.* (eds.) (2000) *Weidianzi Jishu [Microelectronic Technology: Genie of Information Equipment]*, Beijing: Guofanggonye Chubanshe.

Huang, Joyce (2002) "Venture Capitalists Question Plan," *Taipei Times*, 5 September.

—— (2003) "MOEA Levies Fines on 4 Companies for Investing in China," *Taipei Times*, 19 January.

Hughes, Christopher W. (2000) "Globalisation and Security in the Asia-Pacific: An Initial Investigation," Working Paper, Centre for the Study of Globalisation and Regionalisation, University of Warwick.

Humphrey, D. *et al.* (2000) "An Avionics Guide to Uprating of Electronic Parts," *IEEE Transactions on Components and Packaging Technologies*, 23(3): 595–9.

Hung, Chih-Young *et al.* (2006) "Global Industrial Migration: The Case of the Integrated Circuit Industry," *International Journal of Technology and Globalisation*, 2(3/4): 362–76.

Huntington, Samuel P. (1993) "Why International Primacy Matters," *International Security*, 17(4): 68–83.

Hurtarte, Jeorge S. *et al.* (2007) *Understanding Fabless IC Technology*, Amsterdam and London: Elsevier/Newnes.

Hutcheson, G. Dan and Hutcheson, Jerry D. (1996) "Technology and Economics in the Semiconductor Industry," *Scientific American*, 274(1): 40–6.

IBM (2007) "IBM Advancement to Spawn New Generation of Chips," 27 January, available at: http://www–03.ibm.com/press/us/en/pressrelease/20980.wss (accessed 2 October 2007).

IC Insights (2006) "Top 10 Pure-Play Foundries Forecast for 2006: IC Insights Expects a Slowing of Chinese IC Foundries' Marketshare Gains," 3 August, available at: http://www.icinsights.com/news/releases/press20060803.html (accessed 14 December 2007).

—— (2008a) "IC Insights Ranks Top Foundry Suppliers," 5 May, available at: http://www.icinsights.com/news/bulletins/bulletins2008/bulletin20080505.pdf (accessed 28 August 2009).

—— (2008b) "Semiconductor R&D Spending to Reach $49.2b in 2008," 25 June, available at: http://www.icinsights.com/news/bulletins/bulletins2008/bulletin20080625.html (accessed 1 July 2008).

—— (2009) *Special 2009 Study: Global Wafer Capacity Analysis and Forecast*, available at: http://www.icinsights.com/prodsrvs/specialstudies/globalcapacity/gwcs2009broc.pdf (accessed 1 February 2010).

Ietto-Gillies, Grazia (2003) "The Role of Transnational Corporations in the Globalisation Process," in Michie, J. (ed.) *The Handbook of Globalisation*, Cheltenham: Edward Elgar, pp. 139–49.

Industrial Development Bureau (2009) *2009 Industrial Development in Taiwan, R.O.C.*, Taipei: Ministry of Economic Affairs, Republic of China.

Institute for Information Industry (2008) *Study on 2008–2010 Talent Supply and Demand in Taiwan's High-Tech Industries: The Case of the Semiconductor Industry*, Taipei: Institute for Information Industry.

Intel (2007a) "Intel's Transistor Technology Breakthrough Represents Biggest Change to Computer Chips in 40 Years," 27 January, available at: http://www.intel.com/pressroom/archive/releases/20070128comp.htm (accessed 2 October 2007).

—— (2007b) "Intel to Build 300mm Wafer Fabrication Facility in China: Fab 68 in Dalian Is $2.5 Billion Investment," 26 March, available at: http://www.intel.com/pressroom/archive/releases/20070326corp.htm (accessed 26 March 2007).

International Business Strategies (2007) "Analysis of the Relationship between EDA Expenditure and Competitive Positioning of IC Vendors: A Custom Study for EDA Consortium," 20 July, available at: http://www.edac.org/downloads/resources/profitability/HandelJonesReport.pdf (accessed 5 December 2007).

International Technology Roadmap (2007) *International Technology Roadmap for Semiconductors 2007 Edition*, available at: http://www.itrs.net/Links/2007ITRS/ExecSum2007.pdf (accessed 1 January 2008).

Investment Commission (2005a) "MOEA Fined Richard Chang NT$5 Million for His Investment in SMIC," 31 March, Taipei: Ministry of Economic Affairs, Republic of China.

—— (2005b) "Definition of Control and Influence over Companies Set up in Third Areas by Taiwan Investors," 14 June, Taipei: Ministry of Economic Affairs, Republic of China.

—— (2007) *Investment Commission Yearly Report, 2007*, Taipei: Ministry of Economic Affairs, Republic of China.

—— (2008) *Investment Commission Monthly Report, January 2008*, Taipei: Ministry of Economic Affairs, Republic of China.

Jackson, Robert J. and Towle, Philip (2006) *Temptations of Power: The United States in Global Politics after 9/11*, New York: Palgrave Macmillan.

Jen, Chien-Wei (2004) "What's Next for Taiwan's IC Industry," *Solid State Technology*, 47(8), 1 August.

Jencks, Harlan W. (1999) "COSTIND Is Dead, Long Live COSTIND! Restructuring China's Defense Scientific, Technical, and Industrial Sector," in Mulvenon, J. C. and Yang, A. N. D. (eds.) *The People's Liberation Army in the Information Age*, Santa Monica, CA: RAND, pp. 59–77.

Jiang, Zemin (2001) *Lun Kexue Jishu [On Technology]*, Beijing: Zhongyangwenxian Chubanshe.

Jiefangjun Bao (2003) "Army Urged to Be in the Forefront in 'Three Represents' Study,'", *PLA Daily*, 10 July.

Johnston, Alastair Iain (2003) "Is China a Status Quo Power?," *International Security*, 27(4): 5–56.

Jones, Terril Yue (2005) "Spying Case Underscores Rivalry of Asian Chip Firms," *Los Angeles Times*, 3 January.

Jorgenson, Dale W. (2001) "Information Technology and the U.S. Economy," *The American Economic Review*, 91(1): 1–32.

—— (2004) *Economic Growth in the Information Age*, available at: http://www.esri.go.jp/jp/prj-rc/macro/macro15/04–1-P.pdf (accessed 3 October 2007).

Jorgenson, Dale W. and Stiroh, Kevin J. (1999) "Information Technology and Growth," *The American Economic Review*, 89(2): 109–15.

Jorgenson, Dale W. and Wessner, Charles W. (eds.) (2002) *Measuring and Sustaining the New Economy: Report of a Workshop*, Washington, DC: National Academy Press.

Jorgenson, Dale W. *et al.* (2000) "Raising the Speed Limit: U.S. Economic Growth in the Information Age," *Brookings Papers on Economic Activity*, 2000, 1: 125–235.

Jorgenson, Dale *et al.* (2006) "Potential Growth of the US Economy: Will the Productivity Resurgence Continue?," *Business Economics*, 41(1): 7–16.

Kahler, Miles (2004) "Economic Security in an Era of Globalization: Definition and Provision," *The Pacific Review*, 17(4): 485–502.

Kanz, John W. (1991) An Uncertain Shield: U.S. Microelectronics and Foreign Dependencies in a Globalized Industry, PhD Dissertation, Claremont Graduate University.

Karatsu, Hajime (1987) "Significant Differences: A Japanese Perspective on the U.S. Semiconductor Industry," *Defense Electronics*, 19: 91–7.

Kastner, Scott L. (2009) *Political Conflict and Economic Interdependence across the Taiwan Strait and Beyond*, Palo Alto, CA: Stanford University Press.

Keji Ribao (2005) "Minqi Yi Gaoxin Jishu Qieru Jungong Lingyu Qianjing Kanhao [Good Prospect for High-Tech Private Enterprises to Take Part in Defense Production Activities]," *Science and Technology Daily*, 4 June.

Keohane, Robert O. and Nye, Joseph S. (1987) "Review: Power and Interdependence Revisited," *International Organization*, 41(4): 725–53.

—— (2000) "Globalization: What's New? What's Not? (and So What?)," *Foreign Policy*, 118: 104–19.

King, Gary *et al.* (1994) *Designing Social Inquiry: Scientific Inference in Qualitative Research*, Princeton, NJ: Princeton University Press.

—— (2004) "The Importance of Research Design," in Brady, H. E. and Collier, D. (eds.) *Rethinking Social Inquiry: Diverse Tools, Shared Standards*, Lanham, MD: Rowman & Littlefield, pp. 181–92.

Kirshner, Jonathan (ed.) (2006a) *Globalization and National Security*, London: Routledge.

—— (2006b) "Globalization and National Security," in Kirshner, J. (ed.) *Globalization and National Security*, London: Routledge, pp. 1–33.

Klare, Michael T. (2004) *Blood and Oil: The Dangers and Consequences of America's Growing Petroleum Dependency*, New York: Metropolitan Books.

—— (2007) "The Changing Calculus of Conflict?," *Security Studies*, 16(4): 583–97.

Knapp, Kenneth J. and Boulton, William R. (2006) "Cyber-Warfare Threatens Corporations: Expansion into Commercial Environments," *Information Systems Management*, 23(2): 76–87.

Koh, Philip (2004) "India Has Great Potential for Semiconductor Investment," *Gartner Dataquest Research Brief*, 30 November.

Kong, Xianlun (ed.) (2003) *Junyong Biaozhunhua [Military Standardization]*, Beijing: National Defense Industry Press.

Krause, Keith and Williams, Michael C. (1996) "Broadening the Agenda of Security Studies: Politics and Methods," *Mershon International Studies Review*, 40(2): 229–54.

Krugman, Paul (1995) "Growing World Trade: Causes and Consequences," *Brookings Papers on Economic Activity*, 1995, 1: 327–62.

Lammers, David (2004) "Texas Instruments Collects on Split Fab Strategy Bet," *Electronic Engineering Times*, 17 May.

——— (2009) "Freescale to Close Fabs in Japan and France," *Semiconductor International*, 22 April.

Lapedus, Mark (2003) "U.S. Tool Makers at a Disadvantage in China, Says SEMI," *Silicon Strategies*, 18 July.

——— (2005) "Updated: China's Foundry Industry to Slow in 2005," *Electronic Engineering Times*, 14 June.

——— (2006a) "SMIC Tops Chartered in '05 Foundry Rankings," *Electronic Engineering Times*, 25 March.

——— (2006b) "Half of China's Fab Ventures Will Fail," *Electronic Engineering Times*, 10 July.

——— (2007) "U.S. Reduces Export Controls for SMIC," *Electronic Engineering Times*, 19 October.

——— (2008) "Red Alert: China Churns out Chip Fabs," *Electronic Engineering Times*, 1 February.

Leachman, Robert C. and Leachman, Chien H. (2004) "Globalization of Semiconductors: Do Real Men Have Fabs, or Virtual Fabs?," in Kenney, M. and Florida, R. (eds.) *Locating Global Advantage: Industry Dynamics in the World Economy*, Palo Alto, CA: Stanford University Press, pp. 203–31.

Lecuyer, Christophe (2006) *Making Silicon Valley: Innovation and the Growth of High Tech, 1930–1970*, Cambridge, MA: The MIT Press.

Lee, Chyungly (2006) "Taiwan's Economic Security: Confronting the Dual Trends of Globalization and Governance," in Nesadurai, H. E. S. (ed.) *Globalisation and Economic Security in East Asia: Governance and Institutions*, New York: Routledge, pp. 126–45.

Lee, Dao-Cheng (2004) "Mainland China's IC Design Sector Still Small in Scale," *Commercial Times*, 3 June.

Lee, Raymond M. (1993) *Doing Research on Sensitive Topics*, London: Sage.

Lee, Thomas H. (2002) "A Vertical Leap for Microchips," *Scientific American*, 286(1): 50–7.

Lei, Bin (2003) " 'Shenwu Chongqing Ye You Fen – Yuqi Jiang Fu Jing Canjia Qinggonghui [Chongqing Enterprises Contributed to Shenzhou V – Sichuan Enterprises Heading for Beijing to Take Part in the Celebration Banquet]," *Chongqing Shangbao [Chongqing Economic Times]*, 7 November.

Lemon, Sumner (2004) "China's Arca Breaks into Global CPU Market," *IDG News Service*, 17 February.

Leng, Tse-Kang (2002) "Economic Globalization and IT Talent Flows across the Taiwan Strait: The Taipei/Shanghai/Silicon Valley Triangle," *Asian Survey*, 42(2): 230–50.

Lennox, Duncan (2008) "Thesis," Email (22 December 2008).

Leopold, George (2007) "Updated Four Chip Makers in China Added to US Export List," *Electronic Engineering Times*, 19 October.

Lewis, John Wilson and Xue, Litai (1994) *China's Strategic Seapower: The Politics of Force Modernization in the Nuclear Age*, Palo Alto, CA: Stanford University Press.

Lewis, William W. *et al.* (2002) "What's Right with the US Economy," *The McKinsey Quarterly*, 1: 30–40.

Li, Heng (2002) "China Develops New 32-Bit Micro Processor -'Shenwei I'," *People's Daily*, 21 November.

Li, Jie (1993) "Jisuanji Yu Xiandai Zhanzheng [Computer and Modern War]," *Xiandai junshi [CONMILIT]*, 16(12): 15–18.

Lieber, Charles M. (2001) "The Incredible Shrinking Circuit," *Scientific American*, 285(3): 50–3.

Lieberman, Joseph I. (2003) "White Paper: National Security Aspects of the Global Migration of the U.S. Semiconductor Industry," 15 November, United States Senate Armed Forces Committee, available at: http://www.fas.org/irp/congress/2003_cr/s060503.html (accessed 15 October 2004).

Lin, Miao-Jung (2002) "TSMC Cleared of Breaking Wafer Export-Control Laws," *Taipei Times*, 24 December.

Linden, Greg and Somaya, Deepak (2000) "Systems-on-a-Chip Integration in the Semiconductor Industry: Industry Structure and Firm Strategies," *Industrial and Corporate Change*, 12(3): 545–76.

Ling, Yongshun and Wang, Xiaoyuan (2005) *Wuzi Zhuangbei De Xinxihua [Informationalization of Weaponry Systems]*, Beijing: Jiefangjun Chubanshe.

Lippman, Walter (1943) *U.S. Foreign Policy: Shield of the Republic*, Boston, MA: Little, Brown and Company.

Lipsey, Richard G. *et al.* (1998) "What Requires Explanation," in Helpman, E. (ed.) *General Purpose Technologies and Economic Growth*, Cambridge, MA: The MIT Press, pp. 15–54.

Liu, Chenghai (2004) "Liu Chenghai Fubuzhang Zai Junyong Dianzi Yuanqijian Biaozhunhua Jishu Weiyuanhui Chengli Huiyi Shang De Jianghua [Speech by Liu During the Inauguration Ceremony of the Military Electronic Parts and Components Standardization Technology Commission]," *Junyong biaozhunhua [Military Standardization]*, 6: 2–3.

Liu, Lei (2004) "Nide Tongxingzheng Hezai? 'Mincanjun' Zhi Buwanquan Pojieban [Where Is Your Travel Pass? Initial Solution to 'the Participation of Civilian Firms in Defense Productions']," *Zhongguo Junzhuanmin [Defense Industry Conversion in China]*, 4, 35–8.

Liu, Yin *et al.* (eds.) (1986) *Dangdai Zhongguo De Dianzi Gongye [Contemporary China's Electronics Industry]*, Beijing: China Social Sciences Press.

Livingston, Henry (2007) "Avoiding Counterfeit Electronic Components," *IEEE Transactions on Components and Packaging Technologies*, 30(1): 187–9.

Lorber, Azriel (2002) *Misguided Weapons: Technological Failure and Surprise on the Battlefield*, Washington, DC: Brassey's.

Lorell, Mark A. *et al.* (2000) *Cheaper, Faster, Better? Commercial Approaches to Weapons Acquisition*, Santa Monica, CA: RAND.

—— (2001) *Reforming Mil-Specs: The Navy Experience with Military Specifications and Standards Reform*, Santa Monica, CA: RAND.

Lou, Qinjian (ed.) (2003) *Zhongguo Dianzixinxi Chanye Fazhan Moshi Yanjiu [Study of the Development Model of the Chinese Electronics and Information Industries]*, Beijing: Zhongguo Jingji Chubanshe.

Lu, Xinkui (2001) "Xinxi Jishu Zai Xiandai Zhanzheng Zhong De Zuoyong He Diwei [The Role of Information Technology in Modern Wars]," *Guofang Keji Gongye [Journal of Science Technology and Industry for National Defense]*, 1: 16–19.

Luo, Haoping *et al.* (1992) "Junyong ASIC Jiqi Duice [Military ASIC and Its Strategy]," *Bandaoti Jishu [Semiconductor Technology]*, 6(3): 36–59.

Luttwak, Edward N. (1990) "From Geopolitics to Geo-Economics: Logic of Conflict, Grammar of Commerce," *The National Interest*, 20: 17–23.

Ma, Weiye (2003) *Quanqiuhua Shidai De Guojia Anquan [National Security in the Age of Globalization]*, Wuhan: Hubei Jiaoyu Chubanshe.

McCormack, Richard (2004) "$600 Million over 10 Years for IBM's 'Trusted Foundry'; Chip Industry's Shift Overseas Elicits National Security Agency, Defense Department Response," *Manufacturing and Technology News*, 3 February.

McGrath, Dylan (2007) "Fabless Firms Grabbed 20% of IC Revenue in '06," *Electronic Engineering Times*, 17 January.

McHale, John (1996) "Motorola Transfers Military Chip Component Technologies to Lansdale," *Military & Aerospace Electronics*, 10(10): 3.

Macher, Jeffrey T. *et al.* (1998) "Reversal of Fortune? The Recovery of the U.S. Semiconductor Industry," *California Management Review*, 41(1): 107–36.

—— (2000) "Semiconductors," in Mowery, D. C. (ed.) *U.S. Industry in 2000: Studies in Competitive Performance*, Washington, DC: National Academy Press, pp. 245–86.

—— (2007) "The 'Non-Globalization' of Innovation in the Semiconductor Industry," *California Management Review*, 50(1): 1–26.

Mackenzie, Donald (1990) *Inventing Accuracy: A Historical Sociology of Nuclear Missile Guidance*, Cambridge, MA: The MIT Press.

Maggio, Edward J. and Coleman, Kevin (2007) "The Threat of Electronic Warfare 2007," *The Journal of Counterterrorism & Homeland Security International*, 13(3): 24–8.

Maher, Michael C. (2003) "Can COTS Products Be Used in Radiation Environments?," *COTS Journal*, December, available at: http://www.cotsjournalonline.com/articles/view/100089 (accessed 31 December 2003).

Manners, David (2006) "Xilinx Plans R&D in China Next Year," *Electronics Weekly*, 15 November.

Manners, David and Makimoto, Tsugio (1995) *Living with the Chip*, London: Chapman & Hall.

Mansfield, Edward D. and Pollins, Brian (eds.) (2003) *Economic Interdependence and International Conflict: New Perspectives on an Enduring Debate*, Ann Arbor, MI: University of Michigan Press.

Market Intelligence Center (2002) *Retaining the Root in Taiwan, While Going Ahead with Global Positioning: Opening up 8-Inch Wafer Foundry Investment in Mainland China and Its Impact on Taiwan*, Taipei: Market Intelligence Center.

Martin, David (2003) "U.S. Drops 'E-Bomb' on Iraqi TV: First Known Use of Experimental Weapon," 25 March, CBS News, available at: http://www.cbsnews.com/stories/2003/03/25/iraq/main546081.shtml (accessed 24 October 2007).

Mathews, John A. and Cho, Dong-Sung (2000) *Tiger Technology: The Creation of a Semiconductor Industry in East Asia*, Cambridge: Cambridge University Press.

Meaney, Constance Squires (1994) "State Policy and the Development of Taiwan's Semiconductor Industry," in Aberbach, J. D., *et al.* (eds.) *The Role of the State in Taiwan's Development*, Armonk, NY: M.E. Shape, pp. 170–92.

Mearsheimer, John J. (2001) *The Tragedy of Great Power Politics*, London: W.W. Norton & Company.

—— (2005) "E. H. Carr Vs. Idealism: The Battle Rages On," *International Relations*, 19(2): 139–52.

Medeiros, Evan S. *et al.* (2005) *A New Direction for China's Defense Industry*, Santa Monica, CA: RAND.

Mediatek Inc. (2005) *Annual Report 2004*, 30 April, available at: http://www.mediatek. com/upload/files/1584745e3ec079c0ea1322da00c02359.pdf (accessed 8 August 2006).

—— (2006) *Annual Report 2005*, 30 April, available at: http://www.mediatek.com/upload/files/52bf32b125ea3c7546fac14e9fe40e6e.pdf (accessed 4 April 2007).

Ministry of Economic Affairs (2007) *Taiwan's Listed Companies with Approved Investment in Mainland China*, Taipei: Ministry of Economic Affairs, Republic of China.

Ministry of Information Industry (2000) *China: Summary of the Tenth Five-Year Plan (2001–2005) – Information Industry*, Beijing: Ministry of Information Industry, People's Republic of China.

Ministry of Science and Technology (2000) "Bright Perspectives for '909 Project'," *Newsletter*, 239, 30 October.

—— (n.d.) *High Tech Research and Development (863) Program*, Beijing: Ministry of Science and Technology, People's Republic of China, available at: http://www.most. gov.cn/English/Programs/863/menu.htm (accessed 1 June 2004).

Misa, Thomas J. (1985) "Military Needs, Commercial Realities, and the Development of the Transistor, 1948–1958," in Smith, M. R. (ed.) *Military Enterprise and Technological Change*, Cambridge, MA: The MIT Press, pp. 253–87.

Mokhoff, Nicolas (2007) "Intel, TSMC Tout Process Technology Advances at IEDM," *Electronic Engineering Times*, 12 December.

Molas-Gallart, J. (1997) "Which Way to Go? Defence Technology and the Diversity of 'Dual-Use' Technology Transfer," *Research Policy*, 26(3): 367–85.

Moore, Gordon E. (1965) "Cramming More Components onto Integrated Circuits," *Electronics*, 38(8): 114–17.

—— (1975) "Progress in Digital Integrated Electronics," *IEEE International Electron Devices Meeting Tech Digest*, 11–13.

Moran, Theodore H. (1990) "The Globalization of America's Defense Industries: Managing the Threat of Foreign Dependence," *International Security*, 15(1): 57–99.

Moreira, Jose E. *et al.* (2007) "The Blue Gene/L Supercomputer: A Hardware and Software Story," *International Journal of Parallel Programming*, 35(3): 181–206.

Morris, Peter Robin (1994) "The Role of the Ministry of Defense (MOD) in Influencing the Commercial Performance of the British Semiconductor Industry," *History and Technology*, 11: 181–93.

Morrus, Michael (1988) *Competing for Control: America's Stake in Microelectronics*, Cambridge, MA: Ballinger.

Mowery, David C. (2000) *U.S. Industry in 2000: Studies in Competitive Performance*, Washington, DC: National Academy Press.

Mulvenon, James C. (1999) "The PLA and Information Warfare," in Mulvenon, J. C. and Yang, A. N. D. (eds.) *The People's Liberation Army in the Information Age*, Santa Monica, CA: RAND, pp. 175–86.

—— (2005) " 'The Digital Triangle': A New Defense-Industrial Paradigm," in Medeiros, E. S., *et al.* (eds.) *A New Direction for China's Defense Industry*, Santa Monica, CA: RAND, pp. 205–51.

Mulvenon, James C. and Yang, Andrew N. D. (eds.) (2002) *The People's Liberation Army as Organization*, Santa Monica, CA: RAND.

Munck, Gerardo L. (1998) "Canons of Research Design in Qualitative Analysis," *Studies in Comparative International Development*, 33(3): 18–45.

Murphy, M. P. (1988) "Air Force Logistics Command's Very High Speed Integrated Circuits (VHSIC) Insertion Program," *Proceedings of the IEEE 1988 National Aerospace and Electronics Conference*, pp. 1409–12.

Nadamuni, Daya (2004) "Electronic Design Growth in India," *Gartner Dataquest Research Brief*, 24 October.

Nanker Group (n.d.) "About Us," available at: http://www.nanker.com/eng/Company.asp (accessed 1 October 2007).

National Bureau of Statistics *et al.* (eds.) (2004) *2004 China Statistics Yearbook on High Technology Industry [2004 Zhongguo Gaokeji Chanye Tongji Nianjian]*, Beijing: China Statistics Press.

Naughton, Barry (1988) "The Third Front: Defence Industrialization in the Chinese Interior," *The China Quarterly*, 115: 351–86.

—— (2004) "The Information Technology Industry and Economic Integrations between China and Taiwan," in Mengin, F. (ed.) *Cyber China: Reshaping National Identities in the Age of Information*, New York: Palgrave Macmillan, pp. 155–84.

NEC Corporation (1997) "NEC Signs Contract for China's Largest Semiconductor Project," 28 May, available at: http://www.nec.co.jp/press/en/9705/2801.html (accessed 24 January 2008).

Nesadurai, Helen E. S. (2006) "Conceptualising Economic Security in an Era of Globalisation: What Does the East Asian Experience Reveal?," in Nesadurai, H. E. S. (ed.) *Globalisation and Economic Security in East Asia: Governance and Institutions*, New York: Routledge, pp. 1–22.

No. 13 Research Institute, China Electronics Technology Corporation (n.d.) *GaAs Qijian Dianlu Chanpin Shouce [GaAs Device Product Manual]*, Shijiazhuang: No. 13 Research Institute, China Electronics Technology Corporation.

No. 214 Research Institute of China North Industries Group Corporation (2007) "Wo Suo BP1001 Deng Wuzhong Jicheng Dianlu Shunli Tongguo Sheji Dingxing [Five Types of ICs Including BP1001 by the Insitute Smoothly Completed Design Finalization Tests]," 4 December, available at: http://www.cngc.com.cn/MemberDetail.aspx?id=120 (accessed 2 February 2008).

No. 771 Research Institute, China Aerospace Times Electronics Corporation (2005) *No. 771 Research Institute, China Aerospace Times Electronics Corporation*, Xian: No. 771 Research Institute, China Aerospace Times Electronics Corporation.

—— (n.d.a) Chanpin Zhinan [Product Guide], Xian: No. 771 Research Institute, China Aerospace Times Electronics Corporation.

—— (n.d.b) "Lishi Jiaobu [Company History]," available at: http://www.lishan.net.cn/hg.asp (accessed 2 February 2008).

Nolan, Peter (2004) *China at the Crossroads*, Cambridge: Polity.

Noyce, Robert N. and Hoff, M. E. (1981) "A History of Microprocessor Development at Intel," *IEEE Micro*, 1(1): 8–21.

Nye, Joseph S. (1990) *Bound to Lead: The Changing Nature of American Power*, New York: Basic Books.

Nye, Joseph S. and Lynn-Jones, S. (1988) "International Security Studies: A Report of A Conference on the State of the Field," *International Security*, 12(4): 5–27.

Nye, Joseph S. and Owens, William (1996) "America's Information Edge," *Foreign Affairs*, 75(2): 20–36.

Nystedt, Dan (2005) "CEO of Chip Maker SMIC Giving up Taiwan Citizenship," *IDG News Service*, 23 August.

Odell, John S. (2001) "Case Study Methods in International Political Economy," *International Studies Perspectives*, 2(2): 161–76.

Office of Force Transformation, Department of Defense (2005) *The Implementation of Network-Centric Warfare*, Washington, DC: Office of Force Transformation, Department of Defense.

Office of the Deputy Under-Secretary of Defense (Industrial Policy) (2004) *Defense Industrial Base Capabilities Study: Command and Control*, June, Department of Defense, available at: http://www.acq.osd.mil/ip/docs/dibc_command-and-control_6–28–2004.pdf (accessed 13 November 2007).

Office of the Governor, The State of Texas (2007) "Texas Semiconductor Industry Report," available at: http://www.texasone.us/site/DocServer/2005TXSemiRpt.pdf?docID=1781 (accessed 1 March 2008).

Office of the Under-Secretary of Defense Acquisition, Technology & Logistics (2005) *Special Technology Area Review on Field Programmable Gate Arrays (FPGAs) for Military Applications*, Washington, DC: Office of the Under Secretary of Defense Acquisition, Technology & Logistics.

O'Hanlon, Michael (2000) *Technological Change and the Future of Warfare*, Washington, DC: The Brookings Institution.

Okimoto, Daniel I. *et al.* (1987) *The Semiconductor for Competition and National Security*, Palo Alto, CA: Northeast Asia-United States Forum on International Policy, Stanford University, CA.

O'Neil, Ken (2004) "RTAX-S: Actel Radiation Tolerant FPGAs," paper presented at Microelectronics Reliability & Qualification Workshop, Manhattan Beach, CA, 7–8 December.

O'Rourke, Ronald (2007) *China Naval Modernization: Implications for U.S. Navy Capabilities – Background and Issues for Congress*, Washington, DC: Congressional Research Service.

Overend, William (1988) "Increase in Espionage Feared U.S.–Soviet Thaw Worries Some Intelligence Officers," *Los Angeles Times*, 1, 21 November.

Panel on Information in Warfare, Committee on Technology for Future Naval Forces, Naval Studies Board, (1997) *Technology for the United States Navy and Marine Corps, 2000–2035 Becoming a 21st-Century Force: Vol. 3: Information in Warfare*, Washington, DC: National Academy Press.

Patel, P. and Pavitt, K. (1991) "Large Firms in the Production of the World's Technology: An Important Case of 'Nonglobalisation'," *Journal of International Business Studies*, 22(1): 1–20.

Pecht, M. (2004) "How China Is Closing the Semiconductor Technology Gap," *IEEE Transactions on Components and Packaging Technologies*, 27(3): 616–19.

Pecht, M. *et al.* (1997) "An Assessment of the Qualified Manufacturer List (QML)," *IEEE Aerospace and Electronic Systems Magazine*, 12(7): 39–42.

Perraton, Jonathan (2003) "The Scope and Implications of Globalisation," in Michie, J. (ed.) *The Handbook of Globalisation*, Cheltenham: Edward Elgar, pp. 37–60.

Perry, William J. (1994) "Specifications and Standards: A New Way of Doing Business," 24 June, Washington, DC: Department of Defense, available at: http://sw-eng.falls-church.va.us/perry94.html (accessed 19 June 2006).

Pillsbury, Michael (2001) "China's Military Strategy toward the U.S.: A View from Open Sources," 2 November, available at: http://www.declarepeace.org.uk/captain/murder_inc/site/strat.pdf (accessed 16 October 2007).

—— (2005) *China's Progress in Technological Competitiveness: The Need for a New Assessment*, Washington, DC: The US China Economic and Security Review Commission.

Pittman, W.C. (2003) "Evolution of the DOD Microwave and Millimeter Wave Monolithic IC Program," *IEEE Technology and Society Magazine*, 22(1): 40–6.

Pretorius, Joelien (2003) "Ethics and International Security in the Information Age," *Defense & Security Analysis*, 19(2): 165–75.

PricewaterhouseCoopers (2006) *China's Impacts on the Semiconductor Industry 2005/Update*, New York: PricewaterhouseCoopers.

—— (2007) *China's Impact on the Semiconductor Industry: 2006/Update*, New York: PricewaterhouseCoopers.

—— (2008) *China's Impact on the Semiconductor Industry: 2008 Update*, New York: PricewaterhouseCoopers.

—— (2010) *Global Reach: China's Impact on the Semiconductor Industry: 2010 Update*, New York: PricewaterhouseCoopers.

Qian, Pingfan (2005) " 'Mincanjun' Shi Tisheng Woguo Guofang Keji Gongye Guoji Jingzhengli De Zhongyao Tujing ['Civilian Participation in Military Production' as the Major Pathway to Enhance the International Competitiveness of Our National Defense Technology Industry]," *Zhongguo Junzhuanmin [Defense Industry Conversion in China]*, 2: 37–8.

Qiao, Liang and Wang, Xiangsui (2002) *Unrestricted Warfare: China's Master Plan to Destroy America*, Panama City, Panama: Pan American Publishing Company.

Qiao, Songlou (1994) "Jisuanji Bingdu Yu Bingduzhan [Computer Virus and Computer Virus War]," *Jiefangjun Bao [PLA Daily]*, 3, 16 September.

Randazzese, Lucien P. (1996) "Semiconductor Subsidies," *Scientific American*, 274(6): 32–5.

Realtek Semiconductor Corp. (2006) *Annual Report 2005*, available at: http://www.realtek.com/investor/annualView.aspx?Langid=1&PNid=1&PFid=4&Level=1 (accessed 6 April 2007).

Reid, T. R. (2001) *The Chip: How Two Americans Invented the Microchip and Launched a Revolution*, New York: Random House.

Reuters (2007) "Creator Predicts Demise of Moore's Law," 19 September, available at: http://news.zdnet.co.uk/hardware/0,1000000091,39289478,00.htm (accessed 1 October 2007).

—— (2008a) "China's Hejian, Elpida Mull Chip Plant – Paper," 18 March, available at: http://in.reuters.com/article/asiaCompanyAndMarkets/idINTP23202420080318 (accessed 8 May 2008).

—— (2008b) "Taiwan's Ma Says Chip Makers Should Be Allowed More China Access," 10 July, available at: http://uk.reuters.com/article/idUKTP11976620080710 (accessed 15 July 2008).

Rincon, A. M. *et al.* (1999) "The Changing Landscape of System-on-a-Chip Design," *Proceedings of the IEEE 1999 International Conference on Custom Integrated Circuits*, 16–19 May, San Diego, CA, USA, pp. 83–90.

Riordan, Michael and Hoddeson, Lillian (1997) *Crystal Fire: The Birth of the Information Age*, New York: W. W. Norton & Company.

Roberts, Tom (2007) "Multicomputer Programming, Cell Be Processor Boost Signal Processing," *COTS Journal*, May, available at: http://www.cotsjournalonline.com/articles/view/100661 (accessed 2 October 2007).

Romm, Joseph J. (1993) *Defining National Security: The Nonmilitary Aspects*, New York: The Council on Foreign Relations.

Ross, Ian M. (1997) "The Foundation of the Silicon Age," *Bell Labs Technical Journal*, 2(4): 3–14.

Ross, Robert S. (2000) "The 1995–96 Taiwan Strait Confrontation: Coercion, Credibility, and the Use of Force," *International Security*, 25(2): 87–123.

Rothstein, Linda (1999) "Smaller and Smaller . . . (Miniature Military Devices)," *Bulletin of the Atomic Scientists*, 55(1): 5–6.

Roy, J. A. *et al.* (2008) "EPIC: Ending Piracy of Integrated Circuits," *Proceedings of the 2008 Conference on Design, Automation and Test in Europe*, 10–14 March, München, Germany, pp. 1069–74.

Rudolph, Christopher (2003) "Globalization and Security: Migration and Evolving Conceptions of Security in Statecraft and Scholarship," *Security Studies*, 13(1): 1–32.

Ruggie, John Gerard (1998) "What Makes the World Hang Together? Neo-Utilitarianism and the Social Constructivist Challenge," *International Organization*, 52(4): 855–85.

Ruttan, Vernon W. (2001) *Technology, Growth, and Development: An Induced Innovation Perspective*, New York: Oxford University Press.

Salt, John (1997) "International Movements of the Highly Skilled," Social Employment and Migration Working Papers, Paris: Directorate for Education, Employment, Labour and Social Affairs, Organisation for Economic Co-operation and Development, available at: http://www.oecd.org/dataoecd/24/32/2383909.pdf (accessed 14 May 2005).

Sample, Ian (2000) "Just a Normal Town," *New Scientist*, 167(2245), 1 July: 20.

Samuels, Richard J. (1994) *Rich Nation, Strong Army: National Security and the Technological Transformation of Japan*, Ithaca, NY: Cornell University Press.

Sandia National Laboratories (1998) "Intel Provides a No-Fee License to US Government: Sandia Labs to Develop Custom, Radiation-Hardened Pentium Processor for Space and Defense Needs," 8 December, available at: http://www.sandia.gov/media/rhp.htm (accessed 30 October 2007).

Saxenian, Annalee (2002) "Transnational Communities and the Evolution of Global Production Networks: The Cases of Taiwan, China and India," *Industry and Innovation*, 9(3): 183–202.

—— (2006) *The New Argonauts: Regional Advantage in a Global Economy*, Cambridge, MA: Harvard University Press.

—— (2007) "Brain Circulation and Regional Innovation: The Silicon Valley–Hsinchu–Shanghai Triangle," in Polenske, K. R. (ed.) *The Economic Geography of Innovation*, Cambridge: Cambridge University Press, pp. 190–212.

Schiesel, Seth (2003) "Taking Aim at an Enemy's Chips," *New York Times*, Section G, 1, 20 February.

Schwartau, Winn (1994a) "Chipping: Silicon-Based Malicious Software," in Schwartau, W. (ed.) *Information Warfare: Cyberterrosism: Protecting Your Personal Security in the Electronic Age*, 2nd edn, New York: Thunder's Mouth Press, pp. 254–64.

—— (1994b) *Information Warfare: Chaos on the Electronic Superhighway*, New York: Thunder's Mouth Press.

—— (2000) *Cyber Shock*, New York: Thunder's Mouth Press.

Scott, Allen and Angel, David (1988) "The Global Assembly Operations of U.S. Semiconductor Firms: A Geographical Analysis," *Environment and Planning*, 20: 1047–67.

Segal, Adam (2006) "Globalization Is a Double-Edged Sword: Globalization and Chinese National Security," in Kirshner, J. (ed.) *Globalization and National Security*, London: Routledge, pp. 293–320.

SEMI (2011) "SEMI Reports 2010 Global Semiconductor Equipment Sales of $39.5 Billion," 8 March, available at: http://www.semi.org/node/36581 (accessed 5 April 2011).

Semiconductor Industry Association (2003) *2003 Annual Databook: Review of Global and U.S. Semiconductor Competitive Trends, 1978–2002*, San Jose, CA: Semiconductor Industry Association.

—— (2006) "World Market Sales & Shares: 1982–2005," available at: http://www.sia-online.org/pre_sta.cfm?ID=179 (accessed 3 April 2006).

—— (2009) "Maintaining America's Competitive Edge: Government Policies Affecting Semiconductor Industry R&D and Manufacturing Activity," available at: http://www.choosetocompete.org/downloads/Competitiveness_White_Paper.pdf (accessed 1 March 2010).

—— (2010) "Global Chip Sales Decline in 2009," available at: http://www.sia-online.org/cs/papers_publications/press_release_detail?pressrelease.id=1707 (accessed 15 March 2010).

Semiconductor International (2007) "SATS Revenue Surpassed $19b in 2006," 28 June.

Sha, Nansheng *et al.* (2001) "Woguo Junmin Liangyong Jishu Fazhan Xianzhuang He Zhengce Jianyi [The Current Development of Dual-Use Technologies in Our Country and Policy Recommendations]," *Guofang Keji Gongye [Journal of Science Technology and Industry for National Defense]*, 9: 32–5.

Shambaugh, David (1999) "PLA Studies Today: A Maturing Field," in Mulvenon, J. C. and Yang, A. N. D. (eds.) *The People's Liberation Army in the Information Age*, Santa Monica, CA: RAND, pp. 7–21.

—— (1999/2000) "China's Military Views the World: Ambivalent Security," *International Security*, 24(3): 52–79.

—— (2004) *Modernizing China's Military: Progress, Problems, and Prospects*, Berkeley, CA: University of California Press.

Shanghai Fudan Microelectronics (n.d.) " 'Shenwei Yihao' Huo Guojia Guofang Keji Yidengjiang ['Shenwei I' Won a First-Class National Defense Technology Award]," available at: http://www.fmsh.com/news_fb43.htm (accessed 2 February 2006).

Shanghai Industrial Holdings Limited (2005) *Annual Report 2004*, available at: http://www.sihl.com.hk/sihl/eng/jsp/show_doc.action?doc_type=0 (accessed 1 March 2006).

Shanghai Stock Exchange Newspaper (2003) *Hangzhou Silan Microelectronics Co. Ltd. Prospectus Part III*, 19 February, available at: http://memo.stock888.net/030219/102,1310,156226,00.shtml (accessed 6 June 2007).

Sharkey, Brian (2007) "Trust in Integrated Circuits Program," Defense Advanced Research Projects Agency, available at: http://www.darpa.mil/MTO/solicitations/baa07–24/Industry_Day_Brief_Final.pdf (accessed 25 September 2008).

Shen, Chang Xiang *et al.* (2007) "Survey of Information Security," *Science in China Series F: Information Sciences*, 50(3): 273–98.

Shenzhen I-Lacs Technology Co., Ltd. (2004) "I-Lacs Motherboard Family Has Got a New Member," 21 April.

—— (2006) "I-Lacs Passes the Navy's Second Party Certification," 7 August.

—— (n.d.) "Development Milestones," available at: http://www.ilacs.cn/plus/list.php?tid=107 (accessed 2 December 2007).

Sheppard, Gregory (1990) "A New Era for Semiconductors and the DOD," *Solid State Technology*, 33(3), 1 March.

Shih, Chintay *et al.* (2007) "Hsinchu, Taiwan: Asia's Pioneering High-Tech Park," in Rowen, H. S., *et al.* (eds.) *Making IT: The Rise of Asia in High-Tech*, Palo Alto, CA: Stanford University Press, pp. 101–22.

Simon, Denis Fred (1997) "Techno-Security in an Age of Globalization," in Simon, D. F. (ed.) *Techno-Security in an Age of Globalization*, New York: M.E. Sharpe, pp. 3–21.

Simon, Denis Fred and Goldman, Merle (eds.) (1989) *Science and Technology in Post-Mao China*, Cambridge, MA: Council on East Asian Studies, Harvard University Press.

Slade, Rob (1992) " 'Desert Storm' Viral Myths," *The Risks Digest: Forum on Risks to the Public in Computers and Related Systems*, 13(6), 24 January.

Slomovic, Anna (1991) *An Analysis of Military and Commercial Microelectronics: Has DoD's R&D Funding Had the Desired Effect*, Santa Monica, CA: RAND.

SMIC (2001a) "SMIC Raises More Than $1 Billion in Series A Round Financing," 10 November, available at: http://www.smics.com/eng/press/media_press_details. php?id=51407 (accessed 8 July 2004).

—— (2001b) "SMIC Grand Opening," 22 November, available at: http://news.smics. com/website/enVersion/Press_Center/popupDetailNews.action?newsId=163 (accessed 3 March 2004).

—— (2001c) "SMIC to Receive SRAM Process Technology Transfer from Toshiba," 20 December, available at: http://www.smics.com/eng/press/media_press_details. php?id=51396 (accessed 5 July 2004).

—— (2001d) "SMIC and Chartered Announce Alliance on 0.18-Micron Technology and Capacity," 21 December, available at: http://www.smics.com/eng/press/media_press_ details.php?id=51390 (accessed 8 July 2004).

—— (2002) "Infineon Signs Major Foundry Agreement with SMIC in China," 9 December, available at: http://www.smics.com/eng/press/media_press_details.php?id=51379 (accessed 24 August 2004).

—— (2003a) "Toshiba to Transfer SRAM Process Technology to SMIC," 9 January, available at: http://news.smics.com/website/enVersion/Press_Center/popupDetailNews. action?newsId=228 (accessed 5 April 2005).

—— (2003b) "Infineon Expands Foundry Agreement with SMIC," 27 March, available at: http://news.smics.com/website/enVersion/Press_Center/popupDetailNews.action? newsId=246 (accessed 8 April 2004).

—— (2003c) "SMIC Raises US$630 Million in Private Placement," 15 September, available at: http://www.smics.com/eng/press/media_press_details.php?id=51364 (accessed 9 July 2004).

—— (2004a) "SMIC Entered into a US$285 Million Loan Agreement with Four Chinese Banks," 18 January, available at: http://www.smics.com/eng/press/media_press_details. php?id=51347 (accessed 5 December 2005).

—— (2004b) "Semiconductor Manufacturing International Corporation Announces Proposed Dual Listing on SEHK and NYSE," 7 March, available at: http://www.smics. com/eng/press/media_press_details.php?id=51343 (accessed 1 March 2006).

—— (2004c) "SMIC Will Vigorously Defend against the Trade Secret Case Filed by TSMC in State Court," 4 June, available at: http://www.smics.com/website/enVersion/ Press_Center/popupDetailNews.action?newsId=605 (accessed 1 July 2005).

—— (2005a) "SMIC Reaches Settlement with TSMC," 30 January, available at: http://www.smics.com/website/enVersion/Press_Center/popupDetailNews.action? newsId=1105 (accessed 1 July 2005).

—— (2005b) *2004 Annual Report*, 7 April, available at: http://www.smics.com/attach-ment/2011012018070017_en.pdf (accessed 5 May 2006).

—— (2005c) "SMIC Beijing Secures Financing for Expansion," 26 May, available at: http://www.smics.com/attachment/2011012016454117_en.pdf (accessed 5 March 2006).

—— (2006a) "Infineon and SMIC Extend Agreement into 90nm Manufacturing," 6 January, available at: http://news.smics.com/website/enVersion/Press_Center/ popupDetailNews.action?newsId=1867 (accessed 6 April 2007).

—— (2006b) *Annual Report 2005*, 2 May, available at: http://www.smics.com/ attachment/2011012015151217_en.pdf (accessed 18 January 2007).

—— (2006c) "SMIC Denies Allegations and Files Cross-Complaint against TSMC," 13 September, available at: http://www.smics.com/eng/press/media_press_details. php?id=51198 (accessed 24 June 2007).

—— (2007a) *Annual Report 2006*, 30 April, available at: http://www.smics.com/ attachment/2011011916123817_en.pdf (accessed 23 May 2008).

—— (2007b) "Qimonda Expands Foundry Agreement with SMIC," 21 August, available at: http://news.smics.com/website/enVersion/Press_Center/popupDetailNews. action?newsId=2606 (accessed 1 March 2008).

—— (2007c) "SMIC Attains Validated End-User Status from U.S. Government," 19 October, available at: http://news.smics.com/website/enVersion/Press_Center/ popupDetailNews.action?newsId=2726 (accessed 27 December 2007).

—— (2007d) "SMIC and IBM Sign Licensing Agreement," 26 December, available at: http://www.smics.com/eng/press/media_press_details.php?id=51162 (accessed 5 June 2008).

—— (2008a) "SMIC Reports 2007 Fourth Quarter Results," 29 January, available at: http://news.smics.com/website/enVersion/Press_Center/popupDetailNews.action? newsId=2865 (accessed 6 May 2008).

—— (2008b) "SMIC Honored with SEMI China Corporate Social Contribution Award; Dr. Richard Chang Receives Industry Excellence and Contribution Award," 18 March, available at: http://www.smics.com/website/enVersion/Press_Center/popupDetail-News.action?newsId=2948 (accessed 30 August 2008).

Smith, Craig S. (2000) "A Chip Plant That Is Full of Symbolism," *New York Times*, 24 November.

Smith, Steve (2005) "The Contested Concept of Security," in Booth, K. (ed.) *Critical Security Studies and World Politics*, London: Lynne Rienner, pp. 27–62.

—— (2006) "The Concept of Security in a Globalizing World," in Patman, R. G. (ed.) *Globalization and Conflict: National Security in a 'New' Strategic Era*, London: Routledge, pp. 33–55.

Solid State Technology (2007a) "Research: Most Global Wafer Capacity in Hands of Few," 28 June.

—— (2007b) "UMC Execs Acquitted in He Jian Case," 29 October.

Song, Michelle (2007) "Exploring Emerging Cities in China for the Semiconductor Industry," *Semiconductor Insights: Asia*, 2: 22–4.

Song, William (2003) *Local Design House Analysis Report*, Shanghai: Global Advanced Packaging Technology.

Song, Zengbing and Niu, Lingjun (2006) "The Civil Enterprises Produce Military Products: Problems and Countermeasures," *Junshi jingji yanjiu [Military Economic Research]*, 12: 16–18.

Spreadtrum Communications Inc. (2008) "Spreadtrum Wins Recognition as 2007 China's Top 10 IC Design Companies," 26 March, available at: http://www.spreadtrum.com/en/ shown.asp?id=128&category=1 (accessed 7 September 2008).

State Council Informatization Office (2005) *Zhongguo Xinxihua Fazhan Baogao [China Informatization Development Report 2005]*, Beijing: Publishing House of Electronics Industry.

Steinfeld, Edward S. (2005) "Cross-Straits Integration and Industrial Catch-Up: How Vulnerable Is the Taiwan Miracle to an Ascendant Mainland?," in Berger, S. and Lester, R. K. (eds.) *Global Taiwan: Building Competitive Strengths in a New International Economy*, New York: M.E. Sharpe, pp. 228–96.

Stowsky, Jay (2004) "Secrets to Shield or Share? New Dilemmas for Military R&D Policy in the Digital Age," *Research Policy*, 33(2): 257–69.

Stuart, Kenneth C. (1992) "VHSIC Technology Insertion into the AP–102 Avionics Processor Family," *IEEE/AIAA 11th Digital Avionics Systems Conference Proceedings*, pp. 183–8.

Sullivan, Laurie (2003) "IP Rights, China Design Capabilities Worry Industry, DOD," *Electronic Engineering Times*, 18 September.

Sung, Claire and Chan, Rodney (2007) "Well-Driven 'Small Car': Q&A with GSMC Executive VP Arthur Kuo," *DigiTimes*, 16 February.

—— (2008) "Hejian and Elpida to Jointly Build 12-Inch Fab in China," *DigiTimes*, 18 March.

Swaine, Michael D. and Runyon, Loren H. (2002) "Ballistic Missiles and Missile Defense in Asia," *NBR Analysis*, 13(3): 1–82.

Taiwan Semiconductor Industry Association (2004) *Overview on Taiwan Semiconductor Industry (2004 Edition)*, Hsinchu: Taiwan Semiconductor Industry Association.

—— (2006) *Overview on Taiwan Semiconductor Industry (2006 Edition)*, Hsinchu: Taiwan Semiconductor Industry Association.

—— (2007) *Overview on Taiwan Semiconductor Industry (2007 Edition)*, Hsinchu: Taiwan Semiconductor Industry Association.

—— (2008a) "TSIA 2007 Statistics on Taiwan IC Industry," 14 March, available at: http://www.tsia.org.tw/Files/NewsFile/200832085316.doc (accessed 2 April 2008).

—— (2008b) *Overview on Taiwan Semiconductor Industry (2008 Edition)*, Hsinchu: Taiwan Semiconductor Industry Association.

—— (2009) *Overview on Taiwan Semiconductor Industry (2009 Edition)*, Hsinchu: Taiwan Semiconductor Industry Association.

Tang, Theresa (2005) "UMC Lays Down Challenge over He Jian," *International Herald Tribune*, 22 June.

Task Force on China's IC Industry (2002) "The Development of Our Nation's IC Industry in Accordance with the Objectives and Requirements of Strategic Industries," *Diaoyanbaogao [Survey Report]*, 1–21 (*neibu* material).

Taylor, Colleen (2006) "Qualcomm Heads Fabless Rankings," *Electronic News*, 26 September.

Tetsuya, Umemoto (1988) "Comprehensive Security and the Evolution of the Japanese Security Posture," in Scalapino, R. A., *et al.* (eds.) *Asian Security Issues: Regional and Global*, Berkeley, CA: University of California, pp. 28–49.

Texas Instruments (1964) "Eighteen Improved Minuteman Type Integrated Circuits Now Available to Entire Industry, TI and Autonetics Announce," 19 March.

—— (n.d.) "QML Information," available at: http://focus.ti.com/hirel/qltyreliab_qml.shtml (accessed 15 December 2007).

The Commission of Science, Technology and Industry for National Defense (2003) "Quanyu Yinfa Wuqi Zhuangbei Keyan Shengchan Xukezheng Guanli Zanxing Banfa Shishi Xize De Tongzhi [The Commission of Science, Technology and Industry for National Defense's Circular Concerning the Detailed Regulations on the Interim Procedures of Scientific and Research Certificate Management of Weaponry]," *Gazette* of the Commission of Science, Technology and Industry for National Defense of the People's Republic of China, 25(1): 1–3.

The State Council Information Office (2004) *2004 Nian Zhongguo De Guofang [White Paper on China's National Defense in 2004]*, Beijing: The State Council Information Office, the People's Republic of China.

The Wassenaar Arrangement on Export Controls for Conventional Arms and Dual-Use Goods and Technologies (2006) "List of Dual-Use Goods and Technologies and Munitions List," 6 December, Vienna, Austria: The Wassenaar Arrangement on Export Controls for Conventional Arms and Dual-Use Goods and Technologies Secretariat, available at: http://www.wassenaar.org/controllists/Previous/2006_OK/WA-LIST%20 %2806%29%201%20-%20for%20web%20site%20&%20WAIS%20-%20not%20for% 20photocopy.pdf (accessed 15 January 2007).

—— (2008) "List of Dual-Use Goods and Technologies and Munitions List," 3 December, Vienna, Austria: The Wassenaar Arrangement on Export Controls for Conventional Arms and Dual-Use Goods and Technologies Secretariat, available at: http://www.wassenaar.org/ controllists/2008/WA-LIST%20%2808%29%201/WA-LIST%20%2808%29%201.pdf (accessed 12 January 2009).

Thryft, Ann R. (2007) "Multicore Processors Drive Next-Gen Defense Systems," *COTS Journal*, May, available at: http://www.cotsjournalonline.com/articles/view/100658 (accessed 2 October 2007).

Tickner, Ann (1995) "Re-Visioning Security," in Booth, K. and Smith, S. (eds.) *International Relations Theory Today*, Cambridge: Polity Press, pp. 175–97.

Tilton, John E. (1971) *International Diffusion of Technology: The Case of Semiconductors*, Washington, DC: The Brookings Institution.

Toohey, Brian (2011) "Counterfeit Semiconductors: A Clear and Present Threat, Testimony of Brian Toohey, President, Semiconductor Industry Association," 8 November, Washington, DC: US Senate, available at: http://www.sia-online.org/clientuploads/ directory/DocumentSIA/Brian%20Toohey%20Testimony%20Final_SASC.pdf (accessed 5 May 2012).

Trusted Access Program Office, National Security Agency (n.d.) "Accredited Suppliers," available at: http://www.nsa.gov/business/tapo_suppliers.cfm (accessed 13 November 2007).

Tsar & Tsai Lex News (2007) "Different Views between the Administrative Law Court and the Ministry of Economic Affairs on SMIC's Investment," available at: http://www. tsartsai.com.tw/TTnews/Tsar%20&%20Tsai%20Lex%20News%20(English)%20 2007–05.htm#Different_Views (accessed 6 July 2007).

TSMC (2003) "TSMC Files Law Suit against SMIC for Patent Infringement and Trade Secret Misappropriation," 22 December, available at: http://www.tsmc.com/ tsmcdotcom/PRListingNewsArchivesAction.do?action=detail&newsid=1483&language= E (accessed 1 August 2004).

—— (2004) "TSMC Response Identifies More SMIC Espionage: More Wrong Doing Documented," 24 March, available at: http://www.tsmc.com/tsmcdotcom/ PRListingNewsAction.do?action=detail&LANG=E&newsid=1498&news date=2004/03/24 (accessed 5 May 2004).

—— (2005a) *A Banner Year: TSMC Annual Report 2004*, available at: http://www.tsmc. com/download/ir/annual_report/pdf/e_all.pdf (accessed 1 November 2006).

—— (2005b) "TSMC Reaches Settlement with SMIC," 30 January, available at: http:// www.tsmc.com/tsmcdotcom/PRListingNewsArchivesAction.do?action=detail&newsid =1593&language=E (accessed 2 February 2005).

—— (2006a) *Annual Report 2005*, available at: http://www.tsmc.com/download/ir/2005_annual_report/pic/E-TSMC-all.pdf (accessed 2 March 2007).

—— (2006b) "TSMC Welcomes Government Approval for 0.18 Micron Generation Technology Transfer to China," 29 December, available at: http://www.tsmc.com/tsmcdotcom/PRListingNewsArchivesAction.do?action=detail&newsid=2019&language=E (accessed 31 December 2006).

—— (2007a) *Annual Report 2006*, available at: http://www.tsmc.com/english/e_investor/e02_annual/2006_annual_report/pdf/tsmc_e_h.pdf (accessed 15 June 2008).

—— (2007b) "TSMC: The IC Foundry Industry Leader," available at: http://www.tsmc.com/download/english/a05_literature/1_Corporate_Overview_Brochure_2007.pdf (accessed 4 January 2008).

—— (2007c) "California Court Issues Order on TSMC Motion for Preliminary Injunction against SMIC," 14 September, available at: http://www.tsmc.com/tsmcdotcom/PRListingNewsArchivesAction.do?action=detail&newsid=2301&language=E (accessed 30 September 2007).

—— (2007d) "TSMC Signs Contract with Atmel to Purchase Eight-Inch Wafer Fabrication Equipment," 9 October, available at: http://www.tsmc.com/tsmcdotcom/PRListingNewsArchivesAction.do# (accessed 15 December 2007).

—— (2007e) "TSMC Ships One-Millionth 12-Inch 90nm Wafer: Fast Ramping Process Reaches Milestone in 4.5 Years," 3 December, available at: http://www.tsmc.com/tsmcdotcom/PRListingNewsArchivesAction.do?action=detail&&newsid=2420&&newsdate=2007/12/03&&language=E (accessed 28 December 2007).

Tu, Chih-Hao (2008) "Schumacher Took Helm at Grace Semiconductor Reviving the Company," *Commercial Times*, 22 February.

Tung, An-Chi (2001) "Taiwan's Semiconductor Industry: What the State Did and Did Not," *Review of Development Economics*, 5(2): 266–88.

UMC (2005) "United Microelectronics Corporation and Subsidiaries Unaudited Consolidated Financial Statements with Review Report of Independent Accountants for the Six-Month Period Ended June 30, 2005," 19 July, available at: http://www.umc.com/english/pdf/05_1H_consolidated-e.pdf (accessed 6 September 2006).

—— (2009) "UMC Board of Directors Important Announcement," 29 April, available at: http://www.umc.com/english/news/2009/20090429–2.asp (accessed 5 May 2009).

United Nations Conference on Trade and Development (2005) *World Investment Report 2005: Transnational Corporations and the Internationalization of R&D*, New York and Geneva: United Nations.

Van Creveld, Martin (1989) *Technology and War: From 2000 BC to the Present*, New York: Free Press.

—— (1966) "International Investment and International Trade in the Product Cycle," The Quarterly Journal of Economics, 80(2): 190-207.

Vernon, Raymond (1971) "Multinational Enterprise and National Security," No. 74, Adelphi Papers, Institute for Strategic Studies, London.

—— 1979) "The Product-Cycle Hypothesis in a New International Environment," *Oxford Bulletin of Economics and Statistics*, 41: 255–67.

—— (1998) *In the Hurricane's Eye: The Troubled Prospects of Multinational Enterprises*, Cambridge, MA: Harvard University Press.

Vogel, Steve (1992) "The Power Behind 'Spin-Ons': The Military Implications of Japan's Commercial Technology," in Sandholtz, W., *et al.* (eds.) *The Highest Stakes: The Economic Foundations of the Next Security System*, New York: Oxford University Press, pp. 55–80.

Wade, Robert (2004) *Governing the Market: Economic Theory and the Role of Government in East Asian Industrialization*, Princeton, NJ: Princeton University Press.

Walko, John (2007) "Uncommon Market: U.K. Chip Making under the Hammer," *Electronic Engineering Times Europe*, 12 December.

Walling, Eileen M. (2001) "High-Power Microwaves and Modern Warfare," in Martel, W. C. (ed.) *The Technological Arsenal: Emerging Defense Capabilities*, Washington, DC: Smithsonian Institution Press, pp. 90–104.

Walt, Stephen M. (1998) "International Relations: One World, Many Theories," *Foreign Policy*, 110: 29–46.

Waltz, Kenneth (1979) *Theory of International Politics*, Reading, MA: Addison-Wesley.

Wang, Jing (2005) "Junyong Dianzi Yuanqijian Biaozhunhua De Gongcheng Yingyong [The Implementation of Military Electronic Component Standardization]," *Junyong biaozhunhua [Military Standardization]*, 1: 17–20.

Wang, Junxue and Xu, Jianhua (2003) "Xinpian Jishuzhan Jizhan Zhenghan [The Fierce Chip Technology Battle on the Rise]," *Jiefangjun Bao [PLA Daily]*, 9 April.

Wang, Lisa (2005) "UMC May Get Stake in He Jian to Placate Prosecutors," *Taipei Times*, 22 March.

—— (2007) "ASE and NXP to Form Joint Venture," *Taipei Times*, 3 February.

Wang, Mo-Yun (2005) "Further Penalties Awaiting Should SMIC Moves to 65nm Process Technology," *Commercial Times*, 25 August.

Wang, Shiguang and Zhang, Xuedong (1989) "Fazhan Junshi Dianzi Cujin Guofang Xiandaihua: Zhongguo Junshi Dianzi Jishu Fazhan De Huigu Yu Zhanwang [The Development of Military Electronics Accelerates Defense Modernization: The Overview and Prospect of the Chinese Military Electronic Technology Development]," in *Zhongguo Jixie Dianzi Gongye Nianjian: Dianzi Juan [Chinese Machinery and Electronics Industry Yearbook: Electronics Industry Volume]* (ed.), Beijing: Publishing House of Electronics Industry, pp. I–15-I–21.

Wang, Shih-Chi (2001a) "Berkeley Alumni Formerly as Campaigners to Defend Diaoyu Island Became Pioneers in China's Semiconductor Industry," *Commercial Times*, 1 April.

—— (2001b) "Mainland's First Homegrown 32-Bit CPU to Be Fabricated by TSMC," *Commercial Times*, 8 November.

Wang, Wenrong (ed.) (2005) *Zhongguo Jundui Disanci Xiandaihua Lungang [On the Third Modernization of the PLA]*, Beijing: Jiefangjun Chubanshe.

Wang, Yaxian (1998) "Tan Dianzi Yuanqijian De Zhiliang Kengzhi [On Quality Control of Electronic Components]," *Junyong biaozhunhua [Military Standardization]*, 4, 38–9.

Wang, Yuan *et al.* (2002) *Zhongguo Zhanlue Jishu Yu Chanye Fazhan [The Development of China's Strategic Technologies and Industries]*, Beijing: Jingji Guanli Chubanshe.

Warner, R. (2001) "Microelectronics: Its Unusual Origin and Personality," *IEEE Transactions on Electron Devices*, 48(11): 2457–67.

Washington, Douglas Waller (1995) "Onward Cyber Soldiers: The U.S. May Soon Wage War by Mouse, Keyboard and Computer Virus. But It Is Vulnerable to the Same Attacks," *TIME*, 146(8), 21 August: 38–44.

Weidenbaum, Murray and Hughes, Samuel (1996) *The Bamboo Network: How Expatriate Chinese Entrepreneurs Are Creating a New Economic Superpower in Asia*, London: Free Press.

Wendt, Alexander (1992) "Anarchy Is What States Make of It: The Social Construction of Power Politics," *International Organization*, 46(2): 391–425.

Wessner, Charles W. (ed.) (2003) *Securing the Future: Regional and National Programs to Support the Semiconductor Industry*, Washington, DC: National Academies Press.

Williams, R. D. (2001) "Is the West's Reliance on Technology the Panacea for Future Conflict or Its Achilles' Heel?," *Defence Studies*, 1(2): 38–56.

Wilson, Clay (2005) *Computer Attack and Cyberterrorism: Vulnerabilities and Policy Issues for Congress*, Washington, DC: Congressional Research Service.

Winokur, P.S. et al. (1999) "Use of COTS Microelectronics in Radiation Environments," *IEEE Transactions on Nuclear Science*, 46(6): 1494–503.

Wolfers, Arnold (1952) "National Security as an Ambiguous Symbol," *Political Science Quarterly*, 67(4): 481–502.

World Bank (1997) *Global Economic Prospects and the Developing Countries*, Washington, DC: World Bank.

World Trade Organization (2007) *International Trade Statistics 2007*, Geneva: WTO Publications.

—— (2008) *International Trade Statistics 2008*, Geneva: WTO Publications.

Wu, Friedrich and Loy, Chua Boon (2004) "Rapid Rise of China's Semiconductor Industry: What Are the Implications for Singapore?," *Thunderbird International Business Review*, 46(2): 109–31.

Wu, Jiang and Li, Ming (2000) "Xinghao Kekaoxing Guanjian Zai Luoshi: Fang Woguo Xinxing Yuancheng Daodan Liu Zong Shejishi [The Key to Nomenclature Reliability Lies in Implementation: Interview with Mr. Liu, Chief Designer of Our Nation's New Long-Range Missile]," *Zhiliang Yu Kekaoxing [Quality and Reliability]*, 2: 11–14.

Wu, Xiandong (1990) "Pilot Studies on the Development Strategy of Military ASICs," *Weidianzixue [Microelectronics]*, 20(6): 1–6.

—— (1991) "Junyong He Zhuanyong Jichengdianlu Shichang Gongyi Sheji Yaosu De Fenxi [An Analysis of Market, Technology and Design Components in Military ICs and ASICs]," *Weidianzixue [Microelectronics]*, 21(2): 1–6.

Wu, Yuguang (2004) "Minyong Gaoxin Jishu Zhuan Junyong Wenti Pouxi [An Analysis of Issues Pertaining to the Military Use of Civilian High Technologies]," *Keji Chengguo Zongheng [Perspectives of Scientific and Technological Achievement]*, 6: 24–6.

Xie, Guang (1985) "Accelerating the Task of Military Standardization to Adjust to the Reform Situation," *Junyong biaozhunhua [Military Standardization]*, 1(1): 6 (*neibu* material).

Xilinx, Inc. (2004) "Xilinx Announces Investment in He Jian Technology Co.," 27 July, available at: http://www.xilinx.com/prs_rls/xil_corp/0490hejian.htm (accessed 6 June 2006).

Xinhua-PRNewswire (2007) "CCID Consulting Analyzes and Forecasts the Global and Chinese Foundry Market," 14 September, available at: http://www.prnewswire.com/cgi-bin/stories.pl?ACCT=104&STORY=/www/story/09–14–2007/0004662758&EDATE= (accessed 14 October 2007).

Xinhuanet (2001) "Zhonghua Renmin Gongheguo Guomin Jingji He Shehui Fazhan Dishige Wunian Jihua Gangyao [Outline of the 10th People's Republic of China's Economic and Social Development Five Year Plan]," 18 October, available at: http://news.xinhuanet.com/zhengfu/2001–10/18/content_51471.htm (accessed 1 February 2008).

—— (2003) "Zhonggong Zhongyang Guanyu Wanshan Shehuizhuyi Shichang Jingji Tizhi Ruogan Wenti De Jueding [Decision of the Chinese Communist Party Committee on Several Issues in Perfecting the Socialist Market Economy]," 21 October, available at:

http://news.xinhuanet.com/newscenter/2003–10/21/content_1135402_9.htm (accessed 2 August 2007).

—— (2007) "Full Text of Hu Jintao's Report at 17th Party Congress," 24 October, available at: http://news.xinhuanet.com/english/2007–10/24/content_6938749_8.htm (accessed 1 December 2007).

Xu, Heping and Fang, Hanting (2001) "Zhongguo CPU Lujing De Xuanze Ji Qi Zhengce Jianyi [China's Choice to Develop CPU and Related Policy Recommendations]," *Diaoyanbaogao [Survey Report]*, 51: 1–14 (*neibu* material).

Xu, Shi-Liu (1991) "Dali Fazhan Woguo Junyong Moni IC Jishu [Accelerating the Development of Military Analog IC Technology in Our Country]," *Dianzi Ruankexue [Electronics Soft Science]*, 4: 1–5.

—— (2004) "Junyong Weidianzi Jishu Fazhan Zhanlue [Random Thoughts on the Development Strategy for Military Microelectronics Technology]," *Weidianzixue [Microelectronics]*, 34(1): 1–6.

Xue, Xiang and Chen, Ting (2004) "Tuozhan 'Minpin Junyong' De Guofang Jianshe Xin Luzi" [Expanding the New Road of the 'Military Use of Civilian Products' in Defense Construction]," *Guofang Keji Gongye [Defense Science & Technology Industry]*, 5: 22–4.

Yan, Cai (2007) "Applied Materials Opening China R&D Center," *Electronic Engineering Times*, 22 March. Online Edition.

Yan, Li (2004) "Cong Minqi Canjun Tan Yujun Yumin [A Discussion of Locating Military Potential in Civilian Capabilities from the Participation of Civilian Firms in Defense Productions]," *Zhongguo Junzhuanmin [Defense Industry Conversion in China]*, 3: 15–18.

Yang, Chuan *et al.* (2005) "GPS Weixing Daohang Jieshouji Rf Dianlu De Sheji [Design of a Radio Frequency Circuit for GPS Receivers]," *Weidianzixue [Microelectronics]*, 35(1): 21–4.

Yang, Chyan and Hung, Shiu-Wan (2003) "Taiwan's Dilemma across the Strait: Lifting the Ban on Semiconductor Investment in China," *Asian Survey*, 4(4): 681–96.

Yang, Yuan and Liao, Rui-Hua (2003) "Nami Dianzi Jishu Zai Junshi Lingyu De Yingyong [Military Application of Nanoelectronic Technology]," *Weina Dianzi Jishu [Micro-nanoelectronic Technology]*, 11: 13–17.

Yang, Yuzhong (1989) "Fazhan Zhong De Guofang Kegongwei Junyong Biaozhunhua Zhongxin [COSTIND Center of Military Standardization under Development]," *Junyong biaozhunhua [Military Standardization]*, 3: 5–6 (*neibu* material).

Ye, Weiping (2003) "Kaifa Junmin Liangyong Jishu [The Development of Military–Civilian Dual-Use Technologies]," *Guofang Keji Gongye [Defense Science & Technology Industry]*, 10: 35–7.

Ye, Yuqing (1992) "Quanyu Junyong Dianzi Yuanqijian Kekaoxing Gongzuo De Yujian [Opinions on the Task to Ensure the Reliability of Electronic Parts and Components for Military End Uses]," *Dianzi Chanpin Kekaoxing Yu Huanjing Shiyan [Electronic Product Reliability and Environmental Testing]*, 1: 7–13.

Yin, Robert K. (1994) *Case Study Research: Design and Methods*, Thousand Oaks, CA: Sage.

Yu, Zhenxing (ed.) (2002) *Junyong Yuanqijian Shiyong Zhiliang Baozheng Zhinan [Quality Assurance Guide on the Usage of Military Components]*, Beijing: Hangkong Gongye Chubanshe (*neibu* material).

Yu, Zonglin (2004) "Minyong Qiye Canyu Guofang Jianshe Shi Jianli Yujun Yumin Chuangxin Jizhi De Zhongyao Neirong [The Participation of Civilian Enterprises in

National Defense Construction as the Main Pillar Supporting the Establishment of an Innovation Mechanism through Locating Military Potential in Civilian Capabilities]," *Zhongguo Junzhuanmin [Defense Industry Conversion in China]*, 4: 8–11.

Yuan, Mingwen (1992) "Ershinian Huigu: Jidianbu Shisansuo: Yanzhi GaAs Mesfet He GaAs IC Gaikuang [Twenty-Year Review of the 13th Research Institute of the Ministry of Mechanical and Electronic Industry: Research and Fabrication of GaAs Mesfet and GaAs IC]," *Bandaoti Jishu [Semiconductor Technology]*, 3: 1–7, 23.

Zhang, Bin (1991) "Guanyu Bandaoti Fenli Qijian Guanche Guojunbiao Jige Wenti De Tantao [Issues Concerning the Implementation of National Military Standards for Semiconductor Discrete Devices]," *Junyong biaozhunhua [Military Standardization]*, 4: 29–30 (*neibu* material).

Zhang, Jiuchun and Zhang, Baichun (2007) "Founding of the Chinese Academy of Sciences' Institute of Computing Technology," *IEEE Annals of the History of Computing*, 29(1): 16–33.

Zhang, Liying and Guo, Jianping (2004) "Ershiyi Shijichu Shijie Keji Zouxiang Ji Woguo Keji Anquan Huanjing Yanjiu [World Technology Trends at the Turn of the 21st Century and the Study of Our Nation's Technological Security Environment]," *Keji jinbu yu duice [Science & Technology Progress and Policy]*, 2(2): 14–16.

Zhang, Shaozhong (2004) "Xinxihua Wuqi Zhuangbei Yu Xinjunshi Biange [Informationalization of Weaponry and Equipment and Revolution in Military Affairs]," in Chinese Academy of Sciences (ed.) *2004 Gaokeji Fazhan Baogao [2004 High Technology Development Report]*, Beijing: Kexue Chubanshe, pp. 302–13.

Zhang, Weichao and Li, Chun (2006) "The Access of Individually-Run Enterprises to National Defense Industry," *Junshi jingji yanjiu [Military Economic Research]*, 1: 41–5.

Zhang, Xiaojun (1995) "Modern National Defense Needs Modern Signal Troops," *National Defense*, 10: 20–1.

Zhang, Xiao-Wen *et al.* (2005) "Guowai Junyong Chaodaguimo Jicheng Dianlu De Yingyong Yu Fazhan Qushi [The Overseas Military VLSI Application and Development Trend]," *Bandaoti Jishu [Semiconductor Technology]*, 30(8): 13–17.

Zhang, Xiaoxiang and Zhang, Yiren (2005) *Xiandai Keji Yu Zhanzheng [Modern Technology and War]*, Beijing: Qinghua Daxue Chubanshe.

Zhang, Xuefeng (2008) "Lien Chan Visited Chongqing Xiyong Microelectronics Industrial Park," *Chongqing Daily*, 1 May.

Zhang, Yaowen (1986) "Junyong Biaozhunhua Gongzuo Huigu Yu Zhanwang [The Review and Prospect of the Task of Military Standardization]," *Junyong biaozhunhua [Military Standardization]*, 4: 2–7 (*neibu* material).

Zhang, Yin (2002) "Jiakuai Fazhan Guochan Guangke Shebei Mianlin De Wenti Yu Jianyi [Problems and Suggestions Pertaining to the Attempt to Accelerate the Indigenous Development of Lithography Equipment]," *Diaoyanbaogao [Survey Report]*, 1–8 (*neibu* material).

Zhao, Qunzeng (2000) "DPA Shi Baozheng Junyong Dianzi Yuanqijian Zhiliang De Zhongyao Cuoshi [DPA: An Important Measure for Keeping Quality of Military Electronic Parts]," *Junyong biaozhunhua [Military Standardization]*, 2: 8–11.

Zheng, Hantao and Li, Ruhong (1989) "Huigu Guofang Gongye Jianshe Sishinian [Reflections on Forty Years of Defense Industrial Construction]," in Nie, L. and Huai, G. (eds.) *Huigu Yu Zhangwang: Xin Zhongguo De Guofang Keji Gongye [Retrospect and Prospect: New China's Defense Science and Technology Industry]*, Beijing: Defense Industry Press, pp. 117–20.

Zheng, Zhongyang (2007) "The Origins and Development of China's Manned Spaceflight Programme," *Space Policy*, 23(3): 167–71.

Zhong, Wenxue (2005) "Lujing Yilai Yu Guofang Keji Gongye Anquan [Path Dependence and Defense Technology Industry Security]," *Zhongguo Junzhuanmin [Defense Industry Conversion in China]*, 6: 21–4.

Zhongguo Junzhuanmin [Defense Industry Conversion in China] (2004) " 'Mincanjun' Lakai Xumu ['Civilian Participation in Military Production' Opens Its Prologue]," 3: 3–4.

Zolper, John C. (2005) "Integrated Microsystems: A Revolution on Five Frontiers," *24th DARPA Systems and Technology Symposium*, 9–11 August, Anaheim, CA, pp. 118–23.

Zweig, David and Bi, Jianhai (2005) "China's Global Hunt for Energy," *Foreign Affairs*, 84(5): 25–38.

Zysman, John (1991) "US Power, Trade and Technology," *International Affairs*, 67(1): 81–106.

Index